S0-AGP-812

INTRODUCTORY EXPERIMENTAL CHEMISTRY

100
ml
90
80
70
60
50

Introductory Experimental Chemistry

Melanie B. Messer

Educational Consultant
Department of Education
San Diego County, California

Richard A. Hoag

Science Instructor
Grand Blanc Community Schools
Grand Blanc, Michigan

Andrew H. Messer

Quality Assurance Engineer
Rohr Industries, Inc.
Chula Vista, California

PRENTICE-HALL, INC.
Englewood Cliffs, New Jersey

SUPPLEMENTARY MATERIAL

Teacher's Manual

Introductory Experimental Chemistry

Melanie B. Messer, Richard A. Hoag, and Andrew H. Messer

ISBN 0-13-501734-3

10 9 8 7 6 5 4 3 2 1

Design: Robert Sugar

Cover photo by Pierre Berger/Photo Researchers Inc.

Title page photo by Hugh Rogers/Monkmeyer

PRENTICE-HALL INTERNATIONAL, INC., London
PRENTICE-HALL OF AUSTRALIA, PTY. LTD., Sydney
PRENTICE-HALL OF CANADA, LTD., Toronto
PRENTICE-HALL OF INDIA PRIVATE LTD., New Delhi
PRENTICE-HALL OF JAPAN, INC., Tokyo

PREFACE

So you want to study chemistry. . . .

With the beginning of this course you will enter a whole new world of experience. This book is a doorway to that world. This course is designed to give you an experience from which you can profit and grow. That is not to say that the course is going to be easy. It will require you to think. You will spend a great deal of your class time in the laboratory; but it will be an enjoyable and profitable time. Sometimes what you do in class and in the laboratory will raise more questions than answers. But *your* answers will help you to understand better the world you live in, and give you some tools to help you plan your future.

We hope you enjoy working with these materials as much as we have enjoyed preparing them.

Melanie B. Messer
Richard A. Hoag
Andrew H. Messer

CONTENTS

George W. Martin/dpi

UNIT 1

Same Old Problem

INVESTIGATION

1

Who says you have to study chemistry? What is chemistry, anyway? Do we ever use it? What can chemistry do for you? Why bother?

Do I have to know a lot of math?

Is it hard?

Will I hurt myself in the laboratory?

Can I use chemistry to get a job?

Let's tackle just one question now. Later you will answer the other questions for yourself. Why study chemistry? You may be studying it for one of several reasons.

a. You need a science to graduate.
b. You need to fill a hole in your program.
c. Last year's science course was fun. You'd like some more.
d. You need chemistry to help you in a job after you finish school. For example, you may want to be a practical nurse, laboratory or dental technician, fireman, beautician, printer, police officer, or dietician. These are some of the jobs in which chemistry is useful.

This course can help you meet all of these needs, if you want it to. What you get out of it depends on what you put into it. The purpose of these investigations is to make it all as worthwhile and as painless as possible.

A. Leap Right In

Some things that you take for granted are really complicated chemicals. And chemistry deals largely with familiar things. Get to work now and look at some of them. At the same time, you can sharpen your powers of observation.

You have all seen table salt and baking soda in your homes. Did you know that a chemist calls salt *sodium chloride?* and

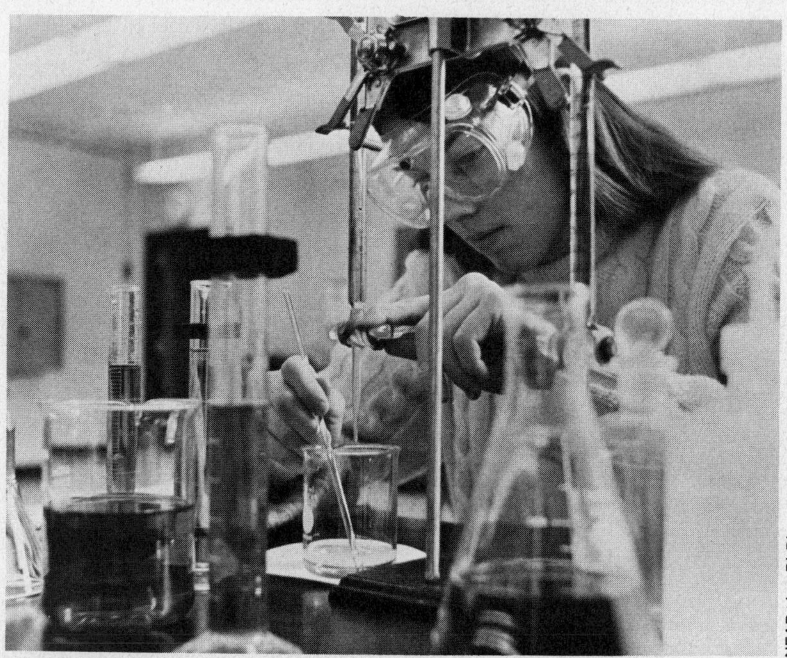

NEAP / Joe Di Dio

baking soda *sodium bicarbonate?* Many other things found in the home are also found in the chemical laboratory. Other examples of household products are sugar, starch, and Epsom salts. Do you know their chemical names?

What you need

Acetic acid
  (vinegar—CH_3COOH)
Citric acid ($C_6H_8O_7$)
Iodine (I_2), 0.1 *M*
  solution
Methyl orange indicator
Sodium bicarbonate
  ($NaHCO_3$)
Starch [$(C_6H_{10}O_5)_n$]
Sucrose ($C_{12}H_{22}O_{11}$)
Dropping bottles, 3
Hand lens
Aluminum foil,
  8 cm × 8 cm, 4
Notebook
Pen/Pencil
Safety goggles
Spatula

What to do

1. In your laboratory notebook, set up a page for this investigation. At the top of the page, write "Unit 1, Investigation 1." Then add today's date.

It is very important to write all observations in a notebook. Otherwise it will be difficult to remember what happened when you need the information later on. Make certain you write down what happens each time you work with chemicals.

a. Put on your safety goggles. This is a must whenever you are working in the laboratory.
b. With a spatula, take a very small sample, about the size of a pea, from one of the bottles of solid substances. Place the sample on a piece of aluminum foil.

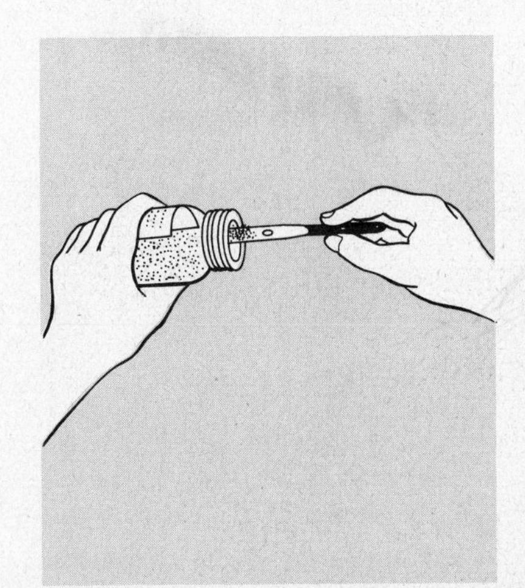

c. Label the foil with the name of the solid.
d. Using a hand lens, look at the sample carefully.

2. What does it look like? (What is its color? shape? texture?) Write the answer in your notebook.

e. Using your spatula, separate the sample into four piles on the foil.

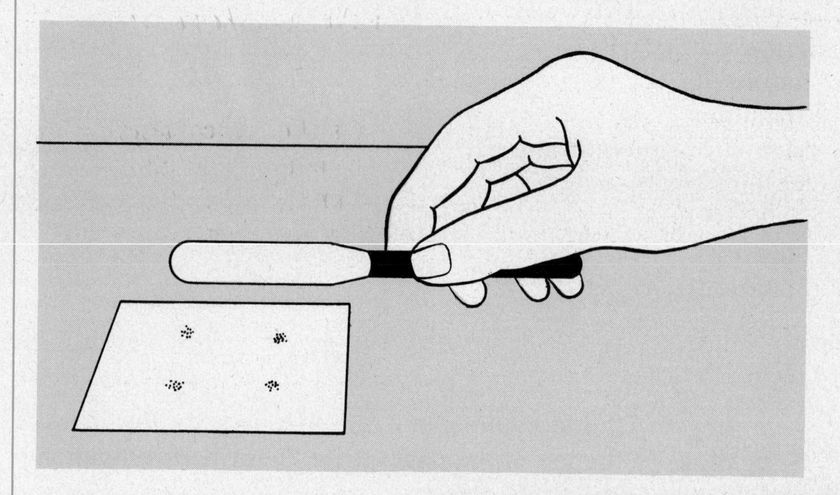

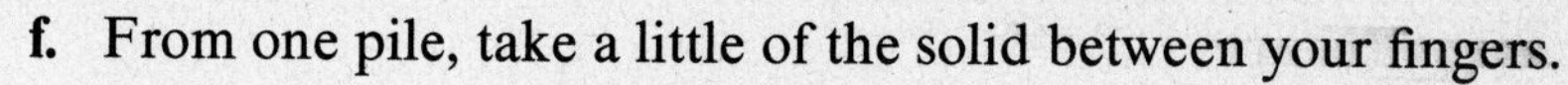

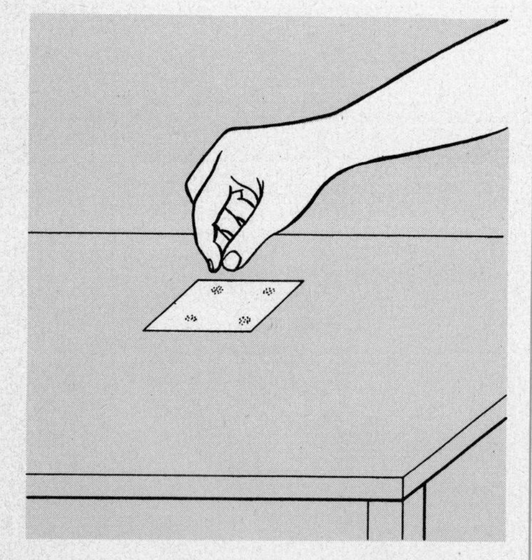

f. From one pile, take a little of the solid between your fingers.

3. What does it feel like—hard, soft, gritty?

g. Rinse and dry your fingers before going on to the next step.
h. Get a dropping bottle of acetic acid. Place a few drops on one of the other piles.

4. What happened?

i. Place a few drops of methyl orange indicator on a third pile.

5. What happened?

j. Place a few drops of iodine solution on the last pile.

6. What happened?

k. Repeat steps **b** through **j** for the other three samples.

7. Answer questions 2–6 for each sample.

B. What Happened?

You have just mixed some substances together. Chemists do this all the time. In some cases nothing happened; this sometimes occurs in chemistry.

But in other cases something did happen. You saw bubbles given off or a color change. Both are signs that a *chemical change* has occurred. In chemistry we call such a change a *chemical reaction.* Simple enough???

You will be doing many different kinds of chemical tests during the year. Sometimes you will use fancy equipment; other times, you will use only your senses. Each time you will be trying to find out about something. That is what chemistry is all about.

INVESTIGATION

2

Round Two

Some day we've got to get organized!

In the first investigation you added acetic acid (vinegar), iodine solution, and methyl orange indicator to different white solids. You learned that color changes or bubbles are signs of chemical reactions. Some of the substances reacted; some did not.

Along the way, you collected a lot of notes and observations. Compare your notes with those of other groups in your class. Each student probably wrote notes in a different way. In chemistry it helps if everyone's notes and observations are as organized as possible. Then anyone who reads them can understand them. Chemists often use *tables* to help them organize their observations. See if a table helps.

A. A Little Organization, Please

What you need

Acetic acid (vinegar—CH_3COOH)
Ammonia water (NH_4OH), 1 *M*
Boraxo ($NaBO_3$)
Chalk ($CaCO_3$)
Methyl red indicator
Plaster of Paris ($CaSO_4$)
Washing soda (Na_2CO_3)

Dropping bottles, 3

Aluminum foil, 8 cm × 8 cm, 4
Notebook
Pen/Pencil
Safety goggles
Spatula
Wax pencil

What to do

1. In your laboratory notebook, set up a page for this investigation. Remember to include today's date. This is always important so that you know when a test was done. Write the answers to all questions in your notebook.

2. Copy the table below into your notebook.

TABLE 1: Results of Tests

	Acetic acid (CH_3COOH)	*Ammonia water (NH_4OH)*	*Methyl red indicator*	*Dissolves in water (Solubility)*
Plaster of Paris ($CaSO_4$) Calcium sulfate				
Washing soda (Na_2CO_3) Sodium carbonate				
Boraxo ($NaBO_3$) Sodium perborate				
Chalk ($CaCO_3$) Calcium carbonate				

a. Put on your safety goggles.
b. With a spatula, take a small sample, about the size of a pea, from one of the bottles of solid substances. Place the sample on a piece of aluminum foil.

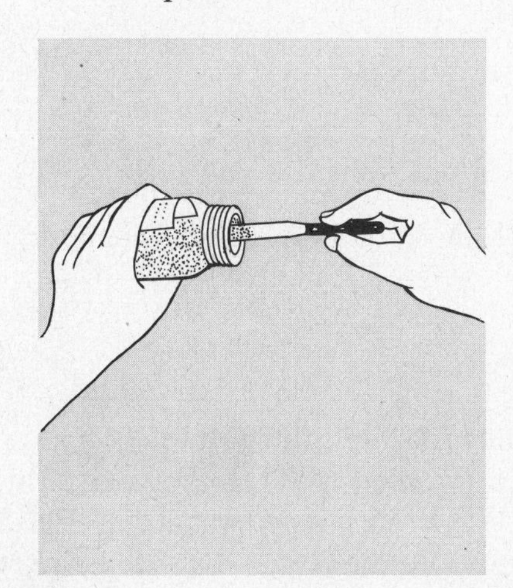

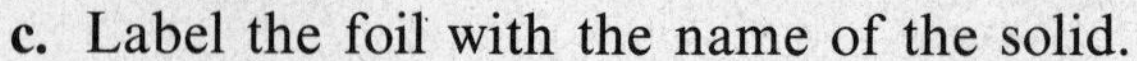

c. Label the foil with the name of the solid.

d. Using your spatula, separate the sample into three piles on the foil.

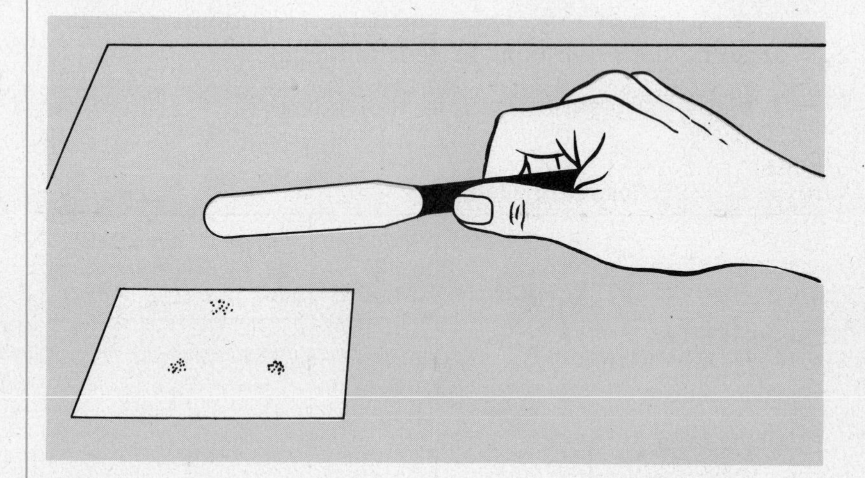

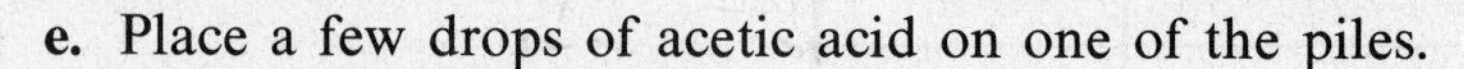

e. Place a few drops of acetic acid on one of the piles.

3. What happened? Record your answer in Table 1.

f. Place a few drops of ammonia water on the second pile.

4. What happened? Record your answer in the table.

g. Place a few drops of methyl red indicator on the last pile.

5. What happened? Record your answer in the table.

h. Repeat steps **b** through **g** using the other three samples.

6. Record your results in the table.

7. What evidence of chemical reactions did you see?

8. Compare your results with those of your classmates, and explain how tables help.

B. The Great Unknown

What if you were given an unknown white solid and told it was one of the eight you have already worked with? Could you tell which one it was by using methyl orange, methyl red, acetic acid, ammonia water, and iodine solution?

You might be able to with some white solids, but not with others. So—here is another tool of the trade used to help identify unknown substances.

What you need

Boraxo ($NaBO_3$)
Chalk ($CaCO_3$)
Plaster of Paris ($CaSO_4$)
Washing soda (Na_2CO_3)

Beaker, 250 ml
Test tubes, 4

Safety goggles
Spatula
Stoppers, solid, 4
Test tube rack
Wax pencil

What to do

a. Carefully fill a test tube $\frac{2}{3}$-full with water.

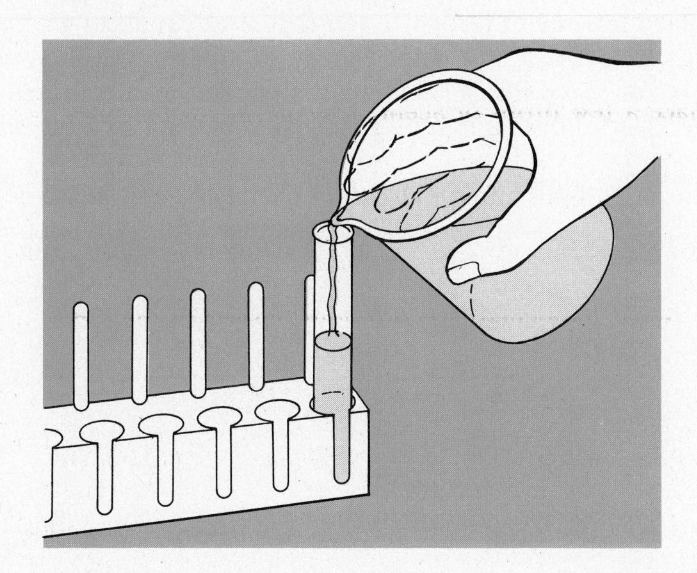

b. With a spatula, add a small quantity of one white solid, about the size of a grain of rice, to the tube.
c. Stopper the tube and shake well.
d. Wait 2–3 minutes. Then shake the tube again.
e. Let the tube sit for 2–3 minutes. Then observe.

9. Record in your table if you can see any solid at the bottom of the tube.

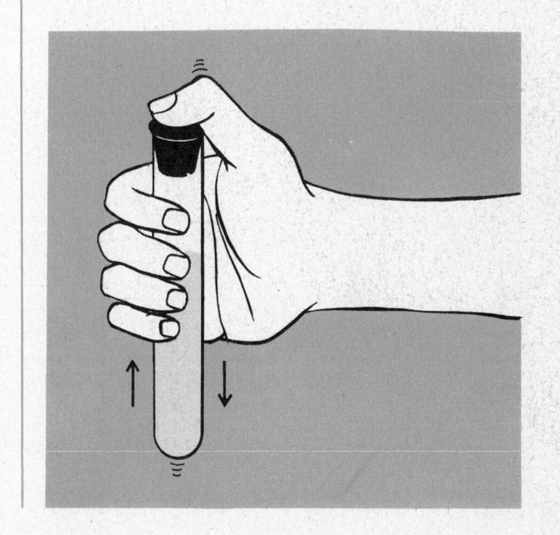

When a solid disappears in a liquid, it is said to be *dissolved.* The ability of a substance to dissolve is called *solubility.*

f. Repeat steps **a** through **e** with the other three white solids.

10. Record the results in the table.

Believe it or not, in the last two investigations you have performed several standard chemical tests. These are the same tests chemists do all the time in a laboratory. You tested for:

a. the appearance of a chemical
b. its texture
c. the reaction of chemicals with an acid (acetic acid)
d. the reaction of chemicals with a base (ammonia water)
e. the reaction of chemicals with indicators (methyl orange, methyl red, iodine)
f. the ability of a chemical to dissolve in a liquid (solubility)

You have learned to recognize some signs of chemical change, such as when there is a color change or a gas given off (bubbling). You have investigated solubility. You also collected a lot of observations, and you organized your data in a table.

Organization is an important part of chemistry, and in the next investigation you will learn how a chemist writes a report.

INVESTIGATION 3

The Report's the Thing

If you had to make a list of materials and equipment found in any laboratory, you would discover something interesting. The list would be different for every laboratory, except for two items—notebook and pen.

Records must be kept, not only of experiments which were successful, but also of experiments which were flops.

Records are figures, notes, charts, and any other information which shows the progress of a laboratory investigation. A record includes all of the details connected with the work. By looking at the records of any of your chemical tests, a classmate should be able to duplicate your work. Or, your teacher should be able to tell exactly how you did your experiment.

Bound notebooks are used to record all data. Scraps of paper should never be used, since they are easily lost. Good records are done in ink, and should be easily read by anyone.

A. The Making of a Report

The way results of experiments are recorded is important. But just as important is the way these results are reported. *Reports* are the results of the work described in the records. A report doesn't have to be long or complicated to be complete. It can consist of a set of figures showing the results of an experiment. It can even be simply a statement that a material has been examined and found to be satisfactory. Reports are for people not familiar with the investigation, whose main concern is for the results. Since the final report depends on the records, it is most important that every laboratory worker keep careful records.

Sometimes industrial reports have to be made on specially prepared forms. These are numbered and dated. All industrial reports should be signed by the chemist or technician doing the test. All of this information serves as a permanent account of *what* happened *when,* and *who* did it. The following is a sample report form used by an industrial laboratory.

R. U. Testing Co. Number _____
2 Accurate Drive
New York, New York

Name of Client: ____________________

Description of Sample: ____________________

Purpose of Test: ____________________

Test Data: ____________________

Summary of Results: ____________________

Conclusion: ____________________

Date: __________ Signature: __________

By the end of this year, you will have a pretty full notebook. You will use it to look up many things, many times. You will use it to check results, to refresh your memory, and to help you plan other experiments.

It helps if everyone in the same laboratory uses the same report form. For this course you will follow this general form:

a. Heading: The heading should include your name, class, the date, Unit number, Investigation number, and Part.
b. Purpose of the Work: This should clearly state what you were trying to investigate.
c. Materials and Chemicals: These should be included only if they are helpful information.
d. Method or Procedure: How did you perform your experiment? State this as briefly as possible.
e. Results Obtained: Clearly and simply show your results.
f. Conclusion: What do the results mean? Try to answer any questions raised in the "Purpose of the Work."

1. Explain the difference between recording and reporting results.

2. Write the Purpose for Unit 1, Investigation 2, Part B.

3. Write a Conclusion for Unit 1, Investigation 2, Part B.

B. Now You See It; Now You Don't

What You Need

Aluminum (Al) foil, 8 cm × 16 cm
Copper(II) chloride ($CuCl_2$), 0.01 *M* solution
Beaker, 250 ml
Hand lens
Forceps
Notebook
Paper towel
Pen/pencil
Safety goggles

What to Do

a. Fill your beaker $\frac{1}{3}$-full with copper chloride solution.
b. Loosely crumple the aluminum foil.

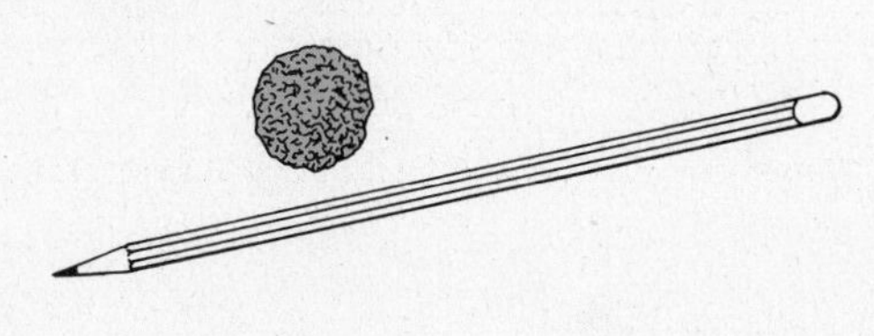

c. Place the loose ball of aluminum foil in the beaker.

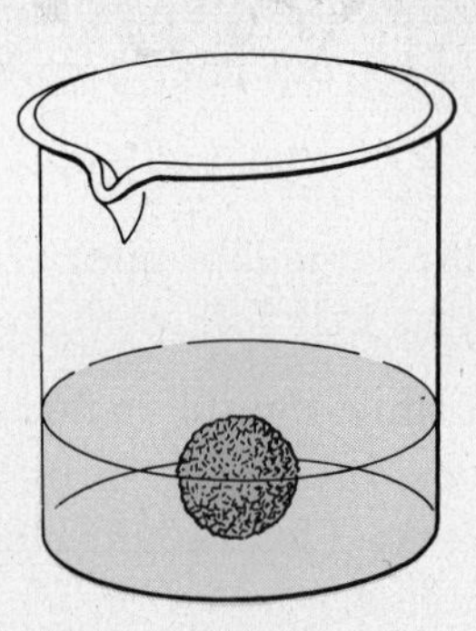

d. Wait 15 minutes. With the forceps carefully remove the foil from the beaker.

e. Place the foil on a paper towel.

f. Examine the aluminum foil carefully.

4. Write a report on this activity. Remember, good reports can be brief. Tell exactly what you did and saw. And, last but not least, try to explain . . . . . . WHAT HAPPENED???

INVESTIGATION 4

Fired Up

You are going to spend a lot of time in the laboratory this year. There are many ways to make it as painless as possible. Obviously, there is a right way and a wrong way to do almost everything in this world. In a chemistry laboratory, the **right** way is the **safest** way. You probably started this course with ten fingers and ten toes. A few simple rules will make sure you have all of them when you leave. While you are learning the rules, look at some additional types of chemical reactions.

A. Light My Fire

Lighting a burner is a good starting point. Two of the most common laboratory gas burners are shown in the diagrams on the next page.

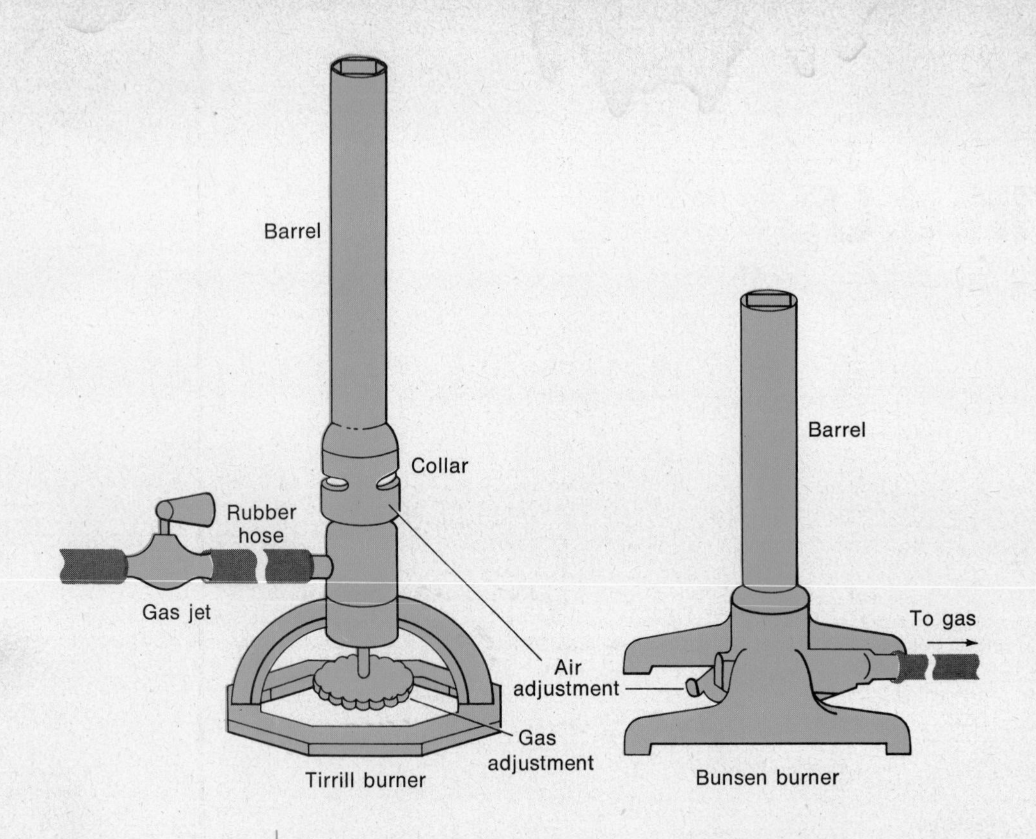

What You Need

Burner with rubber hose
Matches
Safety goggles
Straight pin

What to do

1. Title a new page in your laboratory notebook for this investigation.

a. Carefully study your burner. Compare its parts with those shown in the diagram.
b. Adjust your burner so the air holes are closed.

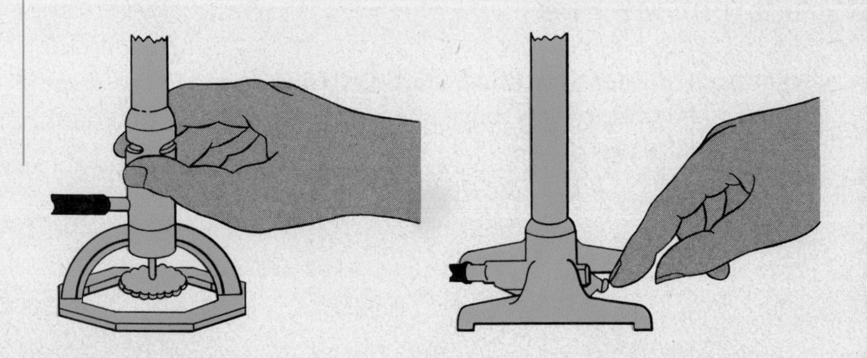

2. What do you think is the purpose of mixing air with the gas?

c. Connect the rubber hose to the gas outlet.
d. Hold a lighted match so that the bottom of the match flame is next to the top of the barrel. The match flame should not be directly over the barrel.

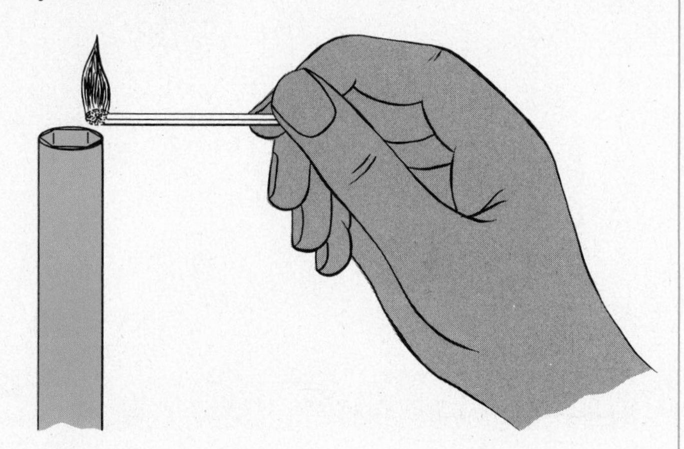

e. Light the burner by opening the gas jet slowly.

3. How do you adjust the amount of gas being fed to your burner?

4. What is the color of the flame?

f. Move the air adjustment to admit air slowly to the flame. Continue to admit air until two well-defined regions appear in the flame.
g. Close the gas jet, turning off the burner. Do not move the air adjustment.
h. Using a straight pin, suspend a paper match in the barrel of the burner. The match head should be within the lower region of the flame.
i. Carefully relight the burner. Do not touch the two matches together.

5. Does the suspended match light?

The blue cone contains unburned gases. The hottest part of the flame is just above the tip of this blue cone.

6. Explain your answer to question 5.

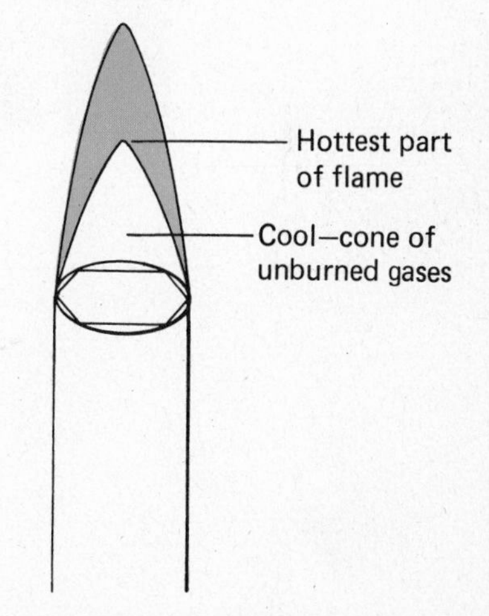

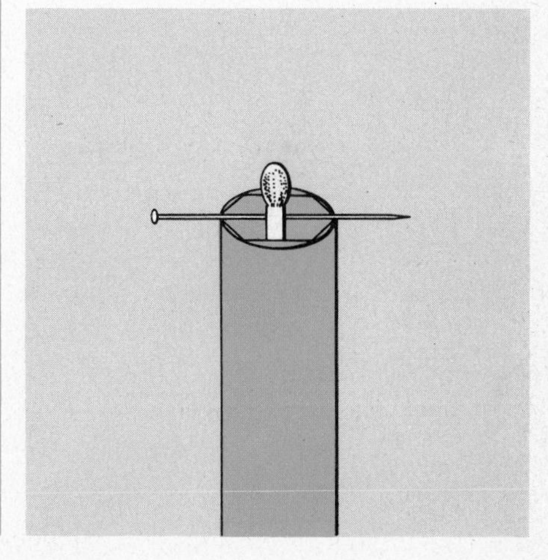

B. Let's Give the Glass a Break

One thing most people working in a laboratory learn is how to work with glass tubing. This skill helps in building both simple and complicated set-ups.

What You Need

Glass tubing, 10 cm	Asbestos pad
Glass tubing, 40 cm	Burner
	Cloth or gloves
	File, triangular
	Matches
	Safety goggles

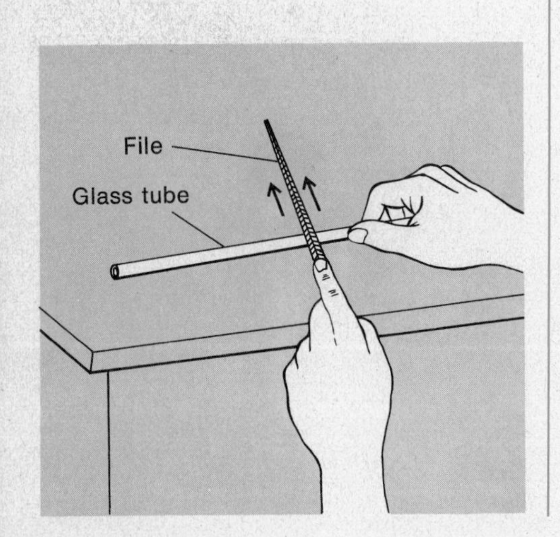

What to Do

a. Place the longer piece of glass tubing flat on your desk. Remember always to wear safety goggles in the lab.

b. Holding the file at right angles to the tube, make a deep scratch in the middle of the tube. Do this by pushing the file away from you. The scratch should be made with one firm, steady stroke.

c. Use a cloth or gloves to protect your hands and hold the tube with both hands. Have your thumbs meet opposite the scratch.

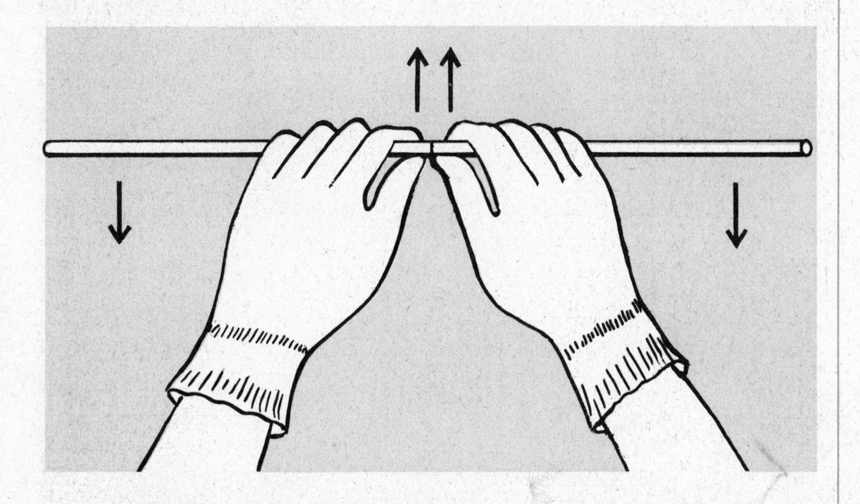

d. Break the glass tubing by pulling the ends toward you, while pushing away with your thumbs.

Look at the cut edges of the glass tubing. They are jagged and sharp and can cause bad cuts. In the laboratory, this problem is solved by smoothing out the edges of the tubing with a flame. This is called *fire polishing*.

e. Light your burner and adjust the air for a hot flame.
f. Hold one piece of the cut glass tubing by one end. Place the other end in the hottest part of the flame.
g. Rotate the tubing as shown in the drawing.

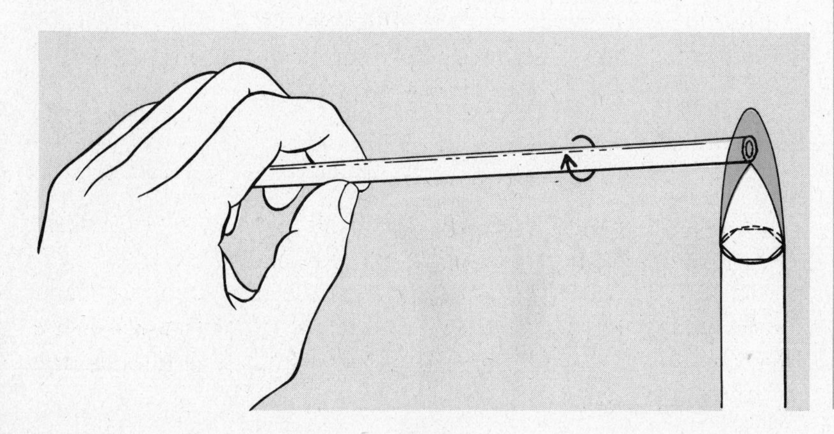

h. When the tubing begins to glow, take it out of the flame. Carefully rest it on the asbestos pad.
i. Allow the tubing to cool at least 5 minutes before picking it up.

7. What has happened to the end of the tubing that was in the flame?

8. Why should glass tubing be fire polished?

j. Repeat steps **f** through **i** for each piece of glass tubing. Fire polish both ends of each.
k. Turn off your burner.

Save all pieces of glass tubing for further use in this investigation.

C. Bend and Stretch, Bend and Stretch

What You Need

Glass tubing from Part B

Asbestos pad
Burner
Flame spreader
Matches
Safety goggles

What to Do

a. Place the flame spreader on the burner.
b. Carefully light the burner.
c. Hold a 20-cm piece of glass tubing in the flame. Hold it off center, as shown below. Slowly rotate the tubing.

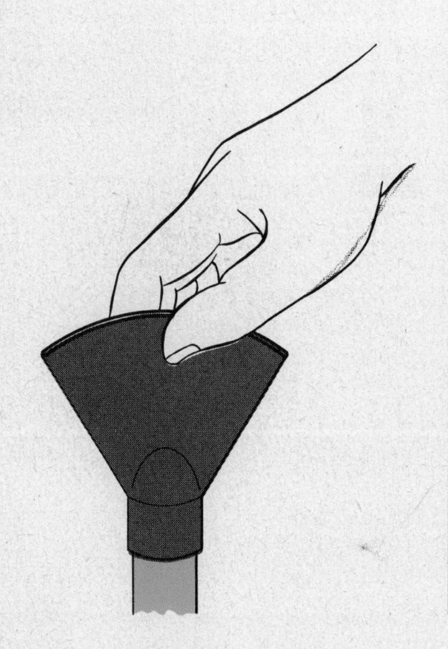

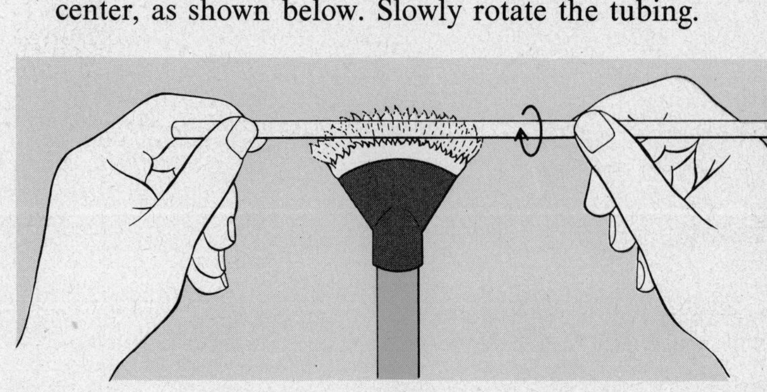

9. What is the purpose of the flame spreader?

d. When the tubing starts to glow and sag, quickly remove it from the flame and bend it into a right angle.

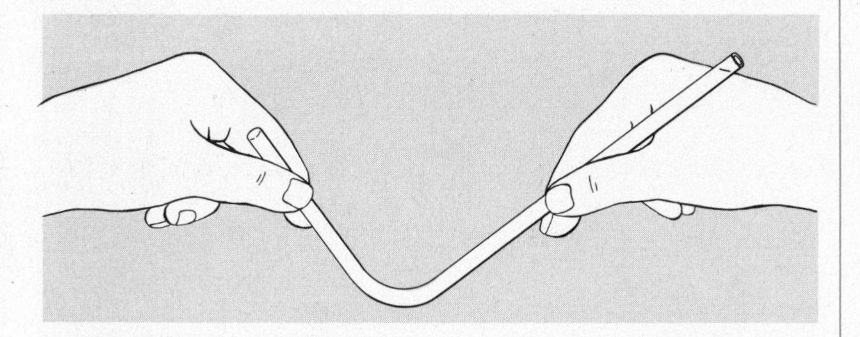

e. Place the tubing on an asbestos pad. Allow it to cool before touching. Save it for Part E.

f. With the other 20-cm piece of glass tubing, repeat steps **c** through **e;** but this time do **not** rotate the tubing.

g. After they are cool, look carefully at both pieces of bent tubing.

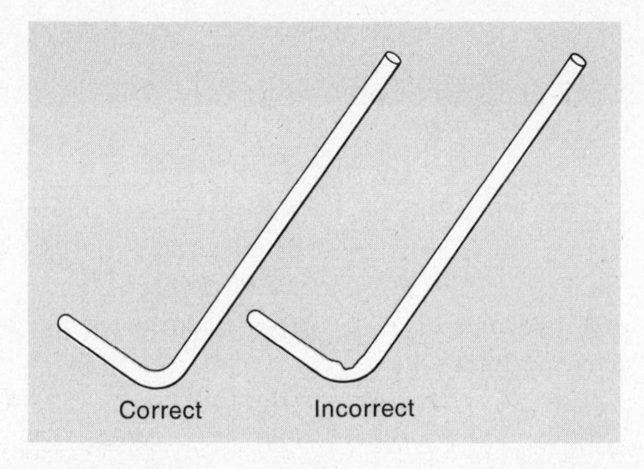

10. Describe the bend you made without rotating the tubing.

11. Describe the bend you made after rotating the tubing.

D. Easy Does It

After bending glass tubing, you may need to put it into a rubber stopper. Chemists often do this in setting up their laboratory equipment.

Museum of Modern Art/Film Stills Archive

What You Need

Beaker
Dropper
Glass tubing, fire polished, from Part C
Cloth or gloves
Safety goggles
Stopper, 2-hole rubber, to fit 250 ml flask

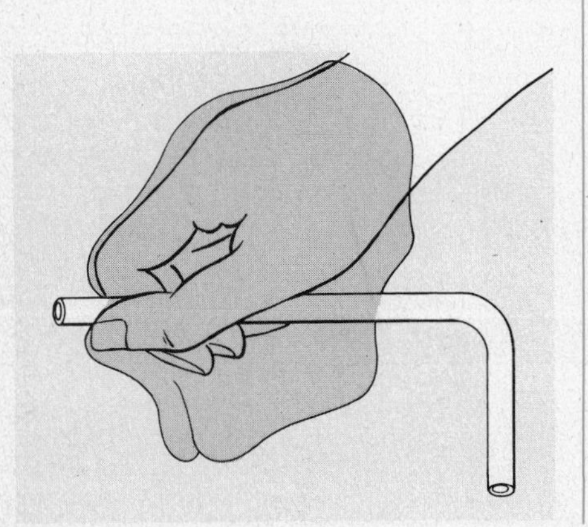

What to Do

a. With a dropper, add several drops of water to one of the stopper holes and to the tip of the long side of your good glass bend.

b. Hold your glass bend close to the end. Always protect your hands with a cloth or gloves when placing glass tubing in stoppers.

c. Very cautiously insert the tubing through the moist hole. **Do not force it!**

d. Save your stopper with tubing for Part E.

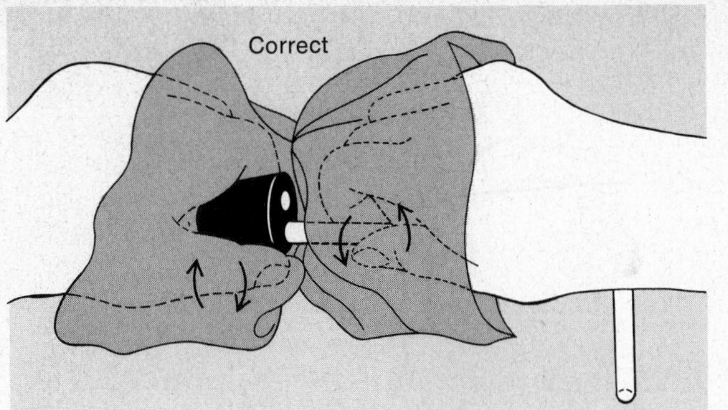

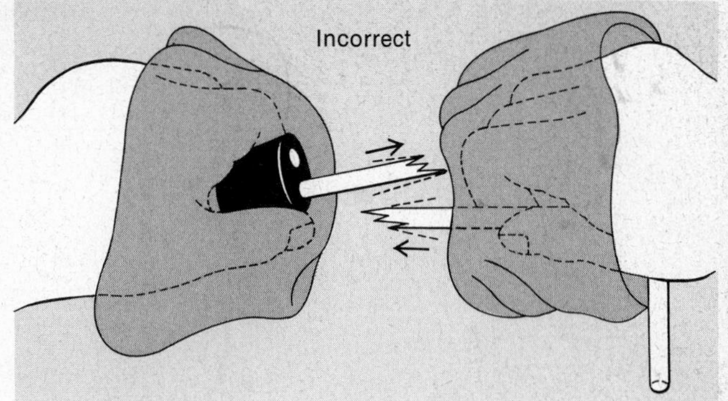

Now that you have some basic skills and tools to work with, it's time to put them to use.

E. Body Chemistry

Carbon dioxide plays an important part in your life. It is normally a colorless, odorless gas. Plants use it to make food. Carbon dioxide can also be written as CO_2, which is pronounced "See-Oh-Two." The air you breathe in contains 0.03% CO_2, but the air you breathe out contains much more. Carbon dioxide is a waste product of your body. Did you ever think of your breath as a chemical?

Try this experiment:

What You Need

Limewater [$Ca(OH)_2$]

Flask, 250 ml

Glass tubing, right angle bend with stopper from Part D

Glass tubing, straight piece from Part B

Cloth or gloves

Safety goggles

What to Do

a. Insert the straight piece of tubing into the open hole of the stopper. **Remember—don't force it!** Only a little piece of the straight tubing should show beneath the stopper. The stopper should look like this.

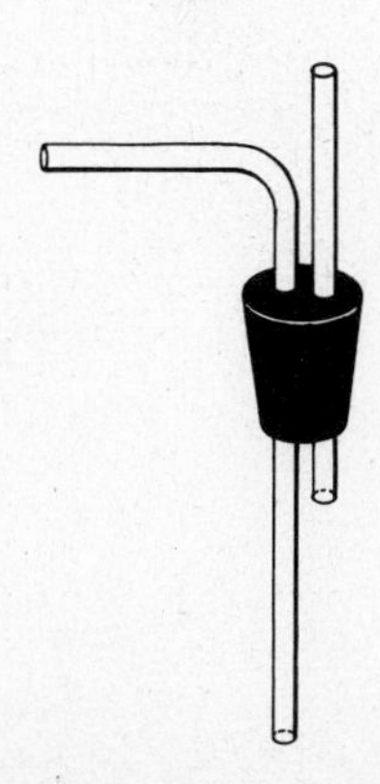

b. Fill the flask $\frac{1}{2}$-full with limewater.

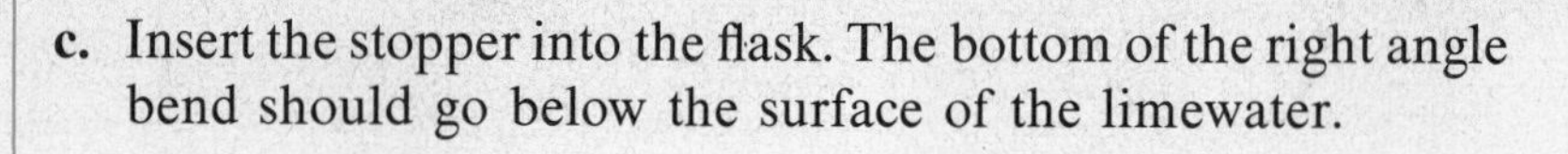

c. Insert the stopper into the flask. The bottom of the right angle bend should go below the surface of the limewater.

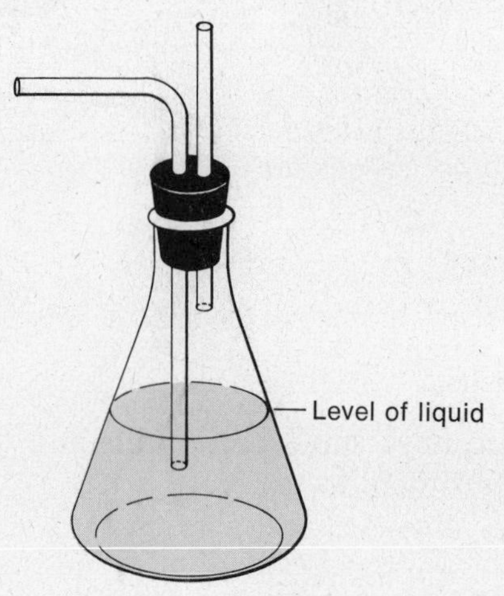

12. Is the limewater cloudy or clear?

d. Blow into the right angle bend.

13. What happened to the limewater?

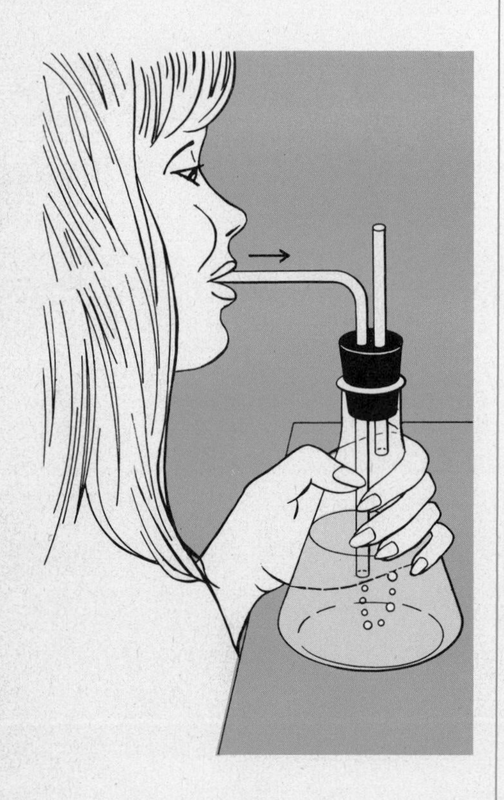

Do you remember the first investigation? You added different liquids to different white powders. Several interesting things happened, which we called *chemical changes* or *chemical reactions.*

14. Do you think a chemical change took place in the limewater?

15. How can you tell?

The white cloud you see in the flask is a solid. It is called a *precipitate.* A precipitate is a solid that settles out of a liquid. This precipitate was formed when you added carbon dioxide to the limewater.

16. Was the precipitate there when you started the experiment?

17. Where did the carbon dioxide come from?

18. What was the evidence that a chemical reaction took place?

Carbon dioxide is normally a gas; but if you get it very cold, it becomes a solid. As a solid, it is called dry ice. Dry ice is used to keep things like ice cream cold.

19. Think of an experiment to prove that dry ice contains carbon dioxide. Write up the procedure for doing the experiment.

You now know at least three ways to show that a chemical change has occurred:

a. Bubbling, which means a gas is given off
b. Color change
c. Precipitate formation

More to follow.

Let's close this investigation with a set of Chemist's Commandments.

1. Always wear safety goggles.
2. Never handle hot glass tubing.
3. When inserting glass tubing into a stopper, use a cloth or gloves to protect your hands.
4. Never point the open end of a test tube at anyone, including yourself.
5. Carefully read every bottle label twice before using the contents.
6. Never use an open flame near a flammable liquid.
7. Never taste any chemical in the laboratory unless the teacher says to.
8. Never pour water into acid.
9. Trust your teacher!
10. Always THINK.

INVESTIGATION

5

Metric —S'il Vous Plaît

One of the biggest communication gaps a science student must cross is learning a new system of measurement. You will not go very far in chemistry before you discover that almost every investigation involves a measurement of some kind. You will have to measure changes in length, temperature, volume, time, and mass. The study of chemistry would be impossible without a suitable system of measurement.

A. In Days of Old

A long time ago, measurement was almost a personal matter. For example, the unit of length was called a cubit. It was the distance from the elbow to the tip of the middle finger. This was great if you knew a shopkeeper with long arms and you were buying. But suppose you had short arms and wanted to sell some cloth. Here is what could happen.

In the 10th century, King Edgar ruled England. He declared that a yard was to be measured from the tip of his nose to the tip of his middle finger, when his arm was stretched out.

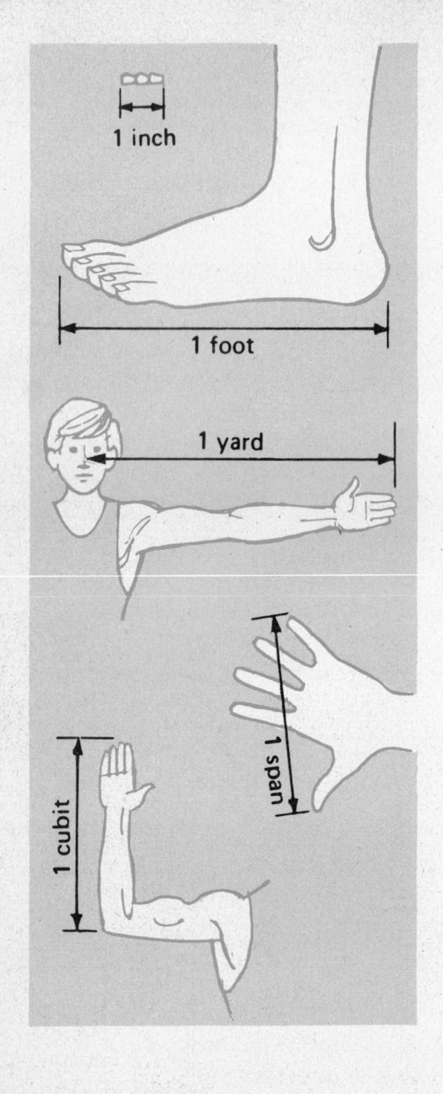

In the 14th century, the inch was determined by King Edward II to be the length of three dry barleycorns laid end to end. Other measurements used through the centuries were the foot and the span of the hand.

1. Do you think these measurements were the same all over the world?

2. Could scientists in various parts of the world in the 14th century compare their work? Why?

A scientific system of measurement didn't come about until the French Revolution in the late 18th century. The new French government formed a committee of scientists to set up a system of measurement. The result was called the *metric system.* It is used today in scientific work throughout the world. It is a *decimal* system—like our money. To change from one unit to another is as easy as changing from dollars to cents. All you do is move the decimal point the correct number of places.

The English system, which we use in our daily lives, presents many disadvantages in scientific measurements. One is that there is no simple numerical relationship between the different units. In the metric system, things are easier. For example, with the metric units for length, the relationship between meter and kilometer is neat and simple:

1,000 meters = 1 kilometer

"Kilo" always means "1,000." In the English system, however, you have to remember feet to yards to miles—Wow!

5,280 feet = 1 mile
1,760 yards = 1 mile

A French, Russian, or Italian student uses the same system in his everyday life that he uses in studying science. He runs the 100 meter dash, drinks a liter of milk a day, and dates a girl whose measurements are 94-60-91 (centimeters, of course). If the temperature is 35 degrees, he thinks about a quick swim after school. This temperature on the Celsius scale equals 95 degrees Fahrenheit.

In the chemistry laboratory, you will use the metric system. The activities in this investigation and the next two will help you get to know this system. The unit of length in the metric system is the *meter.* A meter is a little longer than a yard. The word *meter* is abbreviated as *m.*

$$1 \text{ m} = 39.37 \text{ inches}$$

Yardstick

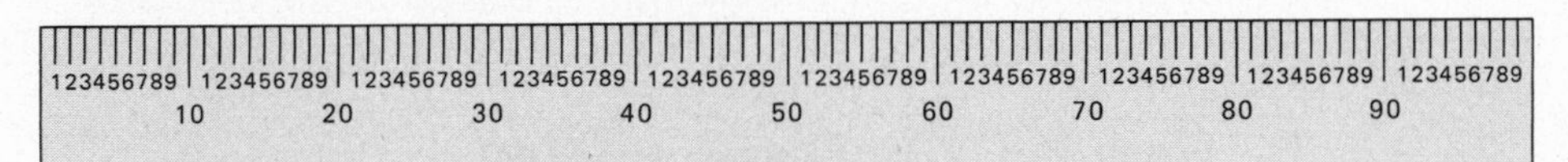

Meter stick

The unit of mass in the metric system is the *gram.* One thousand grams equal one *kilogram. Gram* is abbreviated as *g,* and *kilogram* is abbreviated as *kg.*

$$1 \text{ kg} = 2.2 \text{ pounds}$$

The unit of volume in the metric system is the *liter.* It is abbreviated as the letter *l.* As you can see, it's easy to confuse this abbreviation with the number "1." So, *liter* is usually not abbreviated.

$$1 \text{ liter} = 1.06 \text{ quarts}$$

Here are some tables that show relationships in the metric system. Use them to help you answer question 3.

Kilogram *Pound*

CYNARA

TABLE 1: English and Metric Equivalents

Unit of	*Metric unit*	*Approximate English unit*	*Exact English unit*
Length	meter	little longer than a yard	39.37 inches
Mass	kilogram	little more than 2 pounds	2.2 pounds
Volume	liter	little more than a quart	1.06 quarts

Table 2 shows the prefixes used in the metric system.

TABLE 2: Metric Prefixes

Prefix	*Fractional part of metric unit*
Milli-	1/1,000
Centi-	1/100
Deci-	1/10
Kilo-	1,000

TABLE 3: Metric Conversions

Length
10 millimeters (mm) = 1 centimeter (cm)
100 centimeters (cm) = 1 meter (m)
1,000 meters (m) = 1 kilometer (km)
Mass
1,000 milligrams (mg) = 1 gram (g)
1,000 grams (g) = 1 kilogram (kg)
Volume
1,000 milliliters (ml) = 1 liter (1)

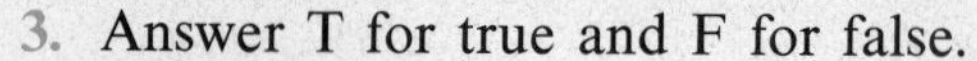

3. Answer T for true and F for false.
   a. 10 mm = 1 km
   b. 100 cm = 1 m
   c. 1,000 mm is less than 1 m
   d. 1 g = 1,000 mg
   e. 1,000 g = 1 kg
   f. 0.001 g is less than 1 mg
   g. 1,000 ml = 1 liter
   h. 0.001 liter is more than 1 ml

It's time to see how the metric system is used in the laboratory.

B. How Do You Measure Up?

What You Need

Alcohol
Ink
Oil
Water

Beakers, 50 ml, 4

Clamp
Filter paper, 4 strips
Paper towels
Pencil, lead
Ring stand
Rod
Ruler, metric
Safety goggles
Tape
Wax pencil

What to Do

a. Label the four beakers with the wax pencil as shown.

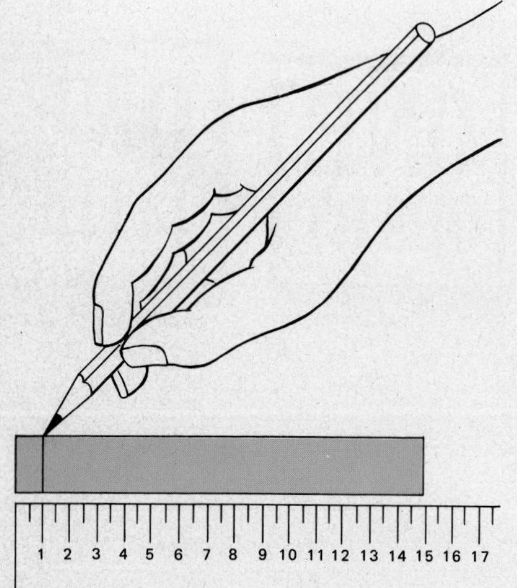

b. Place the beakers in a row next to the ring stand.
c. With a lead pencil, mark each filter paper strip 1 cm from one end.

d. Hang the unmarked ends of the strips from the clamp. Adjust the clamp so the bottoms of the strips just touch the bottoms of the beakers.

e. Pour alcohol into the beaker marked "alcohol" until it just comes up to the mark.

f. Pour ink into the beaker marked "ink" until it just comes up to the mark.

g. Pour oil into the "oil" beaker until it just comes up to the mark.

h. Pour water into the last beaker until it just comes up to the mark.

i. Wait 15 minutes.
j. Carefully remove each filter paper strip and lay it on a paper towel. Mark the name of the liquid on each paper towel.
k. Measure in centimeters the length of the stain above the mark on each strip.

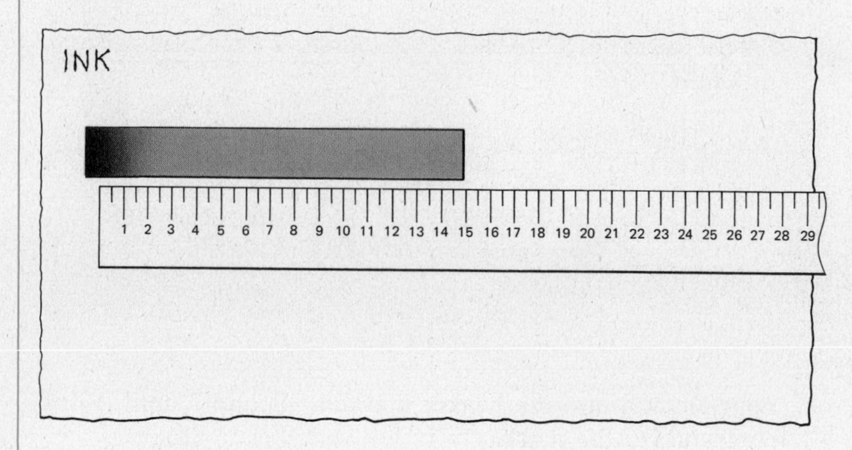

TABLE 4

	Length of stain in cm
Alcohol	
Ink	
Oil	
Water	

4. Record the measurements in a table like Table 4.
5. What is the length of the water stain in millimeters?
6. What is the length of the oil stain in millimeters?
7. How could this method be used to tell substances apart in the laboratory?

C. The Eyes Have It

What You Need

Hydrochloric acid (HCl), 0.1 *M*
Phenolphthalein indicator
Sodium hydroxide (NaOH), 0.1 *M* solution
Bottles, assorted
Dropper
Graduated cylinder, 25 ml
Test tube
Safety goggles
Test tube rack

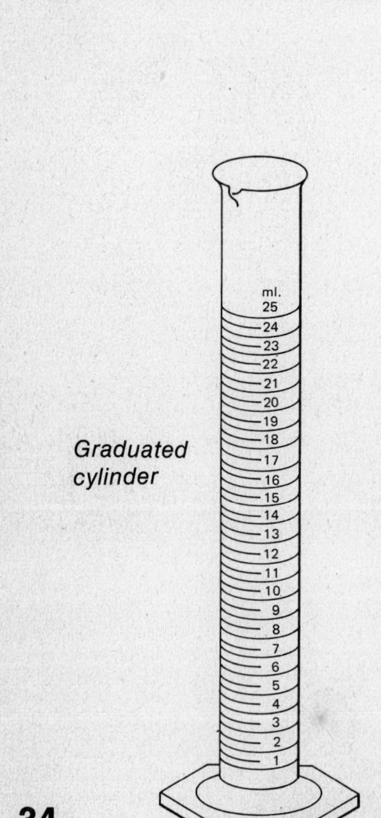

Graduated cylinder

What to Do

a. Carefully examine the graduated cylinder. It is marked in milliliters (ml) and is used to measure the volume of liquids. It is called a "graduate" for short.

b. Fill a test tube with water. Empty it into the graduated cylinder.

c. Put the graduate on the desk top. Bend down until your eyes are on the same level as the water in the graduate.

d. The curved surface at the top of the liquid is called the *meniscus.* Look at the bottom of the meniscus. The line closest to the bottom of the meniscus tells you how many milliliters of liquid are in the graduate.

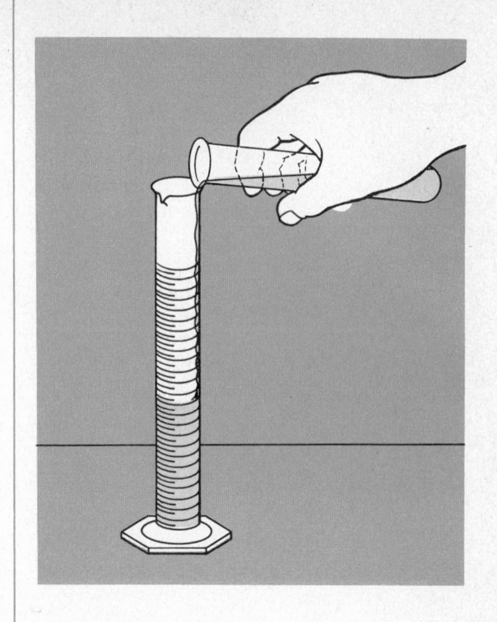

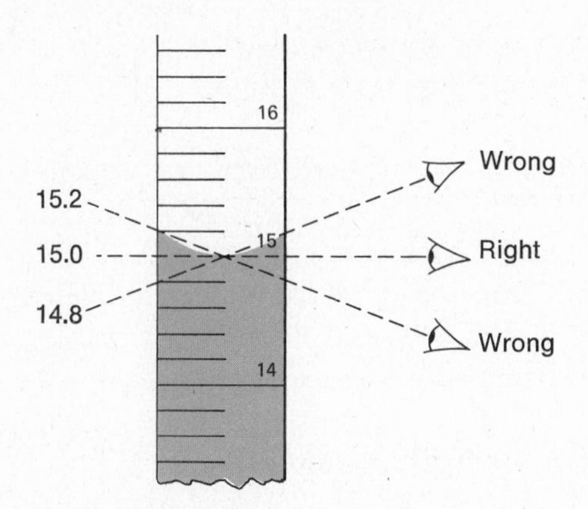

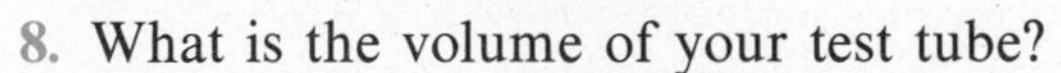

8. What is the volume of your test tube?

Now let's see just how good a guesser you are.

e. Take five or six different bottles from the front desk and arrange them according to how much liquid you think they can hold—from the largest to smallest. Number the bottles from largest to smallest.

f. Estimate the volume of each bottle in milliliters.

9. Record your estimates in a table like the one below.

TABLE 5

Bottle number	*Estimated volume in ml*	*Measured volume in ml*
1		
2		
3		
4		
5		
6		

g. Using the graduate, carefully measure how much liquid each bottle can hold.

10. Record your data in the table. How good are you at guessing?

Now see what happens when you add two chemicals together with an indicator.

h. Fill the test tube about $\frac{1}{4}$-full with sodium hydroxide.
i. Pour the liquid into the graduate.

11. How much sodium hydroxide do you have?

j. Pour the liquid back into the test tube.
k. Add 1 or 2 drops of phenolphthalein indicator to the liquid.
l. Carefully and slowly add enough hydrochloric acid to the test tube so that the color just changes.

12. What is the total amount of liquid in your test tube now?

13. How much hydrochloric acid did you use?

14. What metric unit did you use to express volume?

15. What would be the volume of the two liquids in liters?

D. The Why and Wherefore

Why measure? You measure so that . . .

. . . when you build something, you know how it's going to turn out.

Who's the wise guy who didn't use the metric system?

. . . when you buy new parts for a car, they will fit.

. . . when you read a map, you can tell how far to go.

In this investigation you've learned something about the metric system. You've studied length and volume. In the next investigation you'll find a mass of information about mass.

INVESTIGATION

6

A Pinch of This and a Pinch of That

All people have specific tools which they use in their daily jobs. Can you name the tools of the trade required by the baker? plumber? lawyer? nurse? electrician? mechanic? police officer? safecracker? teacher? grocer? chemist?

A grocer's scale is used to measure the amount of meat, potatoes, or beans. A *balance* in the chemistry laboratory is used to measure the *mass,* or amount, of chemicals. A chemist finds the mass of materials before, as well as after, experiments. He needs to know the mass to answer such questions as "How much chemical do I use in this experiment?" "How much iron is in the steel?" "How much protein is in the bread?" "How much salt is in a sample of ocean water?"

The unit used for mass in the metric system is the *gram.*

$$454 \text{ grams} = 1 \text{ pound}$$

There are many different types of balances used in laboratories. The photographs show some of the most common ones.

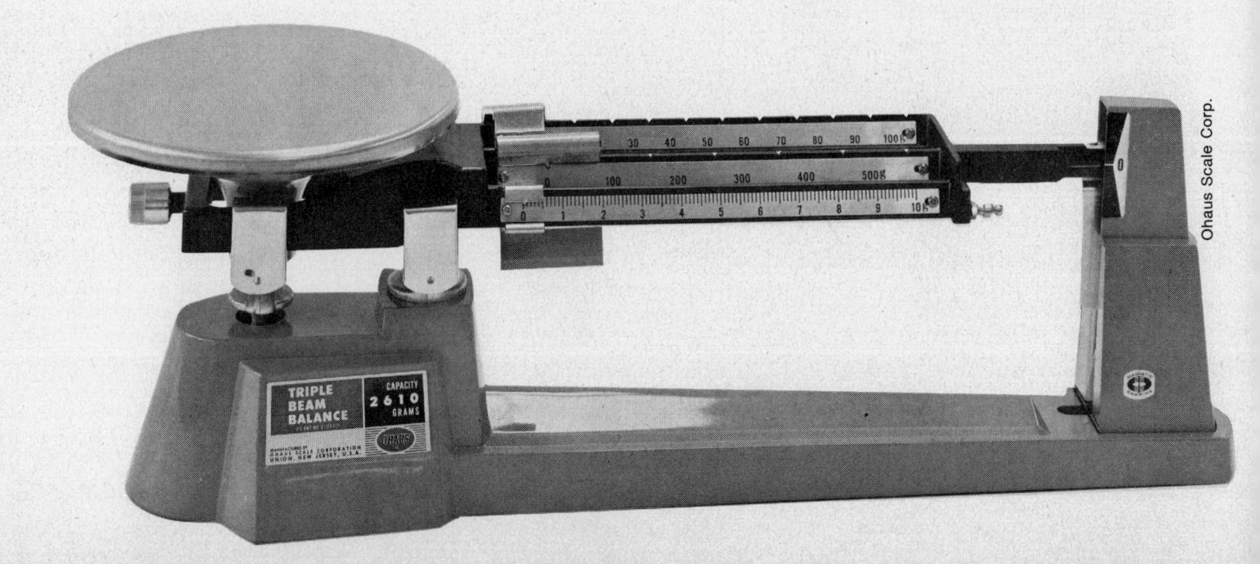

Ohaus Scale Corp.

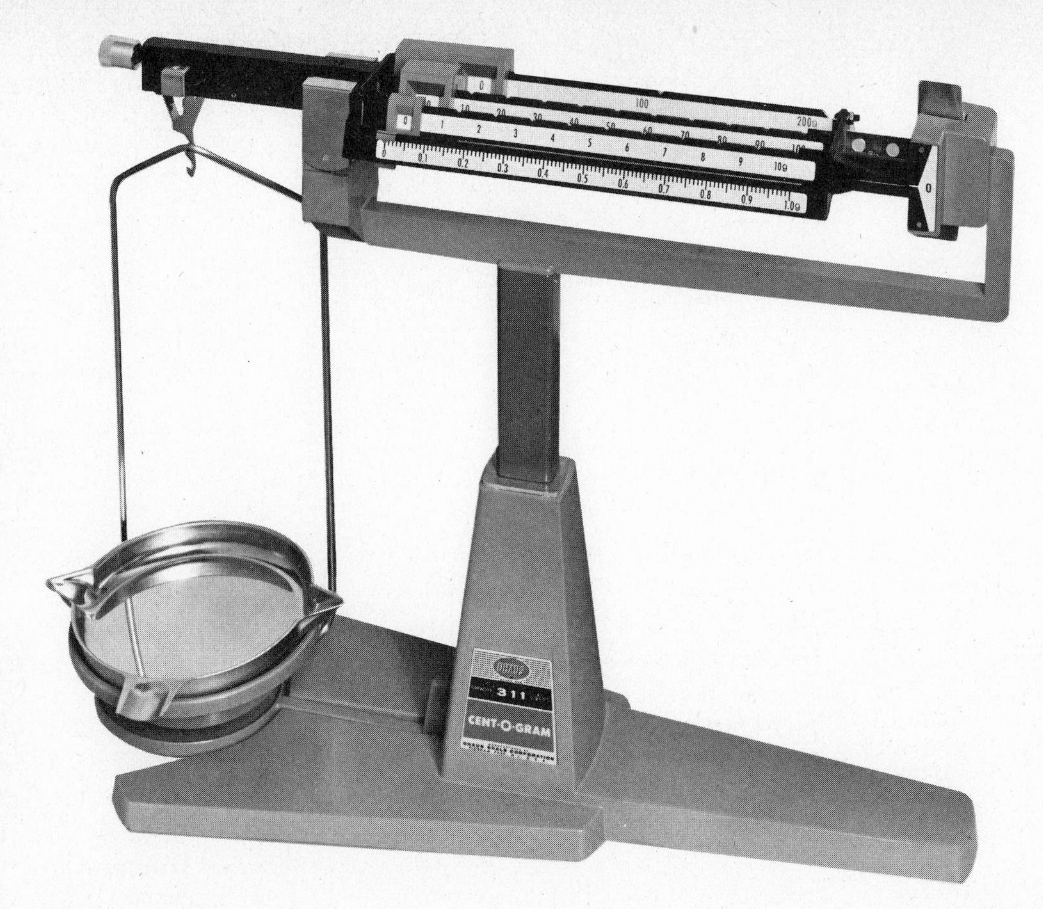

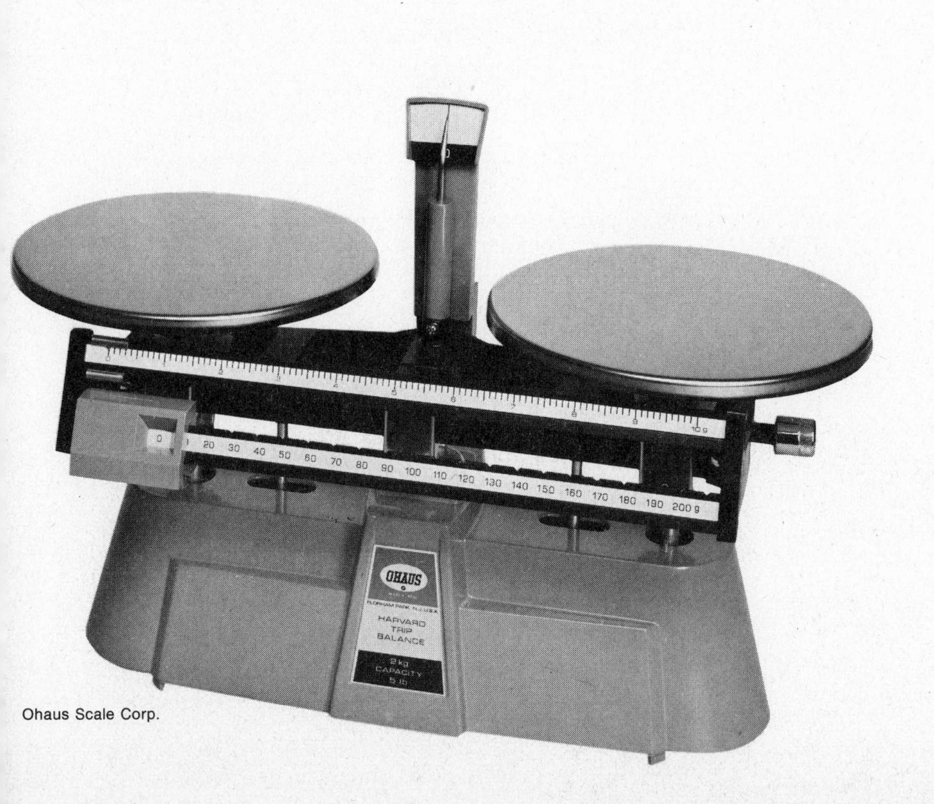

Ohaus Scale Corp.

Mettler Instrument Corp.

One common laboratory balance is the *platform balance.* Different types may have one, two, or three beams. Therefore they are called "single beam," "double beam," or "triple beam" balances. Most of these balances are accurate to the nearest 0.1 gram.

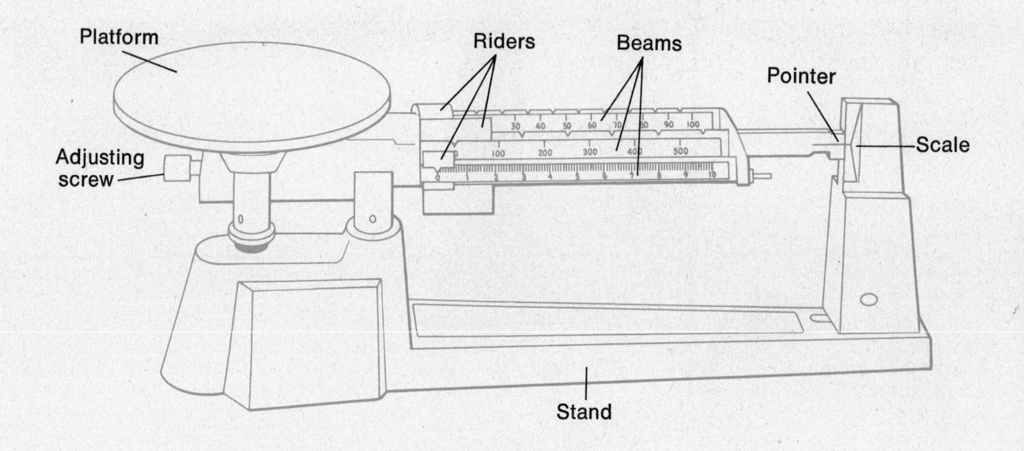

The main parts of the balance are:

a. stand
b. platform or pan
c. beam(s)
d. rider(s)
e. scale with zero point
f. pointer
g. adjusting screw

1. Identify the parts of the balance in this picture.

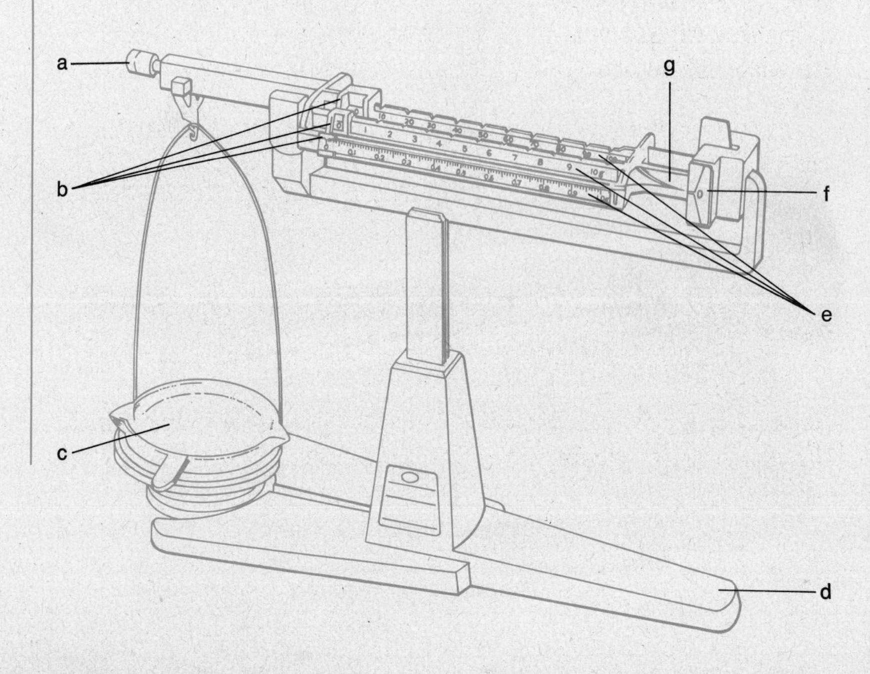

The entrance to many courthouses has a statue of "Blind Justice" holding a double pan balance. Identify the parts of this balance.

All balances have some sort of pointer and an uncalibrated scale. The center of this scale is called the zero point. Before each use, the balance is adjusted so that the pointer comes to rest as close to the zero point as possible.

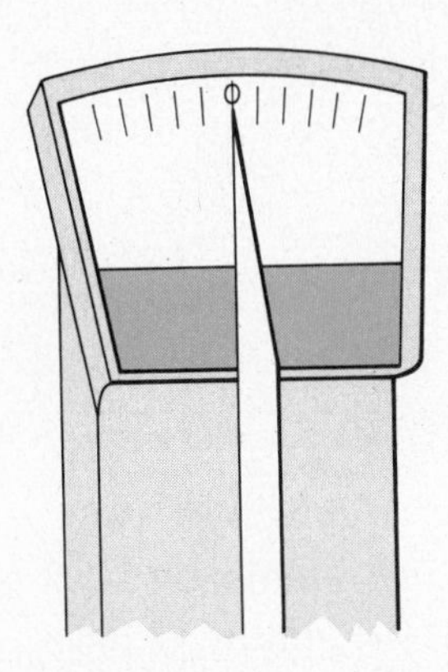

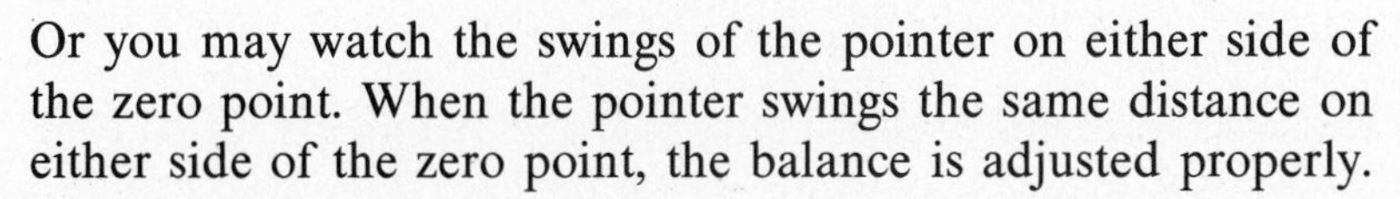

Or you may watch the swings of the pointer on either side of the zero point. When the pointer swings the same distance on either side of the zero point, the balance is adjusted properly.

2. Why must the balance be "zeroed" before use?

A. Rider Readout

One of the first things to learn is how to read a balance. On a double or triple beam balance, each beam represents a different scale. Here are the two beams of a double beam balance:

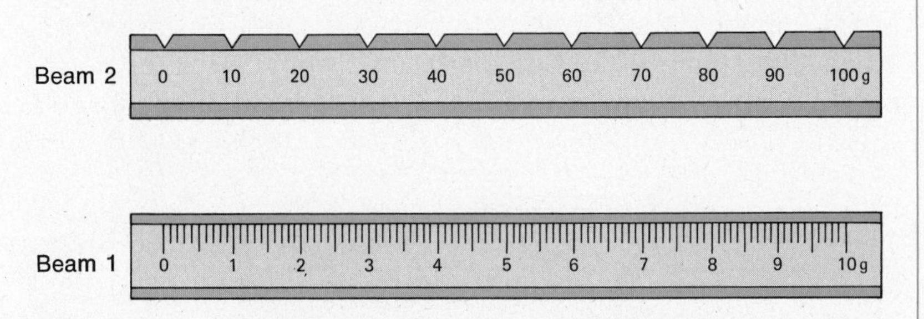

Beam 1 is for masses between 0 and 10 grams. A rider, which slides along the beam, indicates the mass which is balancing the sample on the pan. If the sample is exactly 2 grams, the pointer on the rider would point to the number 2.

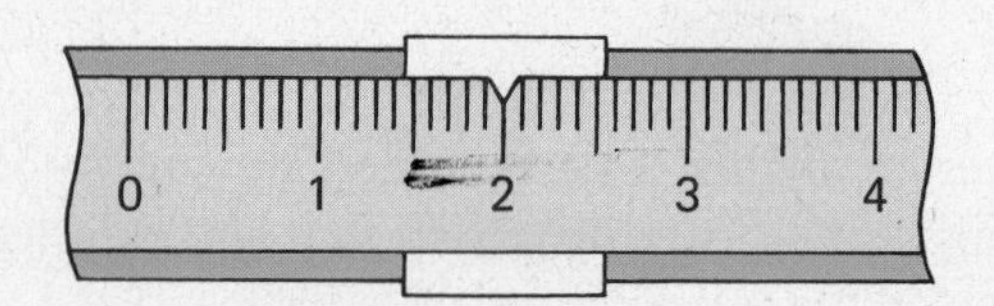

If the rider is five divisions to the right of the number 2, the reading is 2.5 grams. The mass of the sample is 2.5 grams.

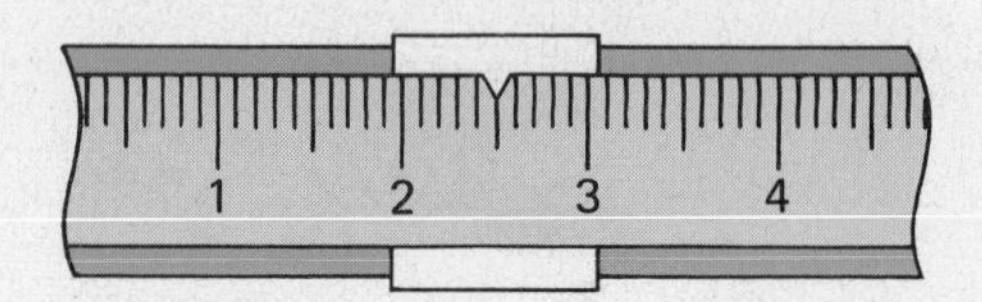

3. Find 5.5 grams, 9.7 grams, and 6.2 grams on a balance or on the diagram at the bottom of page 41. Make sketches of these readings in your notebook.

Beam 2 represents masses between 0 and 100 grams. The beam is notched at each 10-gram division. The rider fits into the notches. If the rider is in the notch for 40, the mass is 40 grams. If it is in the notch for 70, the mass is 70 grams.

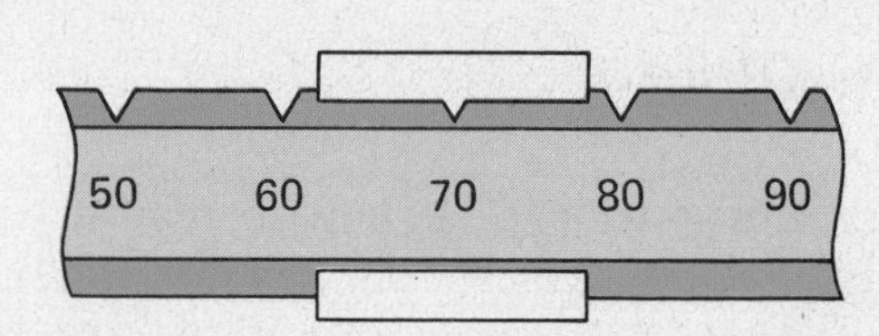

What happens if a sample has a mass of, say, 43 grams? To balance the sample, the rider on beam 2 would be in the notch for 40 grams. The rider on beam 1 would be at 3.0 grams.

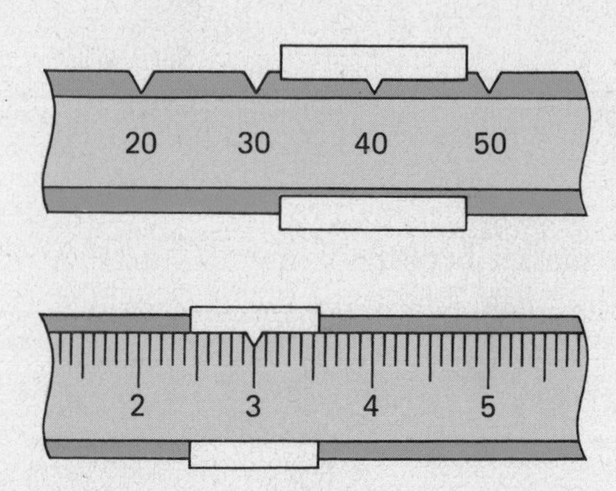

4. Where would the two riders be located to balance a mass of 36 grams?

5. Suppose an object has a mass of 72.3 grams. Where should the rider be on beam 2? Where should the rider be on beam 1?

6. Where would the two riders be located to balance a mass of 80.7 grams?

7. What mass is shown by the riders in the diagram below?

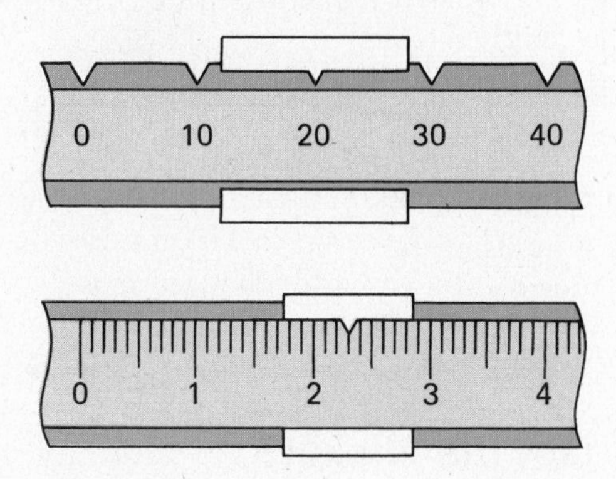

8. What mass is shown below?

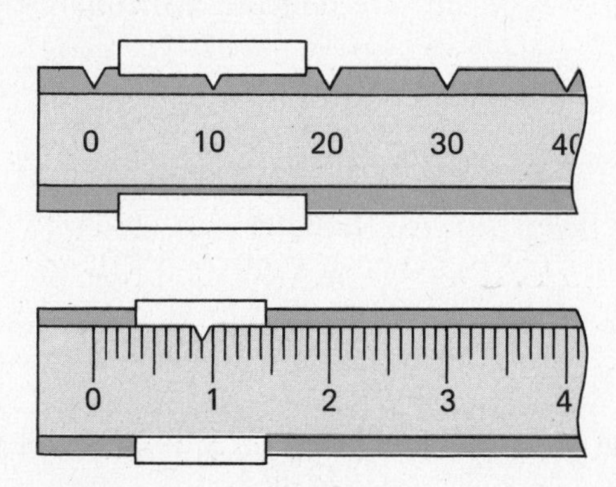

9. What mass is shown below?

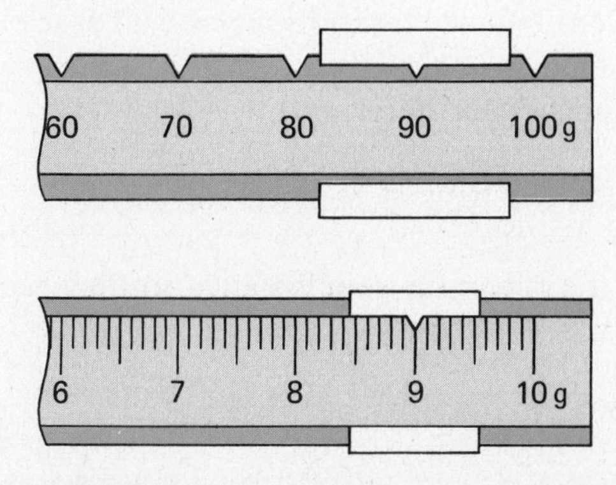

10. What mass is shown below?

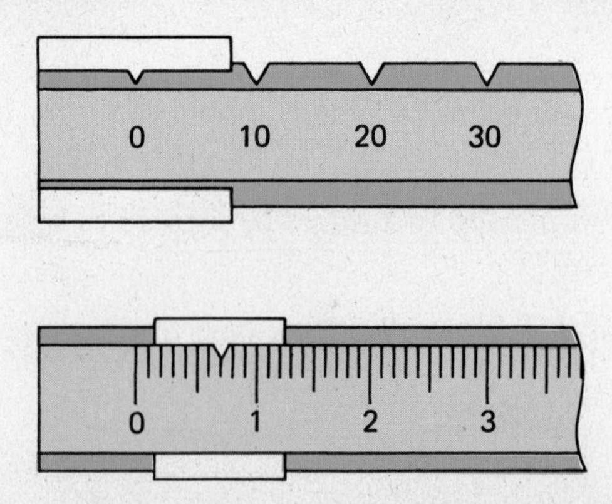

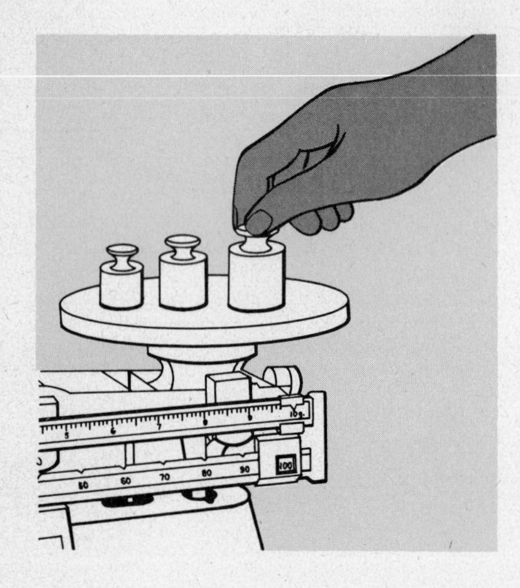

If your balance has two pans, your sample goes on the left-hand pan. You can measure masses up to 110 grams with the riders. For heavier samples, you can add standard masses to the right-hand pan.

B. Who's Got the Nickel?

It's time to get a balance and use it. Always carry the balance with two hands. Balances are delicate instruments.

What You Need

Nickels, 5 Balance

What to Do

a. Make sure your balance is set on a solid, level table.

11. Why should the balance be on a level table?

b. Examine your balance carefully. Locate all of the parts which you have seen in the pictures. If anything seems different, ask your teacher for help.

12. Describe the type of balance you have.

c. Set all of the riders on your balance so they read 0.0.

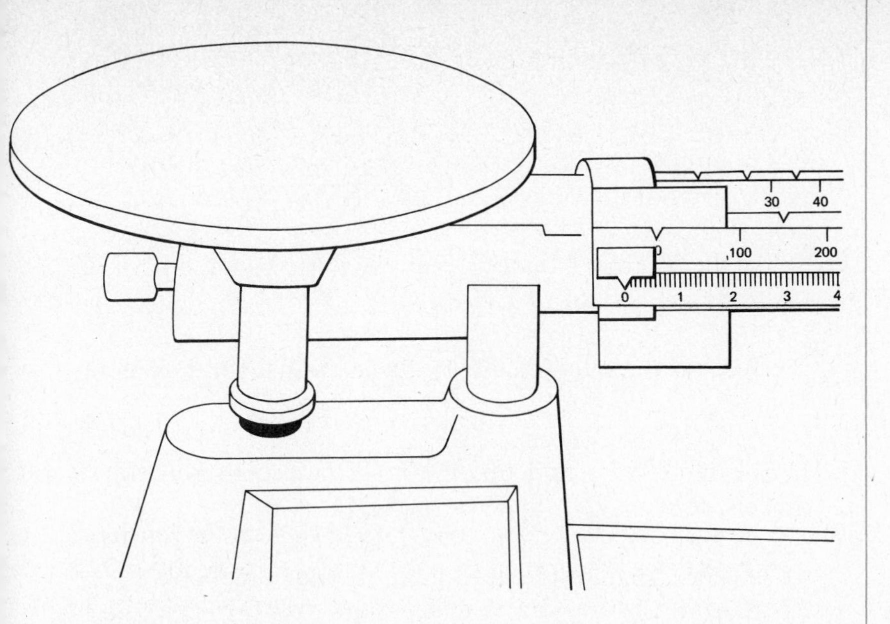

d. Gently touch the pan of your balance. This will set the pointer in motion.

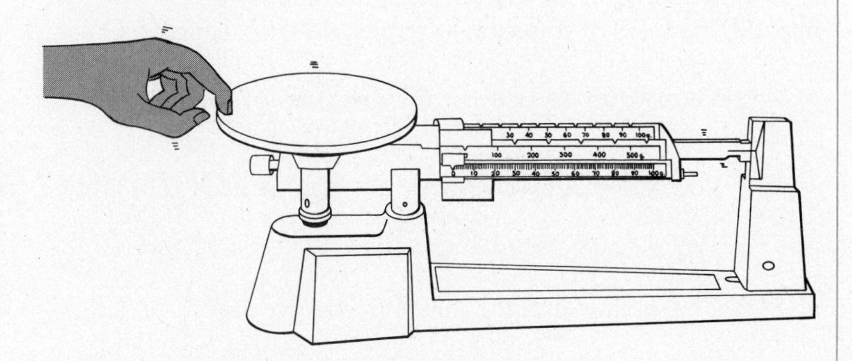

e. Allow the pointer to swing free. Wait until it comes to rest or swings equally on either side of zero. If not, let your teacher know. The balance will have to be adjusted.

13. Why is it important for the balance to indicate zero when no object is on it?

f. Put one nickel on the left pan of the balance.

14. Which way does the pointer move? Why?

g. Slowly move the rider on beam 1 to the right until the pointer begins to move.

15. If the pointer does not move, what does it mean?

h. Carefully adjust the rider until the nickel is balanced. Allow the pointer to swing free until it comes to rest at the zero point; *or* until it swings equal distances on both sides of the zero point.
i. When the nickel is balanced, read the mass from the position of the rider.

16. Record the mass of the nickel in your laboratory notebook.

j. In the same way, find out the mass of two nickels, three nickels, four nickels, and five nickels.

17. Record your results in a table in your notebook.

k. Exchange places with another student. (Do not move the balances.) Take your five nickels with you.
l. Check that your new balance is "zeroed."
m. Find the mass of your same five nickels with the new balance.

18. What value do you get for the mass of the five nickels this time?

19. What value did you get the first time? (Check your table.)

20. How do your two values compare?

n. If you have time, find the mass of your five nickels on a third balance.

21. Does it make a difference if you find the mass of something on different balances?

The comparison of your values tells you the *precision* of your results. Precision refers to how closely the values match. If you make a measurement on an object, and can make other measurements on the same object which agree with the first, the measurement is *precise.*

Accuracy is something else again! It refers to the closeness of a measurement to a standard or to an accepted value. The closer a measurement is to the standard, the more accurate it is. In the United States, the standard for mass is a platinum cylinder kept in the National Bureau of Standards in Washington.

A measurement is precise if it can be reproduced with little or no variation. But it can still be wrong. A basketball player who always hits the front rim of a basket is precise, but not accurate. He is precisely inaccurate!

C. Can a Heavyweight Become a Lightweight?

You've just found the mass of some solid objects. How do you find the mass of a liquid?

What You Need

Potassium iodate (KIO_3), 0.1 *M* solution	Beaker, 100 ml	Balance
Sodium bisulfite ($NaHSO_3$), 0.1 *M* solution	Test tubes, 2	Safety goggles
		Test tube rack
		Wax pencil

What to Do

a. With your wax pencil, mark one test tube **A** and the other **B**.
b. Place both tubes in the beaker.
c. Find the mass of the beaker with test tubes in it.

22. Record the mass in your notebook.

d. Remove the test tubes from the balance. Place them in the test tube rack.
e. Fill tube **A** $\frac{1}{3}$-full with potassium iodate solution.
f. Fill tube **B** $\frac{1}{3}$-full with sodium bisulfite solution.
g. Put both tubes back in the beaker.
h. Find the mass of the test tubes, beaker, and solutions.

23. Record the mass in your notebook.

i. Subtract your answer to question **22** from your answer to question **23.** This gives you the mass of the two solutions.

24. Record the mass of the two solutions.

The mass of the container is called the *tare.* To get the mass of the solutions, you subtracted the *tare* from the mass of the whole set-up.

j. Pour the contents of tube **A** into tube **B**.

25. What happens?

26. Is this a chemical reaction?

k. Put both test tubes back in the beaker.
l. Find the mass again.
m. Subtract the tare.

27. Record the mass of the solution.

28. How do your answers to questions 24 and 27 compare?

29. When a chemical reaction occurs, does the total mass of the substances change?

30. List the ways you can tell that a chemical reaction has taken place.

You're really turning into a chemist! The next investigation will give you a few more tools of the trade.

INVESTIGATION 7

It's All in How You Read It

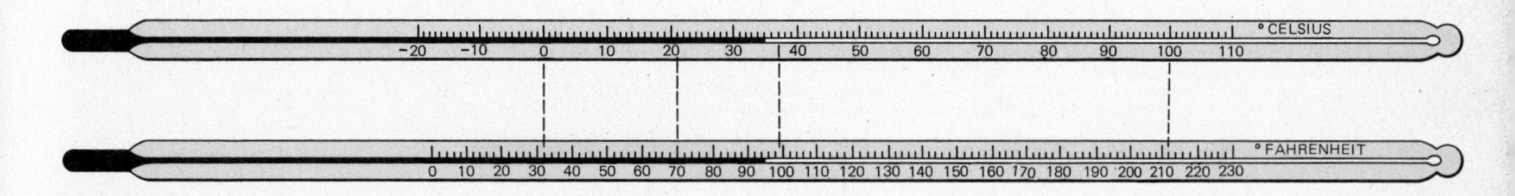

Another tool of the chemistry trade is the *thermometer.* Thermometers are instruments used to measure temperature. There are two types of thermometers in general use. In everyday life we use the Fahrenheit (F) scale. On this scale, water freezes at 32 degrees Fahrenheit (32°F) and boils at 212°F. Normal body temperature is 98.6°F, and comfortable room temperature is around 70°F.

One of us has got to be wrong!

In the laboratory we use the Celsius (C) scale. Look at the diagram below.

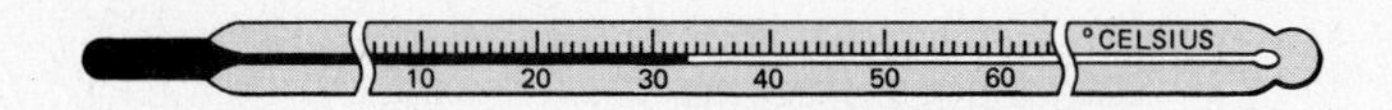

1. What is the temperature shown on the thermometer?

If you said 33, you're wrong. The number 33 by itself means nothing. It could mean 33 apples or 33 desks or 33 degrees Celsius or 33 degrees Fahrenheit. What's missing is the proper *unit.* Any time you give a measurement you must give the unit of measurement as well. You already know the metric units for length, volume, and mass. The units for temperature in the metric system are "degrees Celsius"—abbreviated as °*C*.

Your answer to question 1 should have been 33°C, which is read "33 degrees Celsius."

2. On the Celsius scale, what is:
   a. the freezing point of water?
   b. the boiling point of water?
   c. body temperature?
   d. room temperature?

The marks on a thermometer are usually placed at one-degree intervals. But you can read the thermometer more accurately than that. Look at the following drawing:

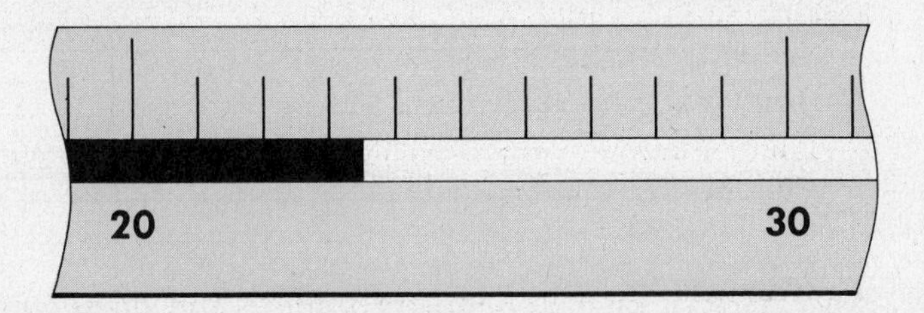

3. What is the temperature shown?

You probably said $23\frac{1}{2}$°C, or 23.5°C. Both are correct, although it is preferable to use decimals rather than fractions in scientific work. With practice, you should be able to estimate most temperatures to one decimal place. Try these:

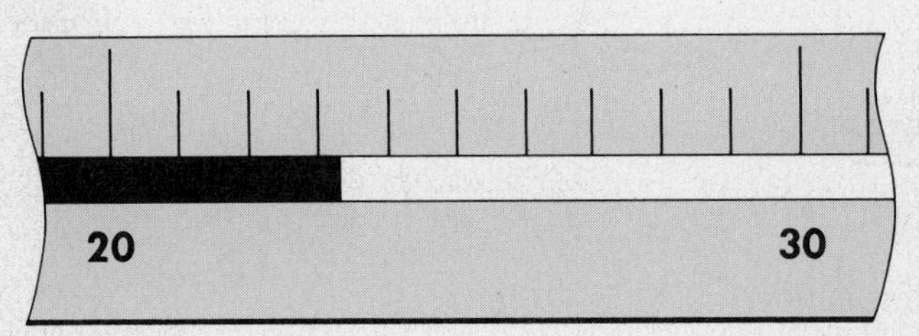

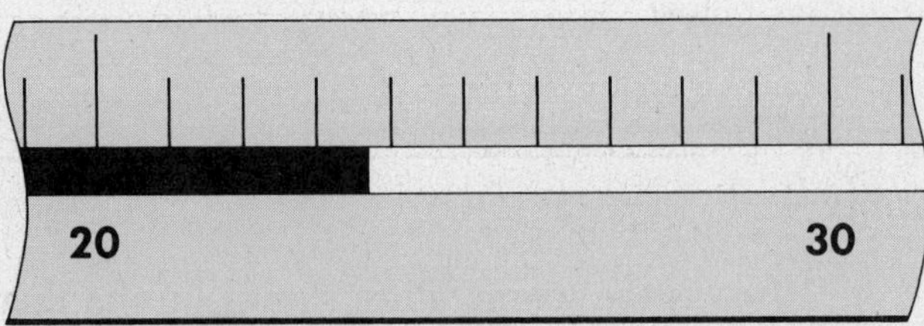

4. What is the temperature shown on the left?
5. What is the temperature shown on the right?

Are you ready for some lab work? Okay, here it is.

A. The Heat's On

What You Need

Sodium hydroxide (NaOH) pellets
Beaker, 50 ml
Graduated cylinder, 25 ml
Stirring rod
Thermometer
Watch glass
Forceps
Safety goggles

What to Do

Let's pause for a few words about handling thermometers

—Thermometers are quite breakable.
—They contain mercury, which is a poison.
—Never use a thermometer as a stirring rod.
—Don't wave a thermometer around or carry it carelessly.
—Thermometers are usually cylindrical. They roll. So don't place a thermometer where it can roll off the desk.

a. Measure out 25 ml of room temperature water in the graduated cylinder. Pour it into the beaker.

6. Copy Table 1 into your notebook.

TABLE 1

Number of pellets of sodium hydroxide	*Temperature in °C*
0	
2	
4	
6	
8	
10	

b. Using your forceps, select 10 sodium hydroxide pellets that are nearly the same size. Place them on the watch glass. **Caution: Sodium hydroxide burns. Do not get any on your skin.**

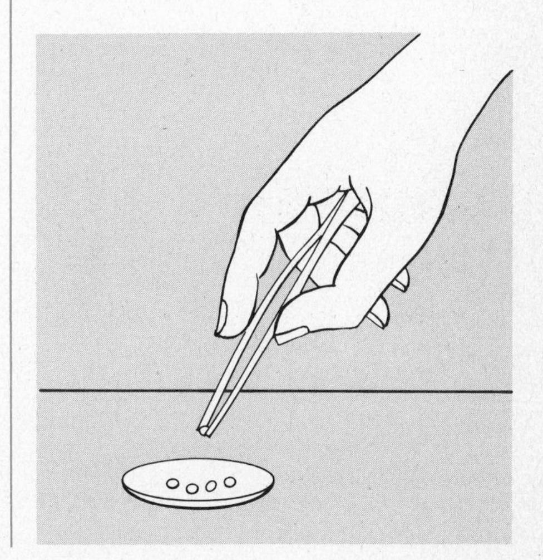

c. Measure the temperature of the water. Always hold the bulb in the center of the water. Do not let it touch the glass. Then remove the thermometer.

7. Record the water temperature in your table.

d. Using the forceps, place two pellets in the beaker.
e. Stir with the stirring rod until the pellets dissolve.
f. Measure the temperature in the beaker. Then remove the thermometer.

8. Enter your results in the table.

g. Add two more pellets to the beaker. Stir until they dissolve.
h. Measure the temperature. Remove the thermometer.

9. Record the temperature in the table.

i. Repeat steps **g** and **h** until you have added a total of 10 pellets to the beaker.

10. Record all your results in the table.

11. As you added sodium hydroxide to the beaker, what happened to the temperature?

12. Do you think a chemical reaction took place in the beaker? Why?

B. The Graphic Approach

Tables are certainly useful in organizing data. But what if you wanted to know what the temperature would be with seven pellets, or nine pellets? Can you tell from your table? Not easily. But if you *graph* your data, it's easier.

A *graph* is often the clearest way to show information about numbers. A graph shows a relationship between two sets of numbers. In the case of your sodium hydroxide experiment, there are two sets of numbers in your table. One set shows the number of pellets of sodium hydroxide; the other set shows temperature.

All graphs have the same basic parts:

a. a horizontal axis, also called the x-axis (the line that runs from left to right)
b. a vertical axis, also called the y-axis (the line that runs up and down)
c. labels for the axes
d. a grid (section divided into squares)
e. a title

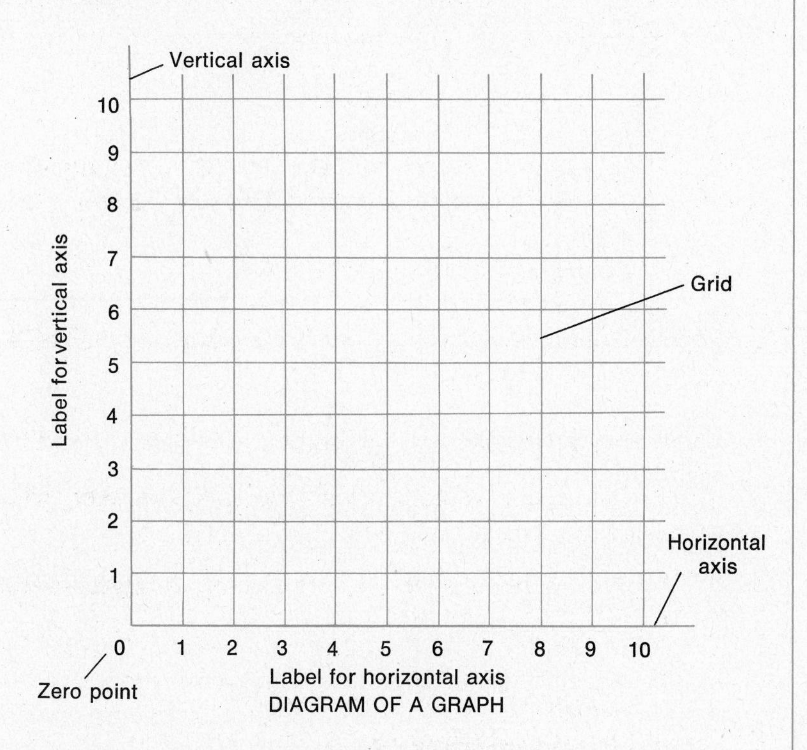

DIAGRAM OF A GRAPH

Now make a graph of the data you collected in Part A.

What You Need

Graph paper
Pencil
Ruler
Tape

What to Do

a. Draw a horizontal axis and a vertical axis on your graph paper.

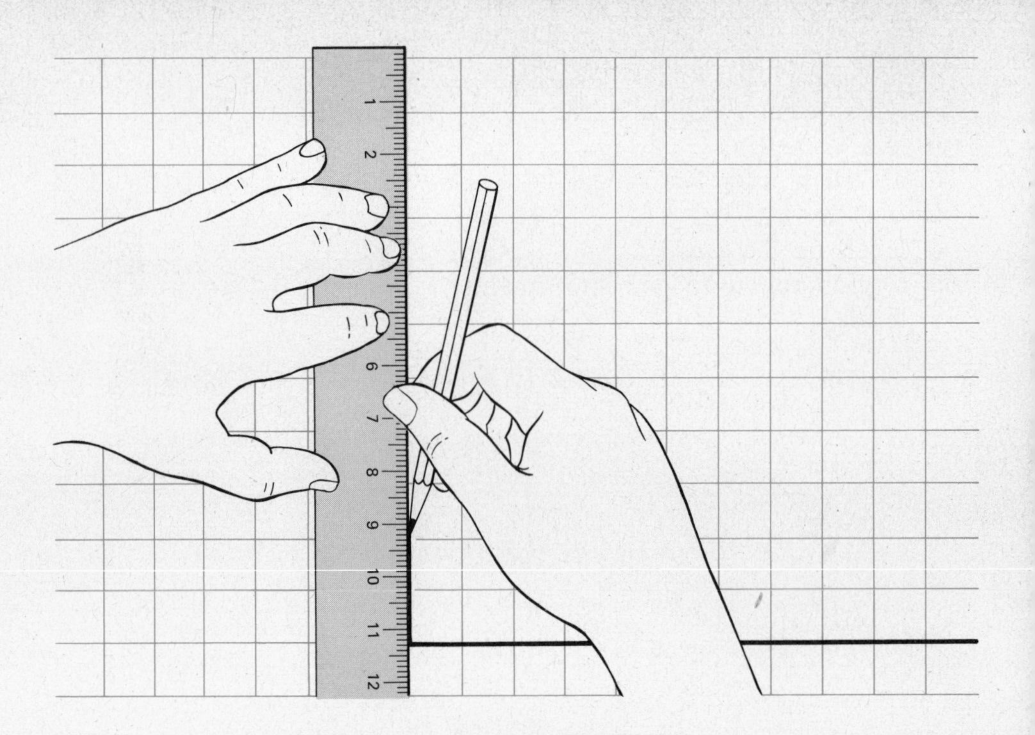

b. Label the horizontal axis "Number of pellets of sodium hydroxide."

c. Write the numbers 0 through 12 along this axis.

d. Label the vertical axis "Temperature in °C."

e. Begin numbering the vertical axis at 15°C. Number up to 40°C.

f. Title your graph.

g. In Table 1, find the temperature which corresponds to "0" sodium hydroxide pellets. Follow the 0-pellet line up on your graph until you come to this temperature and place a small *x* there.

h. Find the temperature which corresponds to 2 pellets. Follow the 2-pellet line up on your graph until you come to the spot that is opposite this temperature. Place a small *x* there.

i. Plot the other four points.

j. Draw a smooth line through your six points.

You now have a *line graph.* The line you drew is called a *curve.*

k. To find out the temperature that correspond to 7 pellets, follow up the 7-pellet line until you come to your curve. From the place where these two lines *intersect,* run your finger left and read the temperature on the vertical axis.

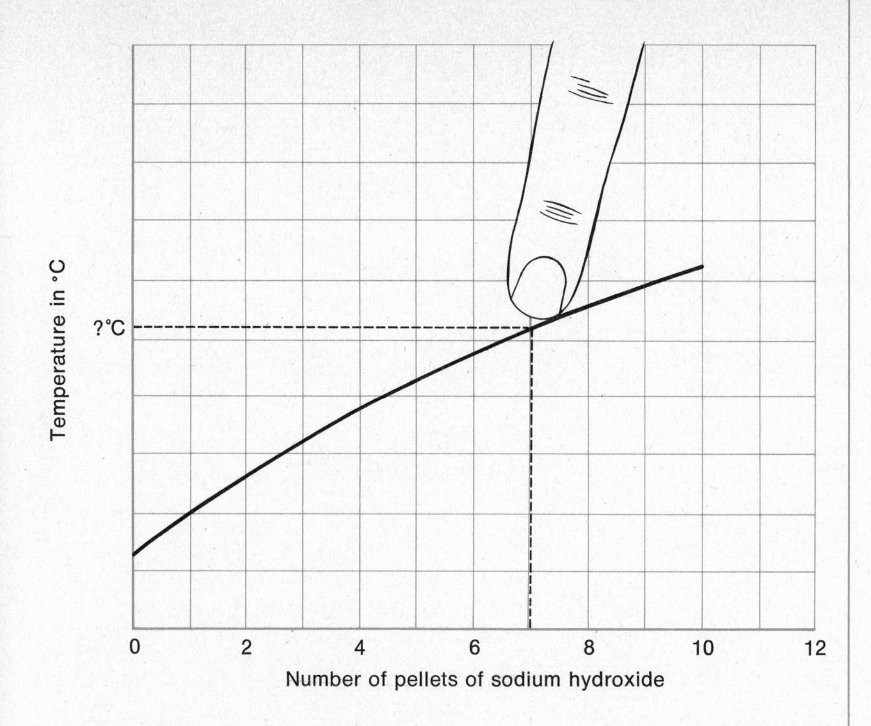

This is called *interpolation*—finding a point between two measured points on a graph.

13. What temperature would correspond to nine pellets in your experiment?

If you get a nice, smooth curve you might extend it slightly beyond the 10-pellet point. You could then *predict* what the temperature would be with 11 pellets. This is called *extrapolation*—finding a point that's outside all your measured points.

14. With tape, fasten your graph into your notebook.

15. Using your graph, what would be the temperature in the beaker if you used 11 pellets?

16. If the temperature in the beaker had been 31°C, how many pellets would you have used?

Sodium hydroxide is a main ingredient of drain cleaners.

17. How do you think drain cleaners work?

Speaking of drain cleaners—check with your teacher about disposing of your sodium hydroxide.

C. One More Time

Do you think all substances will raise the temperature of a beaker of water? Try the experiment in Part A again, this time using ammonium chloride.

What You Need

Ammonium chloride (NH_4Cl)

Beaker, 50 ml
Graduated cylinder, 25 ml
Stirring rod
Thermometer

Balance
Safety goggles
Spatula
Weighing paper

What to Do

a. Place 25 ml of room temperature water in the beaker.

18. Copy Table 2 into your notebook.

TABLE 2

Mass of ammonium chloride in grams	*Temperature in °C*
0	
1	
2	
3	
4	
5	

b. On five pieces of weighing paper, get five 1-gram samples of ammonium chloride.
c. Measure the temperature of the water. Then remove the thermometer.

19. Record the temperature in Table 2.

d. Add 1 gram of ammonium chloride to the beaker. Stir carefully until it dissolves.
e. Measure the temperature. Then remove the thermometer.

20. Record the temperature in the table.

f. Repeat steps **d** and **e** with the other four samples of ammonium chloride.

21. Record your data.

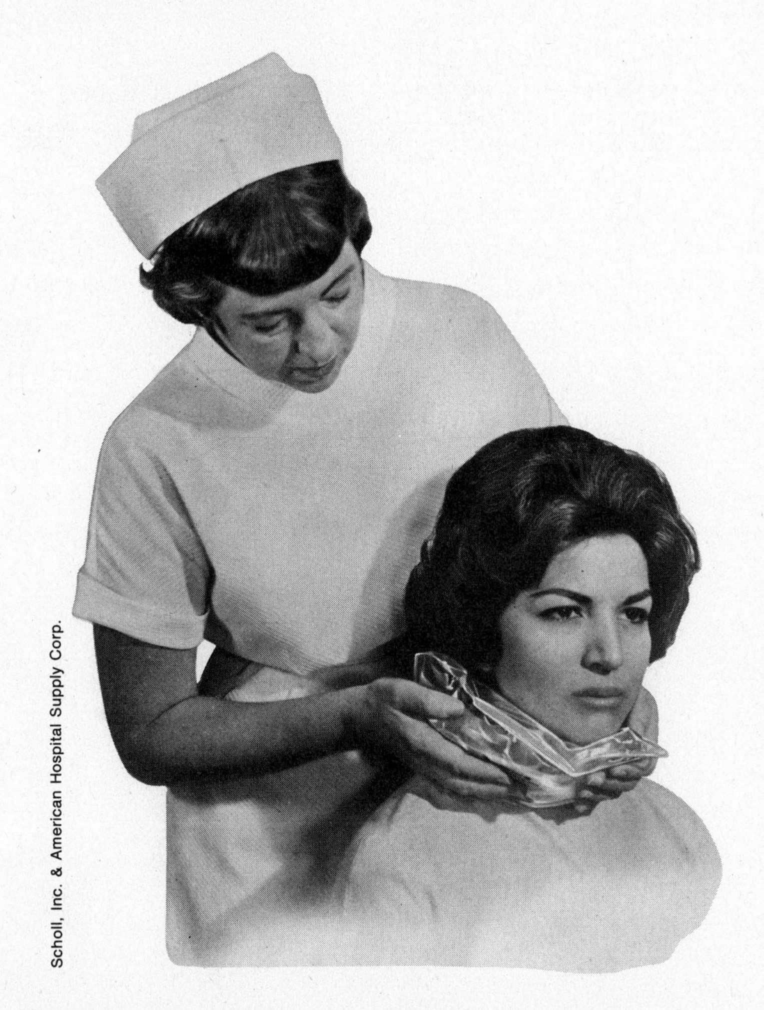

Scholl, Inc. & American Hospital Supply Corp.

Were you surprised that the temperature went down? Ammonium chloride is used in cold packs for people who suffer sprains.

22. Write a report of your ammonium chloride experiment. Include a properly labeled graph.

23. What would have been the temperature with 1.5 grams of ammonium chloride?

24. What would have been the temperature with 6 grams of ammonium chloride?

D. So You Think You're Done

Go back to Investigation 6.

25. Make a graph of your nickel data in Part B.

26. What would be the mass of seven nickels?

You've made a number of graphs in this investigation. You will be using more graphs in the laboratory to give you different kinds of information. But let's back up just a bit before going on.

27. What happened when you mixed
    a. sodium bicarbonate and acetic acid?
    b. carbon dioxide and limewater?
    c. potassium iodate and sodium bisulfite?
    d. sodium hydroxide and water?
    e. ammonium chloride and water?

28. How can you tell if a chemical reaction has occurred?

INVESTIGATION

8

Let's Stroll Down Memory Lane

Do the following activities and see if you can recognize whether any chemical reactions occur. Afterwards, write a report discussing whether or not a chemical reaction has taken place in each activity.

A. Do Your Thing

What You Need

Hydrochloric acid (HCl), 1 *M*
Lead(II) nitrate [$Pb(NO_3)_2$], 0.1 *M* solution
Methyl orange indicator
Potassium chromate (K_2CrO_4), 0.1 *M* solution
Zinc (Zn) metal
Dropper
Graduated cylinder, 10 ml
Stirring rod
Test tubes, 3
Safety goggles
Spatula
Test tube rack
Wax pencil

What to Do

a. With a spatula, put a small piece of zinc, about the size of a pea, into a test tube.

b. With a graduate, add 2–3 ml of hydrochloric acid to the test tube.

1. What happened?

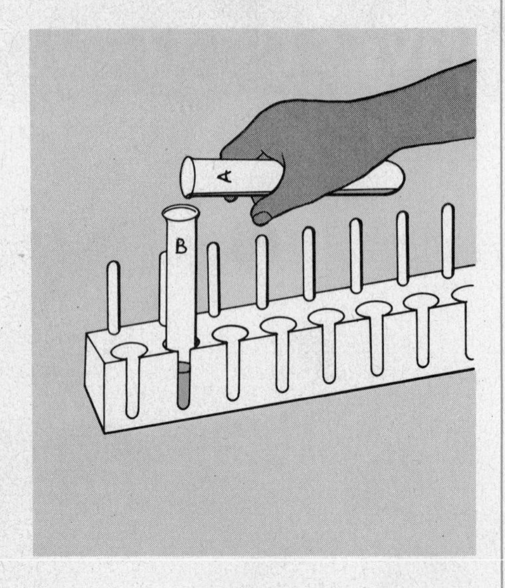

2. What other chemicals that you have worked with reacted in the same way?

c. Label two test tubes **A** and **B**.
d. Add about 2 ml of lead nitrate solution to tube **A** and place it in a test tube rack.
e. Add about 2 ml of potassium chromate solution to tube **B** and place it in the test tube rack.
f. Pour the lead nitrate solution into the potassium chromate solution.

3. Record your observations.

4. What other chemicals that you have worked with have given you a precipitate?

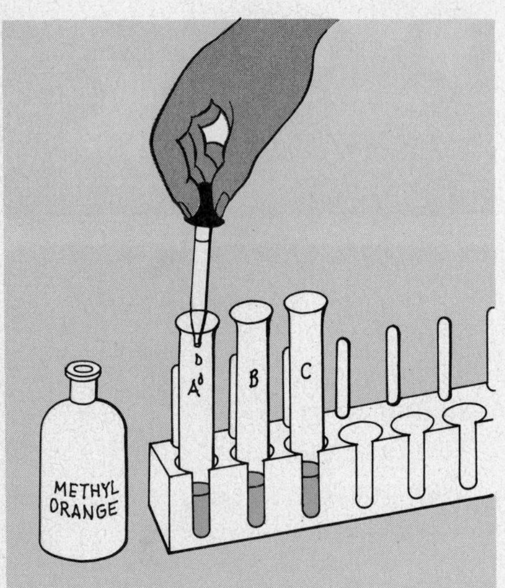

g. Clean your test tubes. Place three tubes in the test tube rack.
h. Label the tubes **A, B,** and **C.**
i. Into tube **A,** pour about 2 ml of hydrochloric acid.
j. Into tube **B,** pour about 2 ml of lead nitrate solution.
k. Into tube **C,** pour about 2 ml of potassium chromate solution.
l. With a dropper, put a few drops of methyl orange indicator into each tube.

5. Record your observations.

B. Onward, Ever Onward

You're practically a pro! Now, to sharpen those lab skills, try these:

What You Need

Ammonium nitrate (NH_4NO_3)	Beaker, 50 ml	Balance
Calcium chloride ($CaCl_2$)	Graduated cylinder, 25 ml	Safety goggles
Lead(II) nitrate [$Pb(NO_3)_2$]	Stirring rod	Spatula
Potassium chromate (K_2CrO_4)	Test tubes, 3	Test tube rack
Zinc (Zn) metal	Thermometer	Wax pencil
		Weighing paper

What to Do

6. Copy Table 1 into your notebook.

TABLE 1: Temperature Changes

	NH_4NO_3	$CaCl_2$
Initial temperature		
Final temperature		
Temperature change		

a. Get 2 grams of ammonium nitrate. Be sure to use weighing paper.
b. Measure 25 ml of water and pour it into a 50 ml beaker.
c. Take the temperature of the water in the beaker. Then remove the thermometer.

7. Record the temperature in Table 1 opposite "Initial temperature" and under "NH_4NO_3."

d. Add the 2 grams of ammonium nitrate to the beaker. Stir with the stirring rod until it dissolves.
e. Measure the temperature again. Then remove and rinse the thermometer.

8. Record the temperature opposite "Final temperature" and under "NH_4NO_3."

9. Subtract and record the temperature change.

f. Repeat steps **a** through **e** with calcium chloride.

10. Record all your results in the table.

11. What chemical reactions have occurred so far in Part B?

g. Place three clean test tubes in the test tube rack.
h. Label the tubes **A, B,** and **C.**
i. Pour 10 ml of water into each test tube.
j. Into tube **A,** put a small piece of zinc metal, the size of a pea.
k. Into tube **B,** put 0.5 grams of lead nitrate.
l. Into tube **C,** put 0.5 grams of potassium chromate.

m. With a clean stirring rod, stir the contents of each test tube.

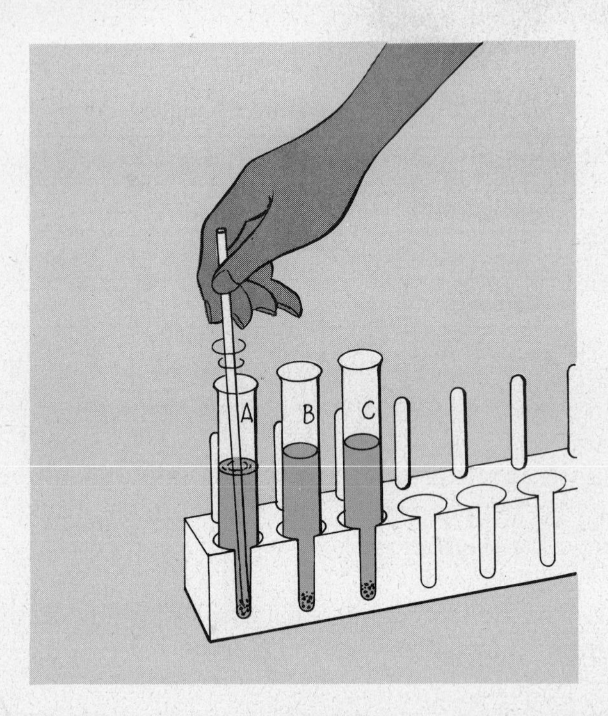

12. Record your observations.

13. If ammonium nitrate had been in test tube **A** instead of zinc, would a chemical reaction have taken place? How do you know?

14. Using all your observations from this investigation, write a report on how to recognize a chemical reaction. Be sure to use specific examples.

It's time to take a breather. Let's take stock of what you have learned so far.

a. Chemistry deals with chemical reactions.
b. Laboratory results are written up as reports.
c. There are special ways to do things safely in the chemistry laboratory.
d. The metric system is used for measurement in science.
e. The balance is used to find the mass of materials.
f. Graphs simplify the interpretation of information.
g. There are a number of ways to recognize a chemical reaction.

. . . and he was such a promising chemistry student!

Now that you've completed the skeleton, it's time to put some flesh on the bones.

Unit 2 will get down to basics and show you why things tick.

Home Activities

Investigation 1

1. Make a list of the chemicals you can find in your:
   (a) kitchen.
   (b) clothes closet.
   (c) laundry.
   (d) work room.

2. Using the classified section of your newspaper, make a list of ten jobs in which a knowledge of chemistry would be helpful. For each job, explain how.

Investigation 2

1. Draw up a table that could be used to show the results of testing salt ($NaCl$), sugar ($C_6H_{12}O_6$), and starch [$(C_6H_{10}O_5)_n$] with acetic acid (CH_3COOH), ammonia water (NH_4OH), and methyl red indicator.

2. Test salt, sugar, and starch for their solubility in water. Record your results in a table.

Investigation 3

1. Read a science fiction story (*e.g., Frankenstein,* by Mary Shelley) and write a report on the scientific experiments described in it.

2. Write a report on an experiment which has as its Purpose, "To fry one egg sunny-side up."

Investigation 4

1. Make a list of household items which serve the same function as laboratory equipment. State which piece of lab equipment is like each item.

2. Make a list of chemical reactions that go on every day in your house.

Investigation 5

1. Measure the perimeter of your kitchen in meters. Calculate the volume of your kitchen in cubic meters.

2. Make a list of items found in your home on which the units are given in the metric system.

Investigation 6

1. Construct a balance.

2. Make a set of standard weights for your balance.

Investigation 7

1. Draw up plans for making a thermometer. How would you calibrate your thermometer?
2. Does adding salt to water change its boiling point? Do an experiment. (Make sure you use the right kind of thermometer. Check with your teacher first.)

Investigation 8

1. Visit a laboratory or shop where some kind of chemical processing goes on and give a report to the class on what goes on there.
2. Interview a chemist or technician and write a description of his or her job.

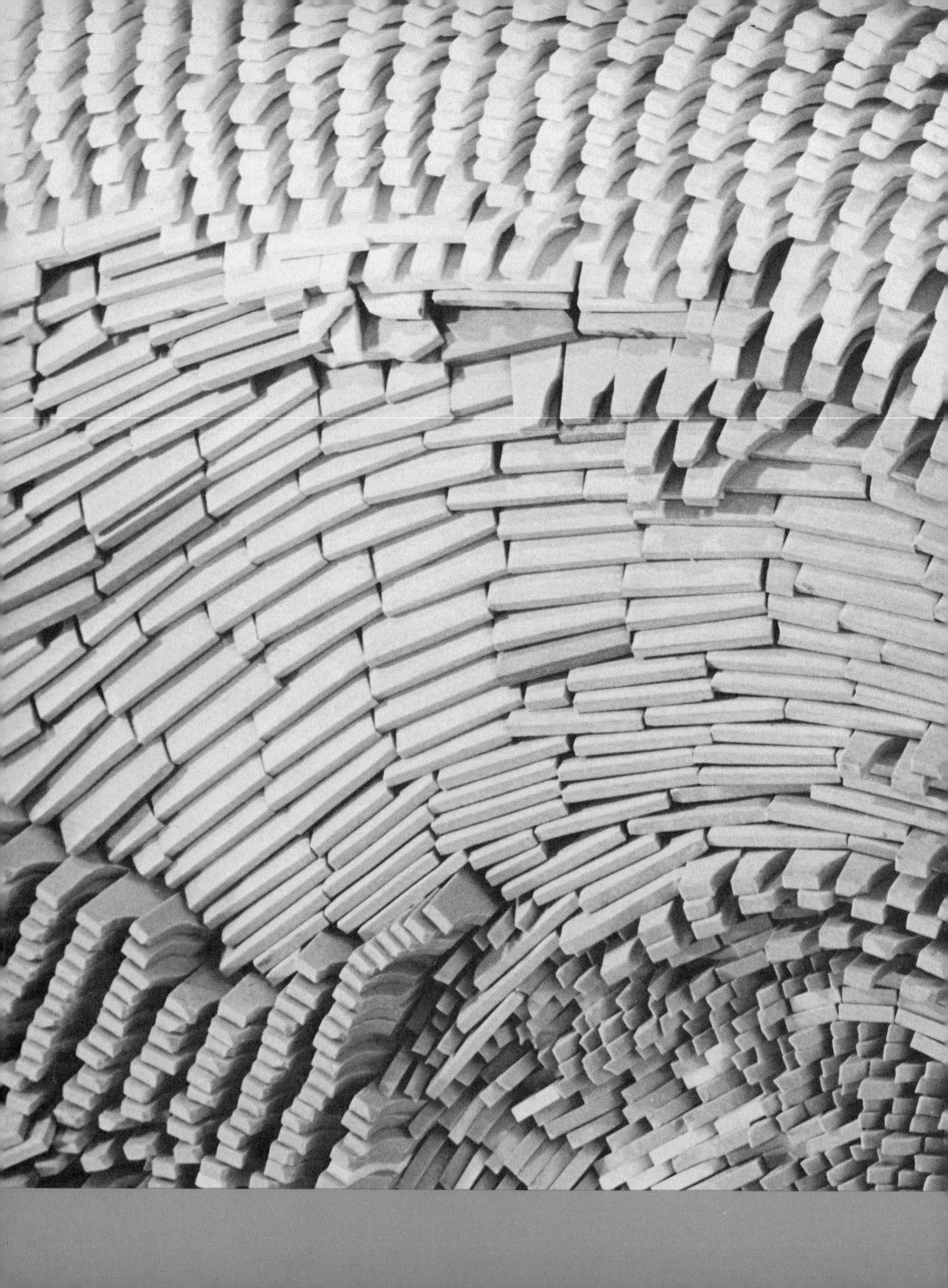

De Wys

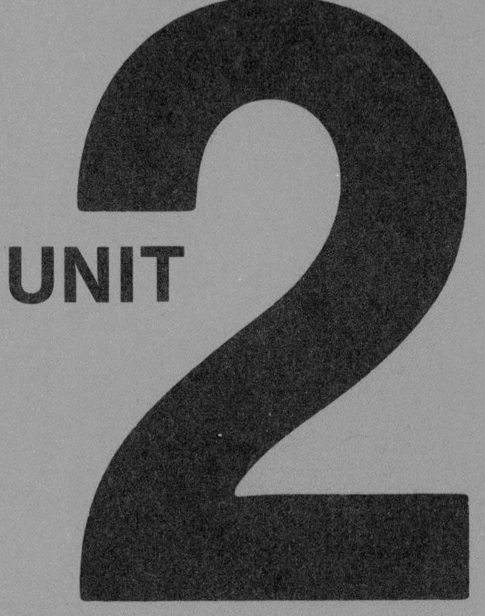

INVESTIGATION 1

Which Property Is Best?

In the first Unit you used many different substances. It was easy to tell that some of the substances were different. However, some looked the same.

1. What are some ways of telling that two substances are different?
2. List some differences between milk and water.
3. Are the metals copper and steel the same? Why?

Sometimes we have to look very closely at objects to tell if they are the same or different. For example, butter and margarine can be the same color. They both melt in warm weather and are both used for the same purpose. But if you tested these two substances you would find that butter comes from cow's milk. Margarine is made from vegetable oils. You might also notice that butter and margarine have slightly different tastes.

Now this is what I call a hot property!

A. Some Like It Hot

In the following experiment you will be given three liquids that look the same. Your job will be to find out whether these liquids are the same or different. To do this you will measure their *boiling points.* A liquid is boiling when bubbles rise to the surface and break. The boiling point is a specific temperature. You will know it has been reached when the temperature stops rising.

What You Need

Liquids **A, B,** & **C**

Graduated cylinder, 10 ml
Test tubes, 3
Thermometer

Boiling chips
Burner
Cork
Matches
Ring Stand
Safety goggles
Test tube clamps, 2
Test tube rack
Wax pencil

What to Do

a. Label three test tubes **A, B,** and **C.**
b. Put 5 ml of liquid **A** into tube **A.**
c. Put 5 ml of liquid **B** into tube **B.**
d. Put 5 ml of liquid **C** into tube **C.**
e. Observe the three liquids.

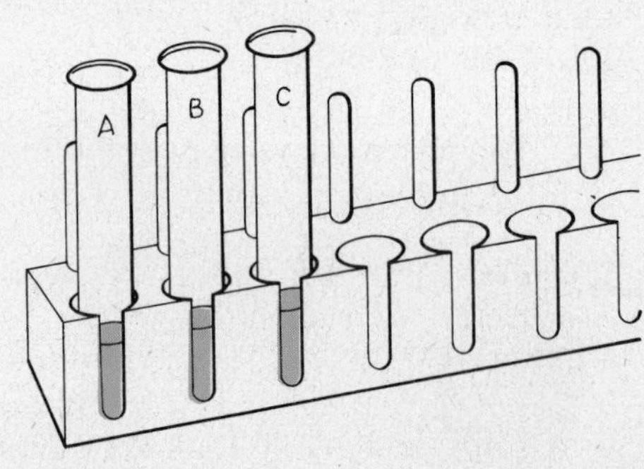

4. Do they look the same?

5. Do you think they are the same or different?

f. Add a boiling chip to liquid **A.**
g. Clamp the test tube to a ring stand. Hold the thermometer in the middle of the liquid.
h. Light the burner. Bring the liquid **slowly** to a boil. Use a **low** flame. Note the drawing below.

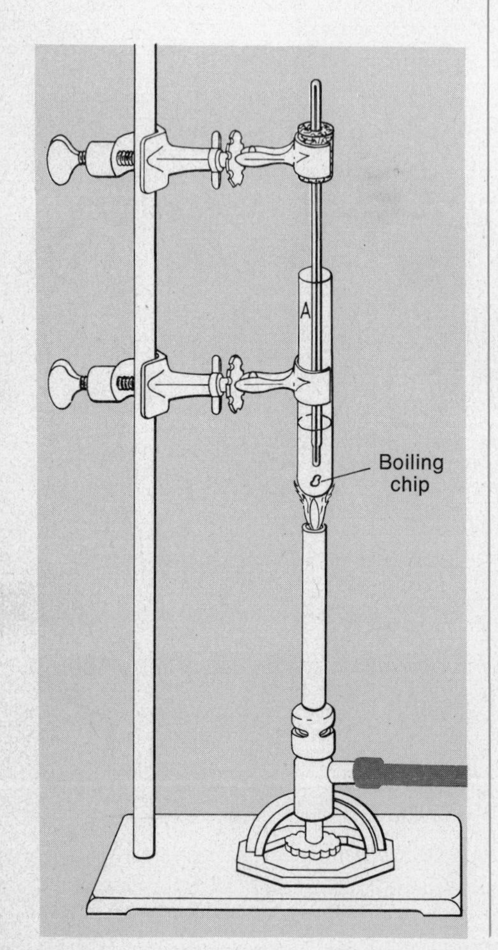

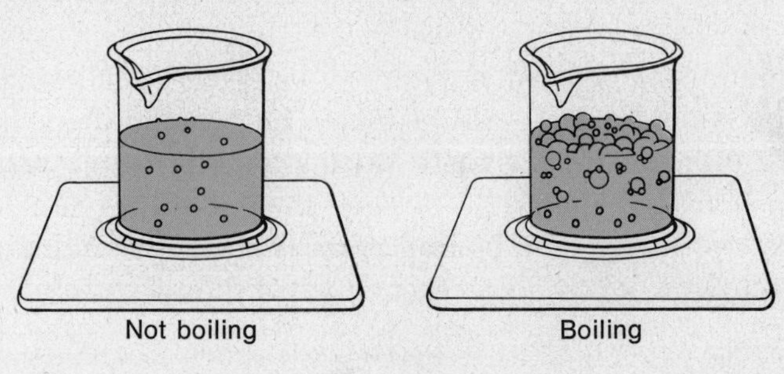

i. Note the temperature when liquid **A** is boiling.

6. What is the boiling point of liquid **A**?

j. Repeat steps **f** through **i** for liquid **B.**
k. Repeat steps **f** through **i** for liquid **C.**
l. Compare your boiling points with those that other students found.

7. Could the three liquids be the same or different; or can't you tell? Why?

B. When It's Hot It's Hot! When It's Not It's Not!

Finding the boiling points of substances can tell you if they are the same or different. Another method of finding out if two substances are the same or different is to determine their *melting points.* Do you think that ice and steel melt at the same temperatures? Try melting two white substances to see if they are the same or different.

What You Need

Solids **A** & **B**

Glass plate
Hand lens

Asbestos pad
Burner
Matches
Paper
Safety goggles
Tongs

What to Do

a. Label two pieces of paper **A** and **B**.
b. Get two or three crystals of solid **A** and place on paper **A**.
c. Get two or three crystals of solid **B** and place on paper **B**.
d. With a hand lens, examine solids **A** and **B**.

8. In your notebook, describe what you see.

9. Do you think the solids are the same or different?

e. Place small, equal-sized pieces of solids **A** and **B** 2 cm apart on a glass plate.
f. Light your burner. Adjust it so that you have a low flame.

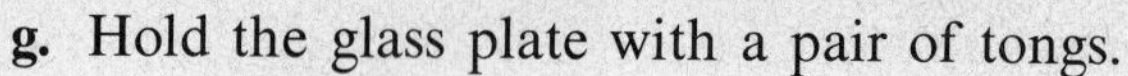

g. Hold the glass plate with a pair of tongs.

h. Heat the plate several cm above the flame by moving it back and forth until the substances melt.

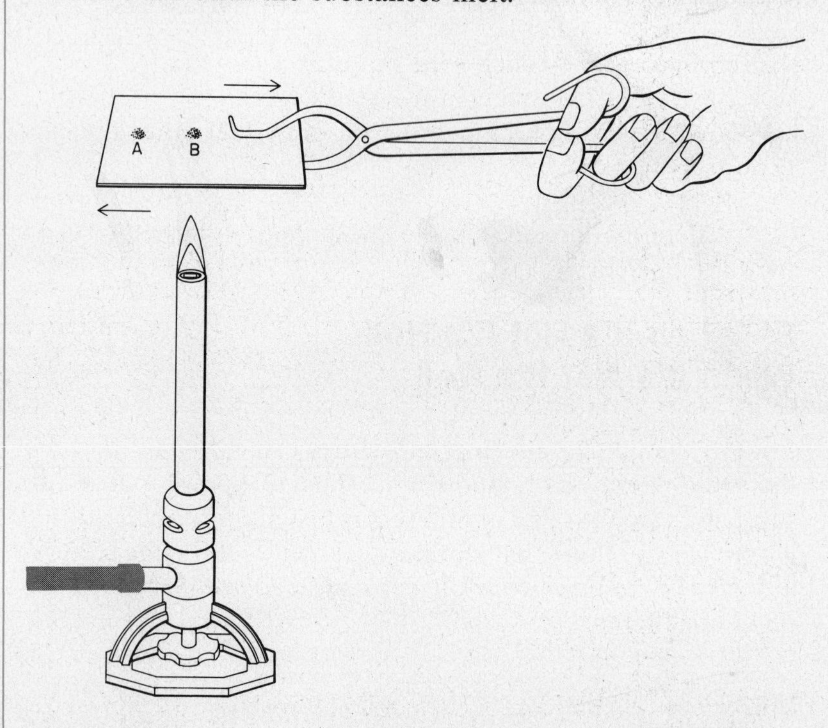

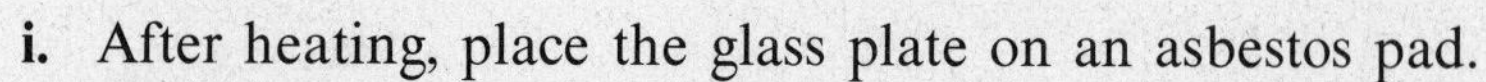

i. After heating, place the glass plate on an asbestos pad.

10. Did both pieces melt at the same time or at different times?

11. If at different times, which melted first—**A** or **B**?

12. Are **A** and **B** the same or different substances?

When you were describing the samples, you probably mentioned their color, odor, and texture. These terms describe their *properties.* Every substance has its own properties, which are determined without changing the substance. Boiling point and melting point are properties. Properties help tell one substance from another.

C. The Old Bag of Feathers Routine

So far you have experimented with the boiling points and melting points of substances. These two properties have helped you find out if substances are the same or different. But the boiling points and melting points are not the only properties of a substance. There are others.

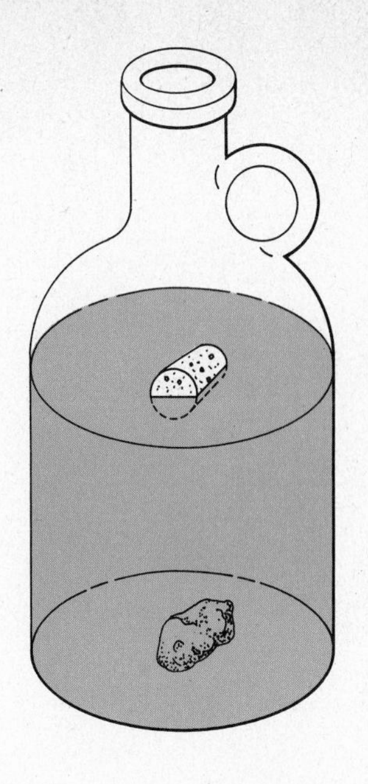

The liquid in the jar shown in the drawing is water.

13. Which object in the drawing is most likely cork?

14. Which object is most likely stone?

A substance that floats on water is less *dense* than water. A substance that sinks in water is more dense than water. It has a greater *density* than water. Another property of any substance is its density. In the following activities you will determine the densities of some solids and liquids.

What You Need

Aluminum (Al) nails	Graduated cylinder, 25 ml	Balance
Lead (Pb) pieces		

What to Do

You will be given different amounts of two solids. The first solid you will work with is aluminum. Some of you will be given one aluminum nail; others two, three, four, or five nails. Your instructor will tell you how many to use. You need to find the mass and volume of your aluminum.

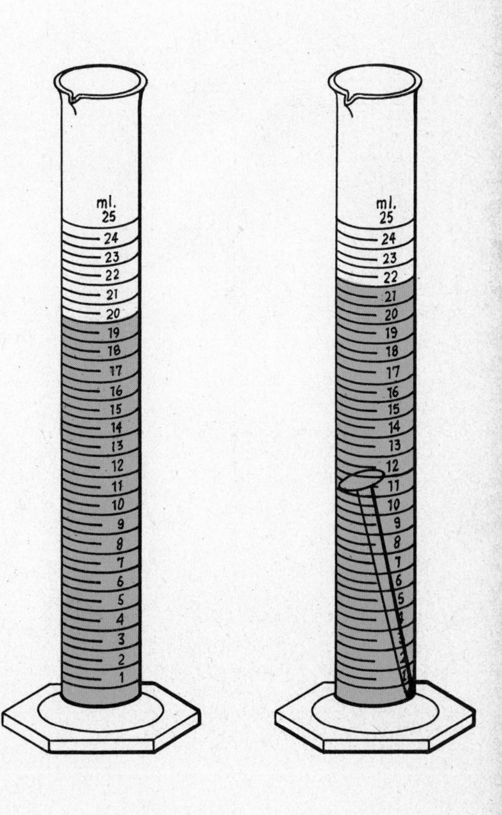

a. Find the mass of your aluminum.

15. Record the mass in your notebook.

To find the volume, you will place the aluminum in water. As you do so you will find that the water level rises. The amount the water rises is equal to the volume of aluminum.

b. Put 20 ml of water into your graduated cylinder.

16. Record the volume of water.

c. Carefully place the aluminum in the graduated cylinder.
d. Observe the level of water now.

17. Record the water level.

18. How much did the water rise when the aluminum was added?

19. What was the volume of the aluminum?

Your second solid is lead.

e. Repeat steps **a** through **d** for lead.
f. Put your data on the board.
g. Plot a graph of the class data as suggested by your instructor.
h. From the class graph, answer the following questions.

20. What would be the mass of 5 ml of aluminum?

21. What would be the mass of 5 ml of lead?

22. Are the masses of 5 ml of aluminum and 5 ml of lead the same or different?

23. What would be the mass of 10 ml of lead? 10 ml of aluminum?

24. What would be the mass of 15 ml of lead? 15 ml of aluminum?

I believe in woman's lib. We'll each take a box!

Two objects with the same mass and volume have the same density. If two objects have the same volume but different masses, then they have different densities.

25. Are the densities of your two metals the same or different?

26. How did you prove that the metals were different?

27. Were your metals more or less dense than water?

28. When an oil slick forms, is the oil more or less dense than water?

29. Is pine wood more or less dense than water?

D. Who's Dense?

Your answer to question **28** tells you that some liquids float on other liquids. Liquids have density. To compare the density of two liquids you have to find their masses and volumes. Your instructor will tell you how much liquid to use.

What You Need

Methanol (CH_3OH)
Water (H_2O)
Graduated cylinder
Balance

What to Do

a. Find the mass of your graduated cylinder.

30. Record the mass in your notebook.

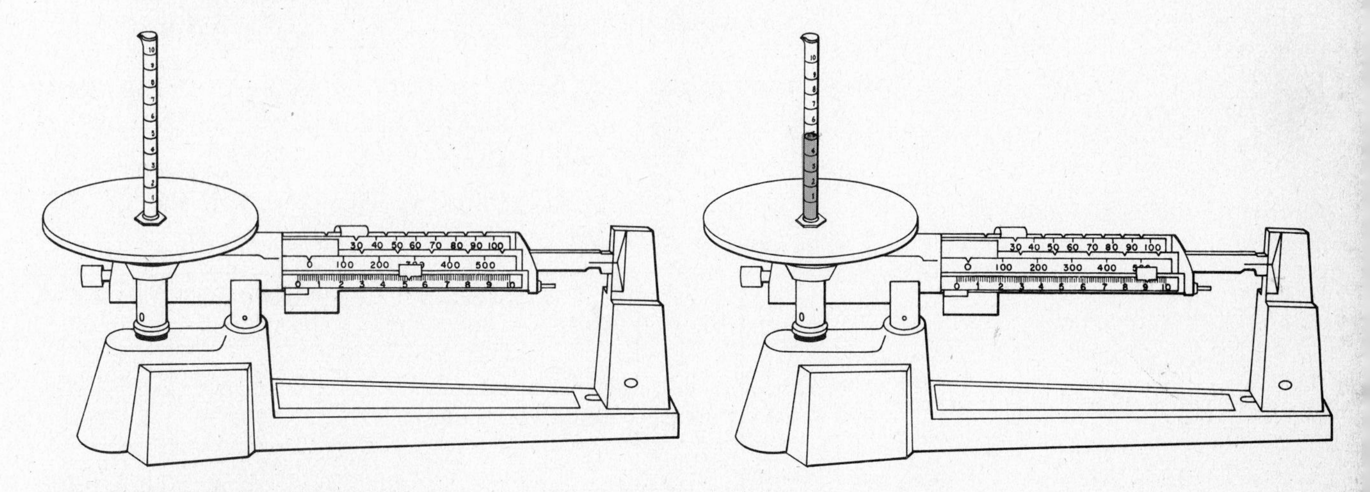

b. Get your sample of methanol.

31. Record the volume.

c. Find the mass of your graduated cylinder with the methanol in it.

32. Record the mass.

33. What is the difference between your answers to questions **30** and **32**?

34. What is the mass of the methanol?

d. Repeat steps **a** through **c** for the volume of water.
e. Put your data on the board.
f. Plot a graph of the class data as suggested by your instructor.
g. From the class graph, answer the following questions.

35. What would be the mass of 5 ml of methanol?

36. What would be the mass of 5 ml of water?

37. Are the masses of 5 ml of methanol and 5 ml of water the same or different?

38. Are the densities the same or different?

39. Are the liquids the same or different?

40. What would be the mass of 20 ml of methanol? 30 ml of water?

41. What would be the volume of 15 grams of methanol? 15 grams of water?

Listed in the table below are some properties that have been tested for three substances. After looking over the table, answer questions **42** and **43**.

TABLE 1

	State	*Color*	*Boiling point in °C*	*Melting point in °C*	*Density in g/ml*
I	solid	white	120	89	0.78
II	solid	white	122	54	0.78
III	solid	white	120.5	88.7	0.78

42. Could any of these substances be the same? Why?

43. Could all three substances be the same? Why or why not?

E. Solubility Power

The boiling point, melting point, and density of a substance are important properties. These properties helped you determine whether two substances were the same or different. The last property you will study is called *solubility*. You looked at solubility briefly at the very beginning of the course. Everyday experience tells about solubility. Salt is soluble in water, sugar in coffee, and gases in soda pop.

What You Need

Solids **A** & **B**	Beaker, 400 ml	Asbestos pad
	Graduated cylinder, 10 ml	Balance
	Stirring rods	Burner
	Test tubes, 2	Matches
		Ring
		Ring stand
		Safety goggles
		Test tube rack
		Wax pencil
		Weighing paper

What to Do

You will try dissolving two white solids in equal amounts of water.

a. Label two pieces of paper **A** and **B**.
b. Find the tare (mass) of each piece of paper separately.

44. Record the masses in your notebook.

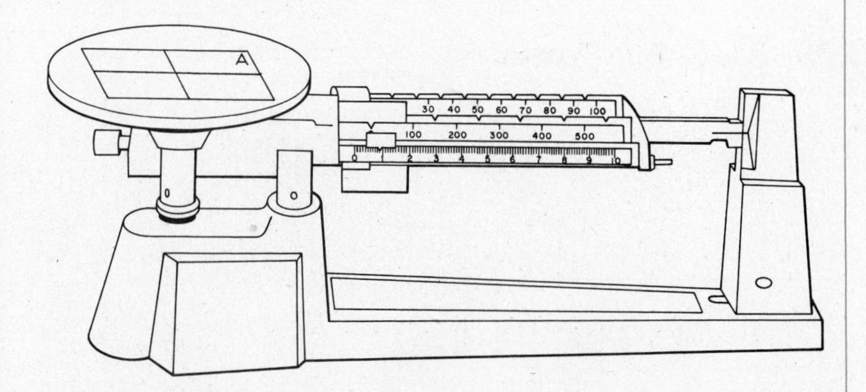

To get 5 grams of a solid, you have to set the balance 5 grams heavier than the tare.

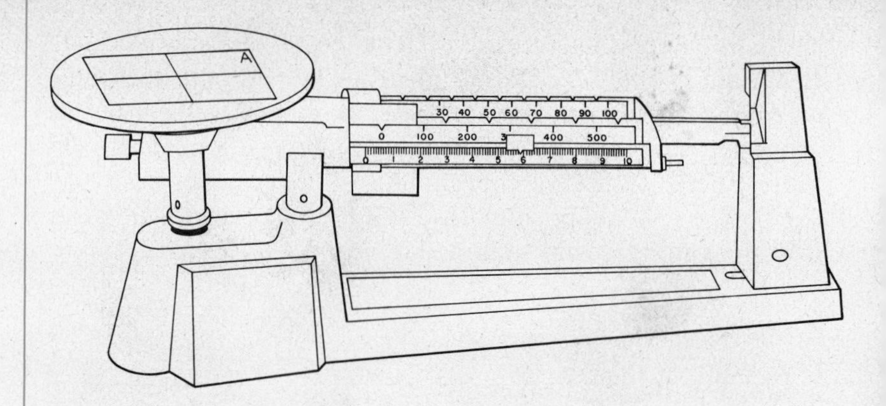

45. To get 5 grams of solid **A**, where would you set the riders on the balance?

c. Get 5 grams of solid **A** on paper **A**.
d. Get 5 grams of solid **B** on paper **B**.

46. Describe the solids.

e. Label two test tubes **A** and **B**.
f. Add 5 ml of water to each tube.
g. Add solid **A** to tube **A** and solid **B** to tube **B**.
h. With a stirring rod, stir the contents of each test tube for 3 minutes.

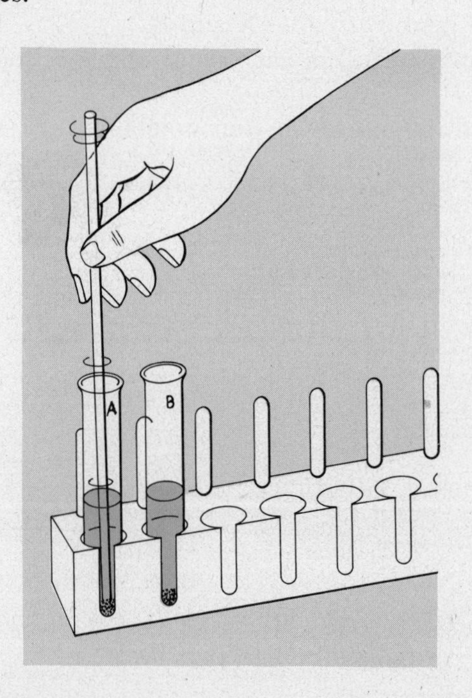

47. Describe what happens.

48. Do the solids dissolve the same or differently?

49. Could the solids be the same substance?

Sometimes warming the liquid increases the solubility of a solid. Try warming the test tubes to see if the solubility is changed.

i. Place both test tubes in 250 ml of water.
j. Heat to boiling.
k. Stir the contents of the test tubes while heating.
l. Observe what happens.

50. Do the solids dissolve the same or differently when heated?

51. Could the solids be the same? Why or why not?

In the last five activities you experimented with various properties. You used these properties to compare substances. These properties helped tell if two substances were the same or different.

52. List the properties studied.

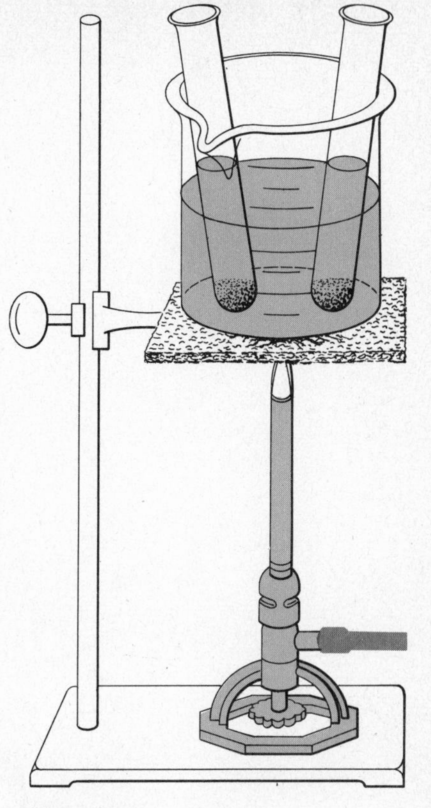

Water bath.

INVESTIGATION

2

There's a Mix-up in the Backfield

Courtesy of General Mills, Inc.

In the next few investigations your knowledge of properties will aid you in separating and identifying substances. You will make a *mixture* of the *pure substances* sand and salt. Then you will separate them. If you have only one kind of material, you have a pure substance. For example, you experimented with lead. If the lead is pure, it's called a pure substance. A pile of salt is a pure substance. When salt is added to water, it is no longer a pure substance. It is then a mixture of substances. The salt water could be separated into salt and water. Which property you have studied will most likely help separate sand and salt?

A. Sandy Salt, Anyone?

What You Need

Salt (NaCl)
Sand (SiO_2)
Sulfur (S)

Beaker, 250 ml
Evaporating dish
Funnel
Graduated cylinder, 25 ml
Hand lens
Stirring rod
Test tubes, 2 (different sizes)

Asbestos pad
Burner
Clay triangle
Filter paper
Matches
Ring
Ring stand
Safety goggles
Spatula
Test tube rack
Weighing paper

What to Do

a. Examine some salt crystals with a hand lens.

1. Describe them in your notebook.

b. Examine some sand crystals with a hand lens.

2. Describe them in your notebook.

c. Mix $\frac{1}{4}$ test tube of sand with $\frac{1}{4}$ test tube of salt in a beaker.

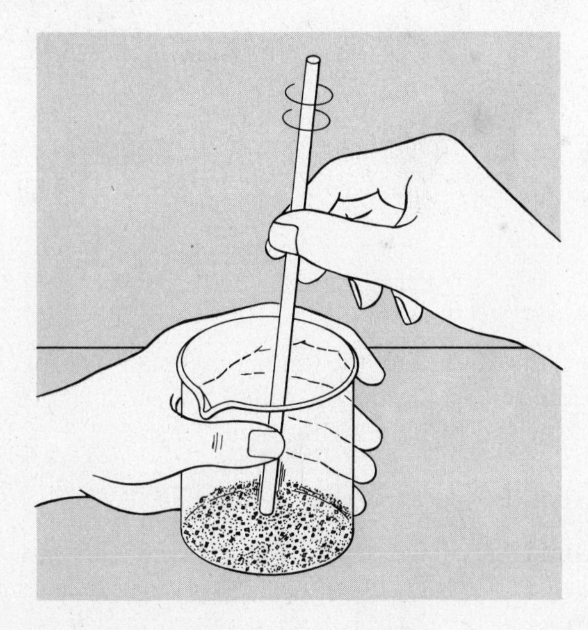

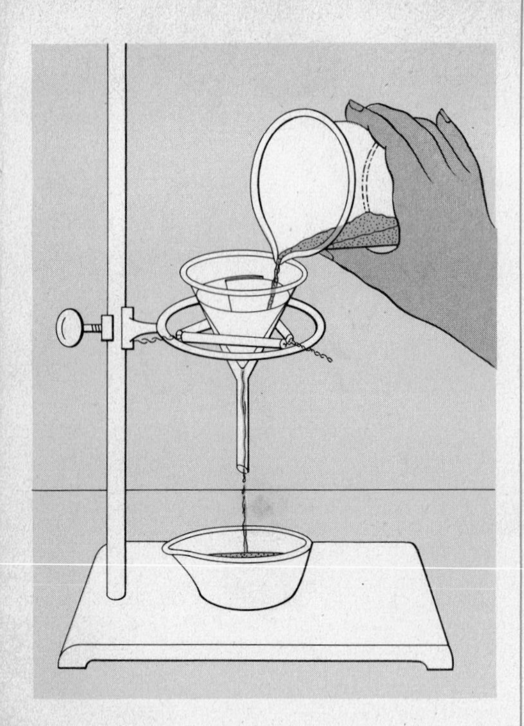

3. Is this mixture a pure substance?

d. Add about 25 ml of water. Stir thoroughly.
e. Filter the liquid into the evaporating dish.
f. Remove the filter paper. Unfold it and place it on a paper towel.
g. With the hand lens, examine the crystals remaining on the filter paper.

4. What has remained on the filter paper?

h. The liquid which went through the filter paper is called the *filtrate.* Boil the filtrate gently until only a small amount of water remains.

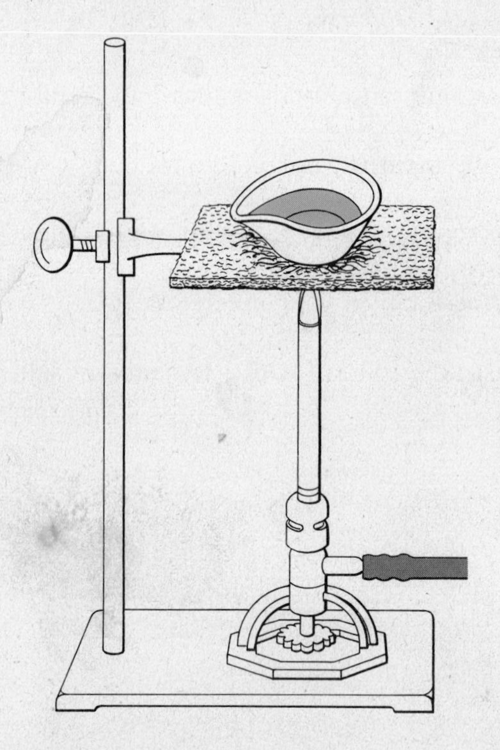

i. Allow the dish and the remaining material to cool. The material left in the dish is called the *residue.*
j. Examine the residue with a hand lens.

5. What do you expect the residue to be—salt or sand?

6. Could you have separated the sand and salt by other means?

7. Check with your classmates. Does everyone have the same amounts of salt and sand?

8. Could you have used different amounts of sand and salt in this experiment? Would your results have been the same?

9. What does this tell you about mixtures?

k. Mix $\frac{1}{2}$ test tube of salt with $\frac{1}{4}$ test tube of sulfur in a beaker.

10. Is the salt and sulfur a mixture or a pure substance?

11. How could you find out?

l. Test your prediction.

12. Record your results in your notebook.

13. How could you separate sand and sulfur?

B. Brighter and Whiter

You just finished separating some mixtures. Now you will make another mixture. Mixtures are very useful. We can find examples of mixtures in gasoline, toothpaste, mouthwash, aftershave lotion, shaving cream, suntan oils, and others. What kind of mixture is toothpaste?

What You Need

Calcium carbonate ($CaCO_3$)
Calcium phosphate [$Ca_3(PO_4)_2$]
Castile soap
Corn syrup
Oil of peppermint
Sugar ($C_{12}H_{22}O_{11}$), powdered
Dropper
Mortar and pestle
Vial, small
Balance
Spatula
Weighing paper

What to Do

You can easily mix the ingredients to make toothpaste.

a. Label four pieces of weighing paper as follows: powdered sugar, castile soap, calcium carbonate, calcium phosphate.

b. Get 8 grams of powdered sugar.
c. Place the powdered sugar in a mortar.
d. Add three drops of oil of peppermint.
e. Mix with a pestle.

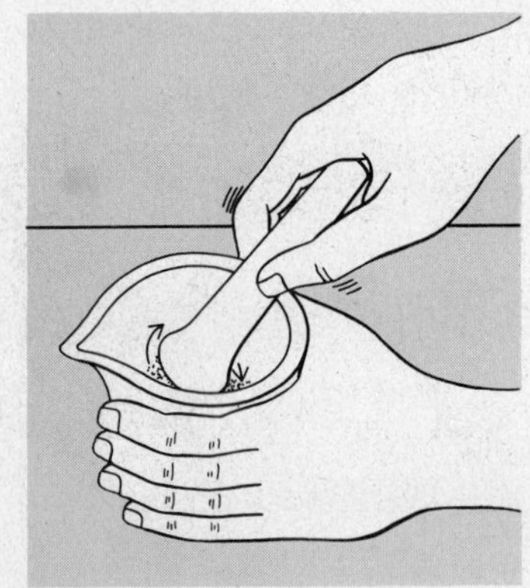

f. Get 3 grams of castile soap.
g. Add to the mortar while mixing.
h. Get 22 grams of calcium carbonate.
i. Add to the mortar while mixing.
j. Get 14 grams of calcium phosphate.
k. Add to the mortar.
l. Mix thoroughly so that there are no lumps.
m. Add enough corn syrup to make a paste.
n. Store in a small vial.

How do we get this toothpaste through the little hole in the tube?

This mixture is called toothpaste. Now that you've learned about mixtures, it's time to take a look at one special kind of mixture—a *solution*.

A. Solve the Solution

What You Need

Acetone (CH_3COCH_3)
Copper(II) sulfate ($CuSO_4$)
Glycerine ($C_3H_8O_3$)
Olive oil
Soda pop, bottle of

Beaker, 250 ml
Graduated cylinder, 10 ml
Mortar and pestle
Test tubes, 2

Asbestos pad
Balance
Bottle opener
Burner
Matches
Ring
Ring stand
Safety goggles
Spatula
Stoppers, solid, 2
Test tube clamp
Test tube rack
Wax pencil
Weighing paper

What to Do

a. On separate pieces of paper, get two 1-gram portions of copper sulfate.
b. In the mortar, grind one portion to a powder with the pestle.
c. Put 8 ml of water into each of two separate test tubes.
d. Add the ground copper sulfate to one test tube. Add the unground portion to the other tube.

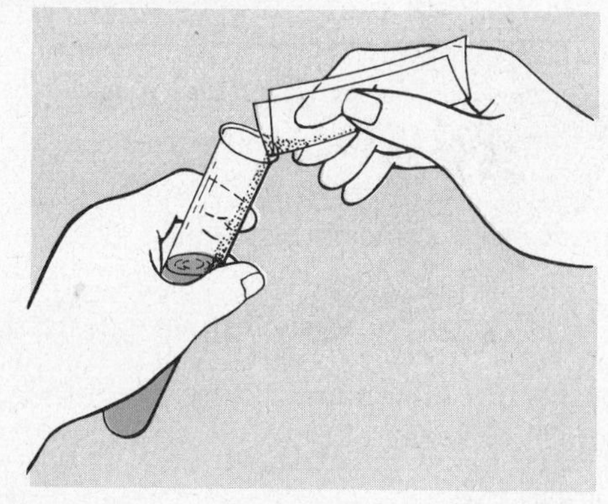

e. Label both tubes.
f. Stopper both tubes securely.
g. Shake both test tubes.

8. Which portion of copper sulfate dissolves more rapidly?

9. Why do you think this is so?

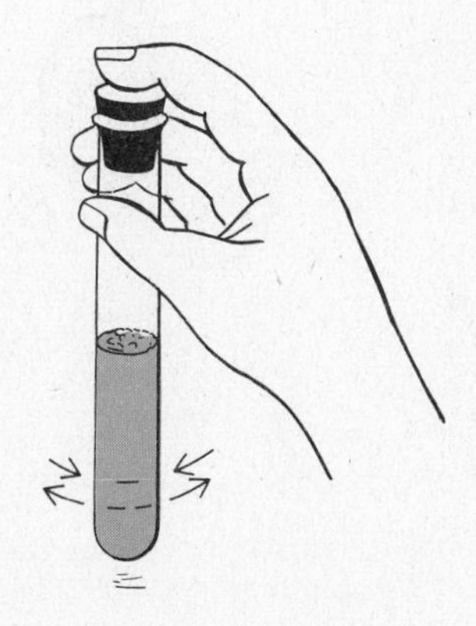

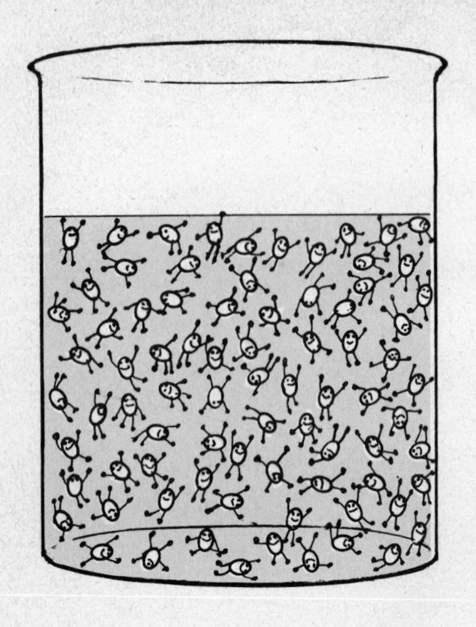

When a material dissolves, the solute particles break up into smaller pieces. These small pieces then scatter throughout the solvent. Since they are very small, they cannot be seen. A true solution is one in which all of the solute has been separated into its smallest pieces and is evenly scattered throughout the solvent.

h. Rinse out the test tubes. Half-fill each with water.
i. Select two small crystals of copper sulfate that are about the same size. Place one crystal in each test tube.
j. Leave one test tube undisturbed.
k. Stopper the other one. Shake the contents until the copper sulfate has dissolved.

10. What has happened to the copper sulfate in the tube that was shaken?

11. Have the crystals in the undisturbed tube dissolved completely?

12. How does shaking affect the rate of solution? Why?

Try dissolving equal amounts of copper sulfate in hot and cold water.

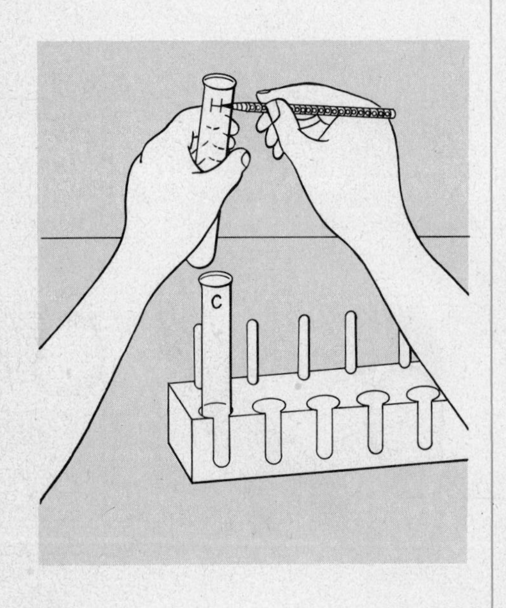

l. Rinse out the test tubes. Then label one tube **C** and the other **H.**
m. Get two 1-gram portions of copper sulfate. Use two pieces of weighing paper.
n. Place one portion in each test tube.
o. Fill tube **C** $\frac{1}{2}$-full with *cold* water.
p. Fill tube **H** $\frac{1}{2}$-full with *hot* water.
q. Stopper both tubes. Then shake them.

13. In which tube does copper sulfate dissolve faster?

14. How does the temperature affect the rate at which copper sulfate dissolves?

A bottle of soda pop contains a solution in which a gas is dissolved in a liquid. In this case, the gas is carbon dioxide (CO_2), which is why soda pop is called a "carbonated beverage."

r. Carefully observe your instructor uncap a bottle of warm soda pop.

15. What happens?

Uncapping the bottle releases the pressure. The bubbles and spray are the gas escaping. This gas was originally dissolved in the liquid, forming a solution. The cap prevented the gas from leaving the solution.

s. Rinse out a test tube.
t. Half-fill the test tube with soda pop. Gently warm it over a flame. **Do not boil.**

16. What do you observe?

17. How is the amount of gas dissolved in a solution affected by an increase in temperature?

How do you remove oil stains from your clothes? Try dissolving oil in different solvents.

u. Rinse out your test tubes. Label one **W** and the other **A.**
v. Fill test tube **W** $\frac{1}{3}$-full with water.
w. Fill tube **A** $\frac{1}{3}$-full with acetone. **(Caution: acetone is very flammable.)**
x. Add a few drops of olive oil to each tube.

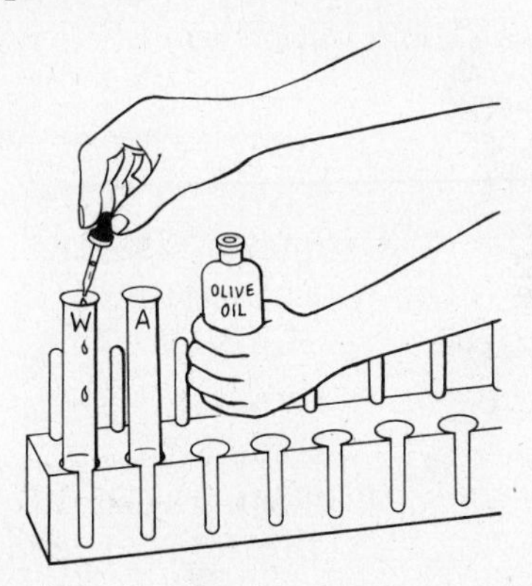

y. Stopper the tubes and shake them.

18. In which solvent is olive oil soluble?

z. Rinse out the test tubes. Then repeat steps **u** through **y** using glycerine instead of olive oil.

19. In which solvent is glycerine soluble?

20. From your work in this investigation, what can be done to dissolve a substance more quickly?

21. What can be done to dissolve a substance more slowly?

B. The Un-Mix

Now you will be given an unknown mixture. Your task is to separate the mixture into as many different substances as you can. Then you should determine some properties of each substance.

The future of the world is in your hands!

What You Need

- Hydrochloric acid (HCl), 1 M
- Methanol (CH_3OH)
- Unknown mixture, 4 g
- Water (H_2O)

- Beaker, 250 ml
- Evaporating dish
- Funnel
- Hand lens
- Stirring rods
- Test tubes, 3

- Asbestos pad
- Balance
- Burner
- Clay triangle
- Filter paper
- Magnet
- Matches
- Ring
- Ring stand
- Safety goggles
- Spatula
- Test tube rack
- Tongs
- Wax pencil
- Weighing paper

What to Do

a. Examine your unknown mixture.

22. Describe your unknown mixture in your notebook.

23. How many different substances do you think are in the mixture?

b. Discuss with your lab partner how you will proceed to separate the mixture and how to test its properties.
c. Then discuss it with your instructor. Determine as many properties of each substance as you can.
d. Make a table to fill in the properties.

24. How many substances were in the mixture?

25. After looking at the properties, and at your past few investigations, what could the substances be?

C. Where You've Been

You started with identifying properties

Mixtures

Properties helped to separate and identify substances in mixtures

You separated pure substances

Can pure substances be broken down or separated more?

Now—Where are you going?
? ? ? ? ? ? ? ? ? ? ? ?

INVESTIGATION

4

Compound It All!

In the previous investigations you separated mixtures into pure substances. You now need to find out if these pure substances can be separated again, producing something new. In Unit 1, you used the pure substances sugar (sucrose) and baking soda (sodium bicarbonate). Will these pure substances separate again? If a pure substance separates, we say it *decomposes.*

A. Don't Compound the Problem

What You Need

Magnesium carbonate ($MgCO_3$)
Sodium bicarbonate ($NaHCO_3$)
Sucrose ($C_{12}H_{22}O_{11}$)
Evaporating dish
Asbestos pad
Burner
Matches
Ring
Ring stand
Safety goggles

What to Do

a. Place a small piece of magnesium carbonate in an evaporating dish.
b. Heat the evaporating dish.
c. Observe carefully.

1. Did the magnesium carbonate change when heated?
2. Was the pure substance composed of more than one thing?

d. Cool and clean the evaporating dish.
e. Repeat steps **a** through **d** for sucrose and sodium bicarbonate.

3. Was there any evidence that sucrose decomposed? What was the evidence?
4. Was there any evidence that sodium bicarbonate decomposed? What was it?

If both mixtures and pure substances can be separated, what is the difference?

B. I'm Forever Blowing Bubbles

In this investigation, the pure substance water will be separated and the products will be examined and tested. Water decomposes into hydrogen and oxygen. Hydrogen and oxygen are elements. An *element* is any substance that cannot be separated into different substances by ordinary means. The *word equation* to represent the decomposition of water is:

$$\text{water} \longrightarrow \text{hydrogen} + \text{oxygen}$$

What You Need

Sodium carbonate (Na_2CO_3), 0.5 *M* solution

Beaker, 250 ml
Graduated cylinder, 25 ml
Test tubes, 2

Alligator clips, 4
D.C. power source (4 dry cells or equivalent
Electrodes, stainless steel
Insulated wire, ends stripped, 2
Matches
Rubber bands, 2
Safety goggles
Wood splints

What to Do

a. Put 200 ml of sodium carbonate solution in a beaker. The sodium carbonate allows the experiment to go faster.
b. From the beaker, fill the two test tubes with the solution.
c. Invert the test tubes into the beaker, as shown by your instructor.
d. Set up the apparatus as shown in the drawing.

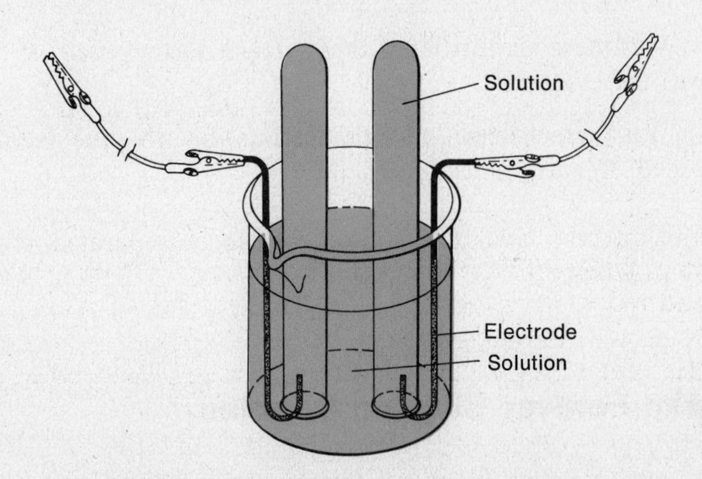

e. Connect the electrodes to your power supply and let run for 10 minutes.

5. What happens to the amount of solution in the test tubes after 3 minutes? 5 minutes? 10 minutes?

6. Does the same amount of gas collect in both test tubes?

Hydrogen gas is coming off the electrode hooked to the negative part of your power supply. Oxygen gas is coming off the electrode hooked to the positive part of your power supply.

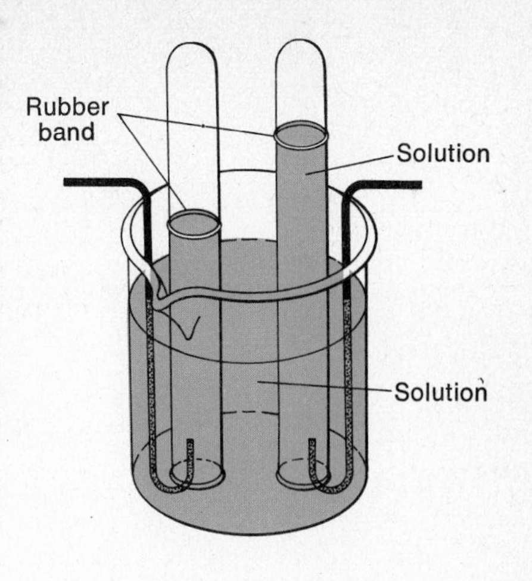

How much gas did you collect? To find out, do the following:

f. Place a rubber band around each test tube at the bottom of the gas level.

7. Record which test tube is connected to the negative terminal of your power supply.

8. Record which test tube is connected to the positive terminal of your power supply.

g. Remove the test tubes and empty them.
h. Fill each test tube with water up to the rubber band.

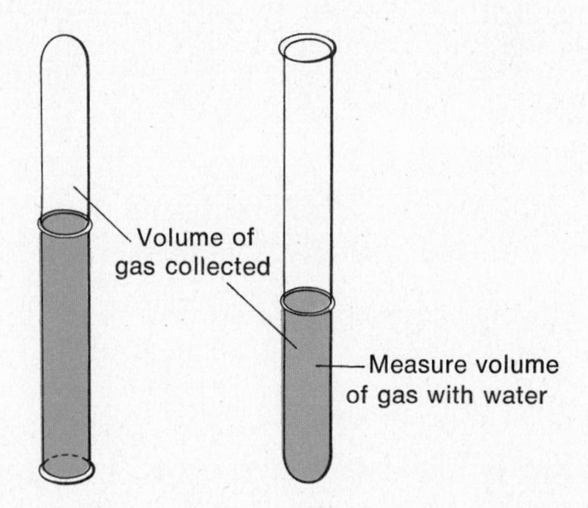

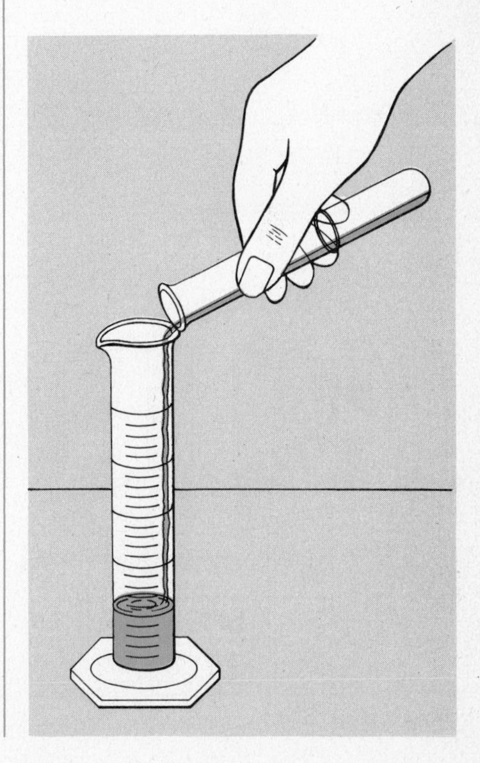

i. Pour the water from one test tube into a graduated cylinder and record its volume. (Same as the volume of gas.)
j. Repeat for the second test tube.
k. Put your data for the volumes of hydrogen and oxygen on the board. Use a table like the one below.

TABLE 1: Volumes of Hydrogen and Oxygen

Volume of hydrogen in ml	*Volume of oxygen in ml*
5	2.5

l. Check the data on the board to answer the following questions.

9. Did everyone collect the same volume of hydrogen?

10. Did everyone collect the same volume of oxygen?

11. Copy into your notebook the statement below which best summarizes the data.
a. The volume of hydrogen was equal to the volume of oxygen.
b. Twice as much hydrogen was collected.
c. Twice as much oxygen was collected.

12. Think back to Investigation 2, where you separated sand and salt. Would all mixtures of sand and salt be the same?

The answers to the last two questions point out the difference between mixtures and compounds. A *compound* is a pure substance containing more than one element. Examples of compounds are: water (H_2O), sugar ($C_{12}H_{22}O_{11}$), and methanol (CH_3OH). Mixtures can be put together any way, but compounds are put together in very special ways. In separating water, you will *always* get twice the volume of hydrogen as oxygen.

13. In a mixture of salt and sand, how many grams of salt can be mixed with 3 grams of sand?

14. In the compound water, how many ml of oxygen would be produced when 10 ml of hydrogen are produced? 20 ml of hydrogen?

If all else fails, follow directions!

Hydrogen is found in water, acids, bases, alcohols, carbohydrates, and petroleum. Oxygen is found in water, alcohol, and minerals. Pure oxygen is used in hospitals, airplanes, and as a rocket fuel.

What are the properties of these two elements that make them so useful?

m. Set up the apparatus you used on page 94. Collect two more test tubes of the gases. Stop collecting when the tube containing hydrogen is filled.

15. In your notebook, describe the hydrogen.

n. Carefully remove the electrode from the test tube of hydrogen. Keep the open end of the test tube **below** the liquid surface.

o. Hold your thumb over the open end of the test tube.

p. Remove the tube of hydrogen from the solution and turn it right side up. Keep the tube closed with your thumb.

q. Have your partner light a wood splint.

r. Put the flaming splint into the tube of hydrogen.

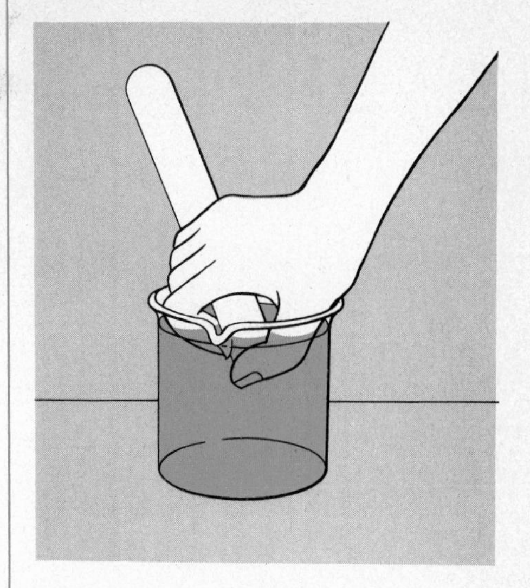

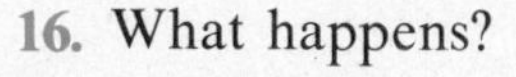

16. What happens?

s. Rinse your thumb thoroughly with water.

t. Repeat steps **q** and **r** with a test tube of air.

17. What happens? Why?

18. Describe the oxygen.

u. Using a *glowing* splint, repeat steps **n** through **r** with the test tube of oxygen.

19. What happens?

Oxygen has the chemical symbol O. It is the most abundant element in the world. Oxygen is 20% of the air we breathe and almost 50% of the earth's crust. The human body is about 67% oxygen.

Hydrogen has the chemical symbol H. It is the lightest element. Because it is lighter than air, hydrogen rises. It was once used to fill weather balloons.

Congratulations!! You have just tested two test tubes of invisible elements that you can't see, taste, or smell!

C. Why Water?

Have you ever given any thought to the importance of water? Without it, there would be no lakes to fish in, no oceans to cross, no beaches to swim at, and no rain for the trees. Industries would come to a halt (why?); food could not be grown; and animals would die of thirst. Although water is colorless, odorless, tasteless, and transparent, we need water to live.

In case you haven't been keeping tabs, we are rapidly running out of water that is fit to use. We could not survive without water, yet we are continually abusing it.

People are busy dumping sewage, industrial chemicals, acid mine drainage, pulp mill wastes, heavy metals, and insecticides into our lakes, streams, rivers, and oceans. What effect does this have? See for yourself.

20. Think for a few minutes. List all the effects of water pollution that you can think of.

We are a world of disposers. We mine something or make something, use it, and then throw it away. What can we do? Whose job is it to stop water pollution?

21. Do you think the people should clean up the mess? Why?

22. Do you think the government should clean up the mess? Why?

23. Do you think industries should clean it up? Why?

Wide World Photos

Water is just one resource. Air, soil, trees, and minerals are all made up of elements. We find these elements in nature as mixtures or as compounds. Water and other resources are being used up or spoiled so that they are not useful.

24. Are the following elements or compounds?

a. water (H_2O)
b. sulfur (S)
c. aluminum (Al)
d. sucrose ($C_{12}H_{22}O_{11}$)
e. methanol (CH_3OH)
f. sodium bicarbonate ($NaHCO_3$)
g. iron (Fe)
h. magnesium sulfate ($MgSO_4$)
i. lead (Pb)
j. iron sulfide (FeS)

In this investigation, you've studied two elements. More will follow!

INVESTIGATION 5

Ever Mix a Compound?

In the previous investigation you separated a compound into its elements. Now try to make a compound out of the two elements, iron and sulfur.

A. What Am I Made of?

What You Need

Hydrochloric acid (HCl), 1 *M*
Iron (Fe) filings
Methanol (CH_3OH)
Sulfur (S)

Beaker, 250 ml
Graduated cylinder, 10 ml
Hand lens
Test tubes, 3 Pyrex

Balance
Burner
Hammer or iron ring
Magnet
Matches
Plastic bag
Safety goggles
Spatula
Test tube holder
Test tube rack
Wax pencil
Weighing paper

What to Do

a. Using your balance, get 4 grams of iron filings on a piece of paper.
b. Get 2 grams of sulfur on another piece of paper.

1. Copy Table 1 into your notebook.

TABLE 1

	Color	Odor	Effect of magnet	Solubility in		
				Water	*Methanol*	*Acid (1 M HCl)*
Row A—Iron						
Row B—Sulfur						
Row C—Iron and sulfur after heating						

2. Describe the color and odor of the iron.

3. How does a magnet affect the iron? Ask your instructor how best to test with a magnet.

4. Record your results in the table in your notebook.

c. Label three test tubes **1, 2,** and **3.**
d. Add 2 ml of water to tube **1.** Add 2 ml of methanol to tube **2.** Add 2 ml of hydrochloric acid to tube **3.**
e. Add one iron filing to each test tube.
f. Observe closely.

5. Record the results in Row A of the table.

g. Wash and dry your three test tubes.

6. Describe the color and odor of the sulfur.

7. How does a magnet affect the sulfur?

h. Repeat steps **d** through **f** for sulfur, using one or two sulfur crystals in each test tube.

8. Record the results in Row B of the table.

i. Mix the remaining iron and sulfur thoroughly on a piece of paper. Use your spatula.
j. Pour half of this mixture into a clean test tube. Save for step **k.**

9. Can you separate the iron and sulfur on the paper? Try.

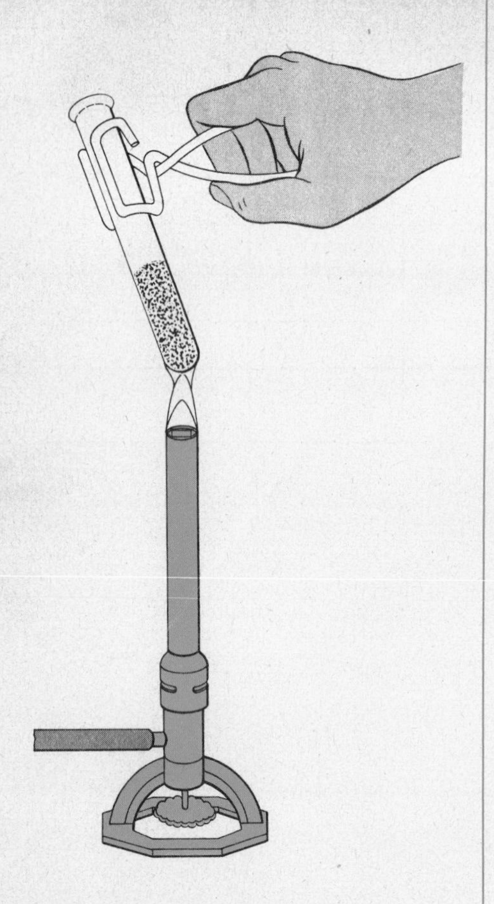

k. Heat the mixture of iron and sulfur in the test tube until the contents begin to glow red.

l. After the red glow stops, dip the hot test tube into a beaker of water. If this doesn't crack the test tube, wrap it in a paper towel and tap with a hammer or an iron ring.

m. Examine the contents.

10. Fill in Row C of the table.

11. Can you separate the contents of the test tube into the two original substances?

12. Is there any evidence that the iron and sulfur changed? What is the evidence?

13. Is the new substance a mixture, a compound, or an element?

The iron and sulfur have reacted to form iron sulfide. The word equation to represent this is:

the element		the element		the compound
Iron	+	Sulfur	⟶	Iron sulfide
(Fe)		(S)		(FeS)

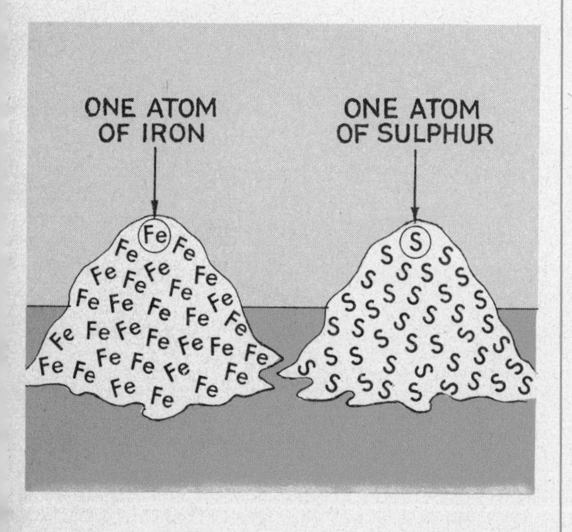

B. Atoms and Molecules

You have worked with several elements: hydrogen, oxygen, iron, and sulfur. Each sample was large enough to work with. But you could have worked with smaller samples. What is the smallest amount of an element you can work with? The smallest part of an element that can be involved in a chemical change is called an *atom.*

The smallest quantity of a compound is a *molecule.*

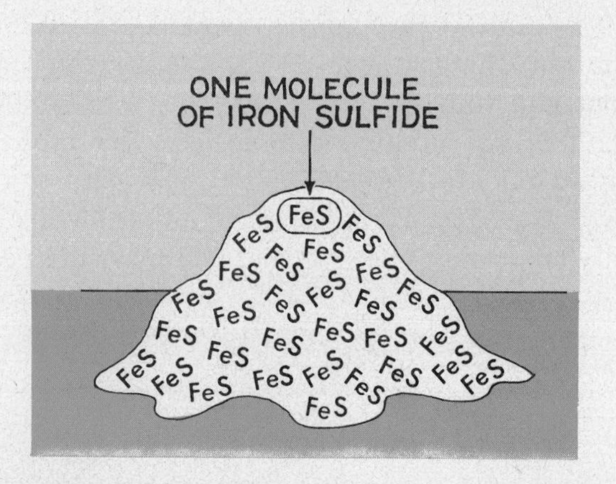

Different atoms can combine to form molecules. An example is the water molecule, H_2O. It contains two atoms of hydrogen (H) and one atom of oxygen (O). When billions of these molecules are together, you have a drop of water.

Another molecule is carbon dioxide (CO_2). One atom of carbon (C) combines with two atoms of oxygen (O) to form a molecule of CO_2.

Try this one on for size: 12 atoms of carbon (C), 22 atoms of hydrogen (H), and 11 atoms of oxygen (O)! They can join up to form a molecule of $C_{12}H_{22}O_{11}$, which is nothing but the sweet stuff called sugar.

14. What do we call the tiny particles which make up molecules?

15. What is the difference between an atom and a molecule?

If you are beginning to get the feeling that everything in this world is made up of elements—and atoms and molecules—you're right!

Your desk is made up of atoms; your food is made up of atoms; water, clothes, cars, books, and yes, even you, are made up of many billions of different atoms. These atoms are atoms of many different elements, and they can combine to form many different molecules. The molecules combine to form the many different materials and objects of our world. These materials all have different properties. You studied properties at the beginning of this Unit.

This is elementary!

C. Everyone is a Model

Atoms and molecules are too small to see. Something larger can be used to represent atoms and molecules. When this is done a *model* of atoms and molecules has been developed.

At the beginning of this investigation you reacted iron (Fe) and sulfur (S) to make iron sulfide (FeS). The word equation is:

$$\underset{(Fe)}{\text{Iron}} + \underset{(S)}{\text{Sulfur}} \longrightarrow \underset{(FeS)}{\text{Iron sulfide}}$$

Let a bolt represent one atom of iron and let a nut represent one atom of sulfur. Therefore, you can rewrite the equation as follows:

$$\underset{(Fe)}{\text{one atom of Iron}} + \underset{(S)}{\text{one atom of Sulfur}} \longrightarrow \underset{(FeS)}{\text{one molecule of Iron sulfide}}$$

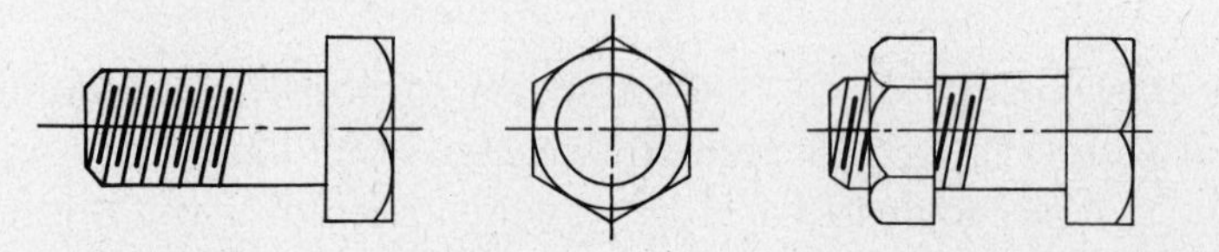

After the reaction, in which iron sulfide was made, you can represent the molecule of iron sulfide by the bolt and nut together.

16. In your notebook, answer the following as a summary of your study of atoms, elements, molecules, compounds, and mixtures.
a. A box of nuts could be used as a model to represent (1) an element, (2) a compound, (3) a mixture.
b. A box of bolts could be used as a model to represent (1) an element, (2) a compound, (3) a mixture.
c. A box of bolts and nuts not combined in any way could be used as a model to represent (1) an element, (2) a compound, (3) a mixture.
d. A box of bolts and nuts combined so that every bolt has one nut on it could be used as a model to represent (1) an element, (2) a compound, (3) a mixture.

e. One bolt could be used as a model to represent (1) a molecule, (2) an atom, (3) a compound, (4) a mixture.

f. One nut could be used as a model to represent (1) a molecule, (2) an atom, (3) a compound, (4) a mixture.

g. One bolt and one nut that are not put together could be used as a model to represent (1) two atoms that are alike, (2) two atoms that are different, (3) a molecule.

h. One bolt with one nut on it could be used as a model to represent (1) a molecule, (2) an atom, (3) an element, (4) a mixture.

INVESTIGATION

6

It's Elementary, Dear Earth

Look around you. What do you see?

You see things—objects—some living, and some non-living. Do you think they have anything in common? As you study chemistry, you can find out for yourself.

A. Don't Pile It On

Suppose you could decompose everything in the universe and sort it into piles. You would have more than 100 separate piles. Each pile would contain a different element.

Some of these piles would contain elements you are quite familiar with. For example, you would find oxygen, hydrogen, copper, and iron. Some piles would contain materials you probably have never heard of. Examples might be ytterbium, americium, einsteinium, and lawrencium.

These materials are all elements. As you have learned, elements are substances that cannot be broken down by ordinary chemical means into simpler substances. For example, you cannot break oxygen down into any other substance. This is because oxygen is an element. You can, however, take a substance like sugar and break it down into carbon, hydrogen, and oxygen. Therefore, sugar is not an element. Do you remember what sugar is? Right! A compound.

At room temperature, most of the elements are solid. Eleven elements are gases, and only two, mercury and bromine, are liquids. During this year, you will become familiar with the most important of the elements.

There are over 100 different elements. Some are very common, while others are extremely rare. Tables 1 and 2 show some of the more common elements found in the earth's crust and in the ocean.

TABLE 1:
Elements in the Earth's Crust

Element	Percent by mass
Oxygen (O)	46.6
Silicon (Si)	27.7
Aluminum (Al)	8.1
Iron (Fe)	5.0
Calcium (Ca)	3.6
Sodium (Na)	2.8
Potassium (K)	2.6
Magnesium (Mg)	2.1
Carbon (C)	0.09

TABLE 2:
Elements in the Ocean

Element	Percent by mass
Oxygen (O)	86.1
Hydrogen (H)	10.6
Chlorine (Cl)	1.9
Sodium (Na)	1.1
Magnesium (Mg)	0.13
Sulfur (S)	0.09
Calcium (Ca)	0.04
Potassium (K)	0.04

Looking at Tables 1 and 2, answer the following questions.

1. Which element is found most often on the earth? in the ocean?

2. Which elements listed can be found in both the earth's crust and the oceans?

The element carbon is an essential part of all living things. Over thousands and thousands of years prehistoric plant and animal

remains were changed into coal, oil, and natural gas. These are used as fuels. Because they came from ancient plants and animals, they are called fossil fuels. Fossil fuels contain carbon.

3. Is carbon a large or a small part of the earth?

4. Are fossil fuels a large or a small part of the earth?

5. Why do we have to worry about the amount of energy available in the world?

B. The Language of Chemistry

As you know by now, each element has a symbol. Chemists label the elements with symbols of one or two letters. For example, hydrogen is H; carbon is C; and magnesium is Mg. Chemical symbols of elements make up an international language.

Before chemistry became a true science, the ancient alchemists used symbols to represent the elements as they understood them.

In 1814, the Swedish chemist Jakob Berzelius proposed a simple system, using letters. This system is still in use today.

Many elements were discovered hundreds of years ago, and their names come from Latin and Greek. This explains why the

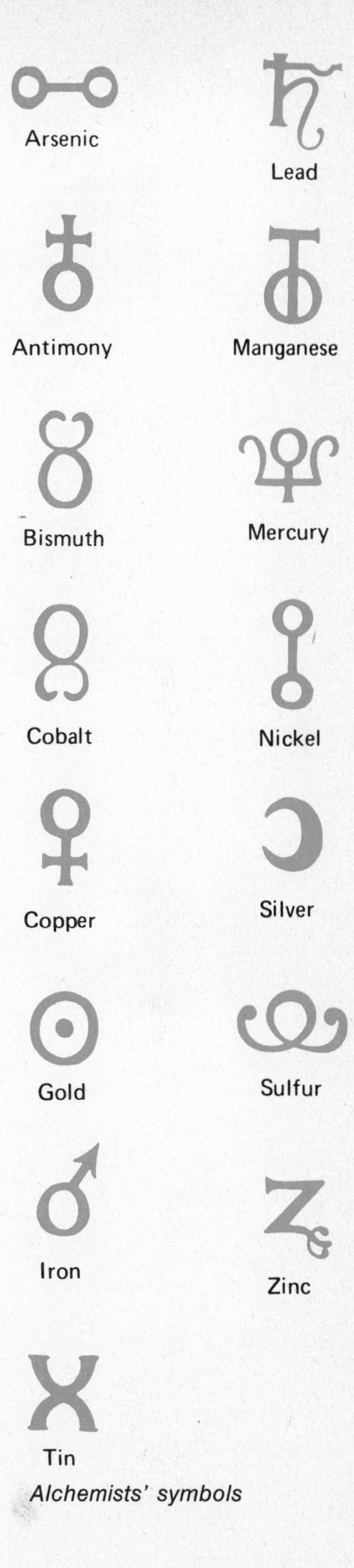

Alchemists' symbols

symbol for copper is Cu (from cuprum), and the symbol for mercury is Hg (from hydrargyrum).

We can give a few simple rules for writing chemical symbols. Don't worry about them; just keep them in mind.

a. When the names of two or more elements begin with the same letter, either one or two letters is used. For example, C stands for carbon; Cl for chlorine; and Cu for copper.

b. When writing a symbol which has two letters, the second letter is *never* capitalized. Copper is Cu, *not* CU; sodium is Na, *not* NA.

Listed on the next page are some common elements. Also listed are some uses of the elements and of the compounds containing them.

6. What elements are used for medical purposes?

7. What elements are used for construction?

8. What elements are used in making paints?

Now that you know something about different elements, it's time to find out how to identify them.

TABLE 3: Uses of Common Elements

Element	*Symbol*	*Uses*
Aluminum	Al	building, foils, outdoor furniture
Barium	Ba	paint, rat poison
Boron	B	antiseptic, water softening, flux, cleaning
Calcium	Ca	limestone, cement, plaster
Carbon	C	petroleum, alcohol, ink, pencils
Chlorine	Cl	swimming pools, bleach, germicide
Chromium	Cr	plating, tanning leather, stainless steel
Cobalt	Co	coloring glass, high temperature alloys
Copper	Cu	brass, bronze, coins
Fluorine	F	etching glass, air conditioning
Iodine	I	table salt, antiseptic
Iron	Fe	steel, wrought iron
Lead	Pb	solder, batteries, X-ray shields
Lithium	Li	flares, paint
Magnesium	Mg	flash bulbs, flares, airplane construction
Mercury	Hg	light switches, thermometers
Nickel	Ni	coins, safes, plating
Sodium	Na	table salt, soap
Sulfur	S	making paper, gunpowder, sulfa drugs
Zinc	Zn	paints, galvanizing

INVESTIGATION 7

The Case of the Missing Elements

The elements are the units from which all other substances are made. Just as the parts of a building can be examined, substances can be *analyzed* in detail. *Analysis* means finding out which elements are present in a given substance. You take the substance apart and study its parts. To do this is to analyze the substance.

A. Let's Analyze

You will first use a flame test to analyze what is in a substance. To perform a flame test, a small amount of the substance is placed in the flame and observed carefully.

What You Need

Copper(II) chloride ($CuCl_2$)
Copper(II) sulfate ($CuSO_4$)
Hydrochloric acid (HCl), 6 *M*
Lithium chloride ($LiCl$)
Lithium nitrate ($LiNO_3$)
Sodium bicarbonate ($NaHCO_3$)
Sodium chloride ($NaCl$)
Strontium chloride ($SrCl_2$)
Strontium nitrate [$Sr(NO_3)_2$]

Beaker, 50 ml
Evaporating dish
Graduated cylinder, 25 ml
Stirring rod

Asbestos pad
Burner
Matches
Safety goggles
Wire loop, platinum or nichrome

What to Do

a. Place 25 ml of hydrochloric acid in a 50 ml beaker.
b. Before testing each substance, clean the wire loop by heating it red hot.
c. Dip the hot wire in the hydrochloric acid.
d. Repeat the heating and dipping until the wire is clean.
e. Get one compound at a time from your instructor. The piece you get should be the size of a pea.
f. Place the compound in the evaporating dish.
g. Heat the wire loop red hot.
h. Place the red hot wire in the compound so that some of the compound will stick to the wire.

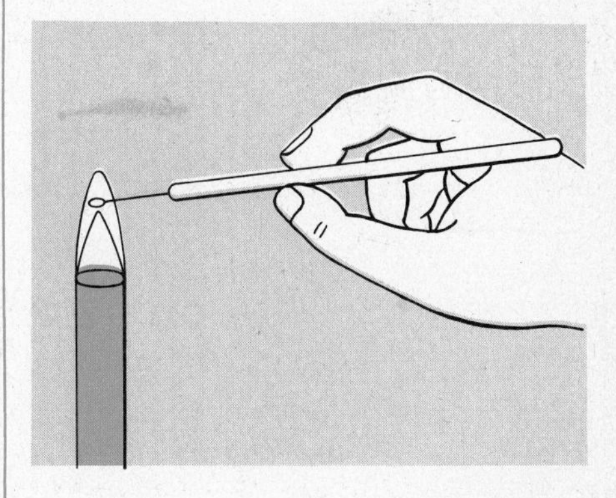

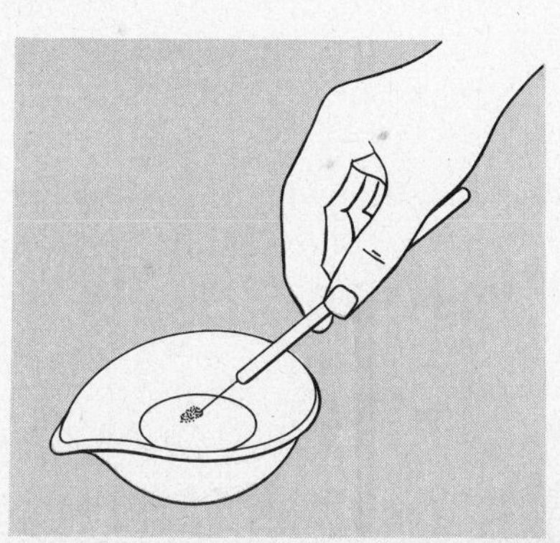

i. Heat the wire in the flame again.
j. Observe the color.

1. Record the color in your notebook. Use a table like the one below.

TABLE 1: Results of Flame Test

Compound	*Elements present in the compound*	*Color of flame*
Copper(II) chloride	Cu, Cl	

k. Repeat steps **b** through **j** for the other compounds listed under **What You Need.**

2. Record your results in the table.

Look at the table to answer the following questions.

3. Were there any compounds that gave the same colors?

4. What compounds gave a green color?

Look at the elements in the compounds that gave a green color.

5. What elements are in these compounds?

6. What element is the same in both compounds?

7. What element must have given you the green color?

8. What compounds gave a bright yellow color?

9. In both compounds, what element is the same?

10. What element must have given you the bright yellow color?

l. Test other materials for elements. You might try chalk dust, baking soda, alum, milk, and salt water. Add the results to your table.

By using the flame test, you were able to analyze substances. You could tell some of the elements in a substance. The color or colors a substance gives in a flame are like fingerprints. Some colors may be similar. But no two elements give exactly the same colors.

11. What compounds could be added to logs and charcoal to give colored flames in fireplaces?

B. Somewhere Over the Rainbow

Have you seen a rainbow recently? Rainbows come from white sunlight. The white sunlight gets separated into many different colors, called a *spectrum.* You can see the same thing by using a prism. The colors of light given off by any glowing object are its spectrum.

You can look more closely at what makes a compound give a green, yellow, or red color in a flame. You and your lab partner can build an instrument to observe spectra. It is called a *spectroscope.*

Museum of Modern Art/Film Stills Archive

What You Need

Cigar or shoe box
Cork borers, #7 or #8 and
#10 or #11
Diffraction grating
Index card
Razor blade, single edge
Ruler, metric
Scissors
Tape

Hole, 15 mm in diameter
25 mm

What to Do

a. In one end of the box, cut a hole 15 mm in diameter. The hole should be 25 mm in from the front edge of the box. It should be centered vertically.

b. At the other end of the box, cut a hole 25 mm in diameter. The center of this hole should line up exactly with the center of the first hole.

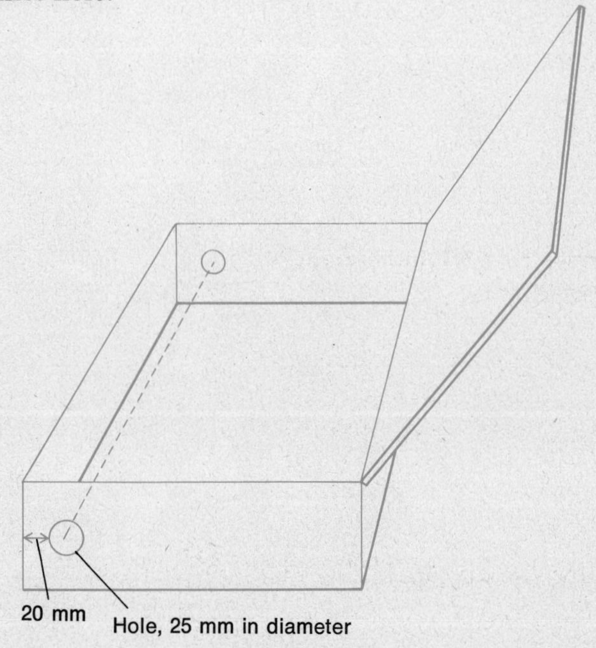

c. Cut a thin slit in an index card. The slit should be about 0.2–0.3 mm wide and 15 mm long. Use a razor blade to make the slit.

d. Trim the card to fit the box.

e. Tape the card over the first hole you cut in the box. (This is the 15 mm hole.) The slit should be centered on the hole and should run up and down.

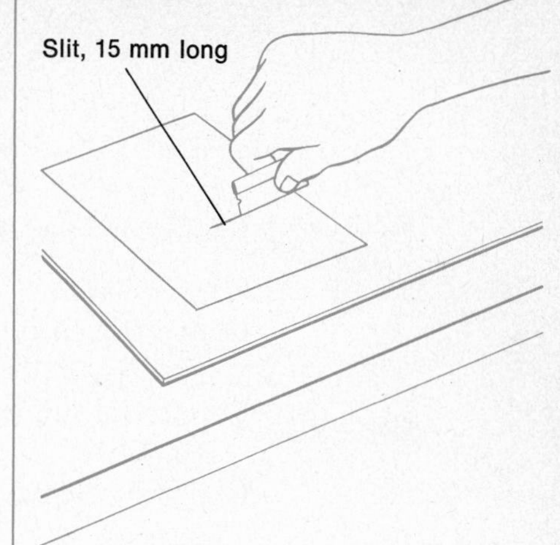

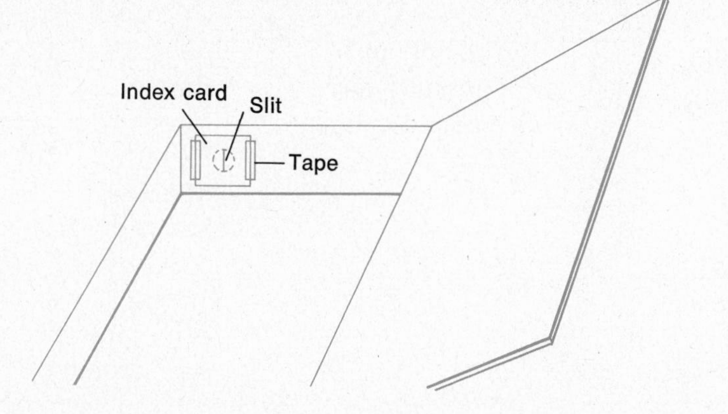

f. Fasten a piece of diffraction grating over the 25 mm hole at the opposite end of the box. It should be fastened to the inside of the box. This is your eyepiece.

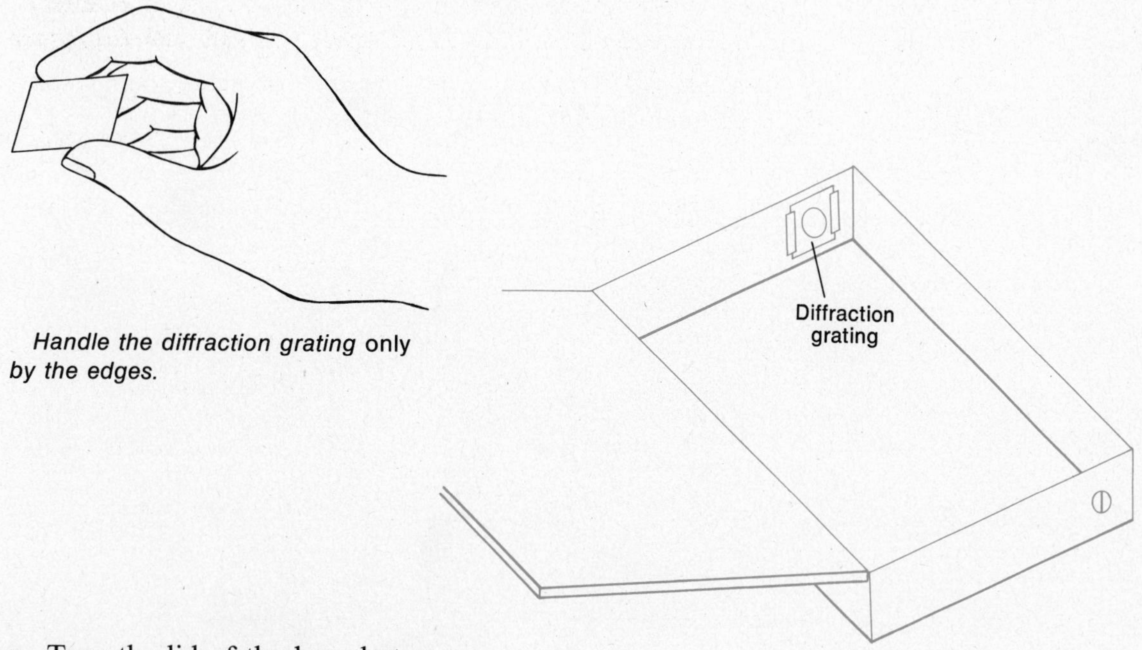

Handle the diffraction grating only *by the edges.*

g. Tape the lid of the box shut.

C. Autumn Leaves

Now you have a spectroscope. What can it do for you? What can it tell you about your world? Will colors look any different through the spectroscope?

What You Need

Copper(II) chloride ($CuCl_2$)
Copper(II) sulfate ($CuSO_4$)
Hydrochloric acid (HCl), 6 *M*
Lithium chloride (LiCl)
Sodium bicarbonate ($NaHCO_3$)
Sodium chloride (NaCl)
Strontium chloride ($SrCl_2$)
Strontium nitrate [$Sr(NO_3)_2$]
Beaker, 50 ml
Evaporating dish
Graduated cylinder, 25 ml
Asbestos pad
Burner
Fluorescent lamp
Light bulb, 100 watt
Matches
Safety goggles
Spectroscope
Tongs
Wire loop, platinum or nichrome

What to Do

a. Point your spectroscope at a fluorescent lamp. You can use a ceiling fixture or a desk lamp. The light should shine through the slit into the spectroscope.

b. Look through the eyepiece. Observe the spectrum.

12. Record in your notebook the colors you see. Use a table like Table 2. Try to draw spectral patterns that look like the spectra observed.

TABLE 2: Colors Observed

Object	*Violet*	*Blue*	*Green*	*Yellow*	*Orange*	*Red*
Fluorescent lamp						
Light bulb, 100 watt						

c. Use your spectroscope to observe the light from an ordinary light bulb.

13. Draw the spectral pattern in Table 2.

You can now check the spectra of the compounds used in Part A.

d. Repeat steps **a** through **k** on pages 113 through 114, but observe the spectra through your spectroscope.

You may also want to observe neon signs, mercury lights, sodium lamps, and any additional light sources your instructor suggests.

14. In Table 2, draw the spectral patterns for all spectra observed.

15. Did any two substances have the same pattern?

16. If they did, did they have the same element in them?

These spectra are called *emission spectra.* They are formed by the energy *given off* (emitted) by atoms when the atoms are heated.

Emission spectra are important in scientific investigations because no two elements have the same spectrum. The spectrum of a given element is as individual as a fingerprint. Scientists can observe the spectra of stars and tell what elements make up a star.

Suppose you had drawn the spectra of all the known elements. Then, if you looked at the spectrum of an unknown sample, you could find what elements were present in it by comparing the spectra. A new spectrum, which didn't match any of the known ones, would mean that a new element must be present. In fact, this is how the elements helium, cesium, and rubidium were first discovered.

D. You're a Colorful Character

Another test to identify the elements in a substance is the borax bead test. The borax bead test is a test for the metals present in a substance.

What You Need

Borax ($Na_2B_4O_7$)
Chromium(III) sulfate [$Cr_2(SO_4)_3$]
Cobalt(II) chloride ($CoCl_2$)
Copper(II) chloride ($CuCl_2$)
Iron(II) chloride ($FeCl_2$)
Manganese dioxide (MnO_2)
Nickel chloride ($NiCl_2$)

Aluminum foil dishes
Asbestos pad
Balance
Burner
Hammer or iron ring
Matches
Safety goggles
Spatula
Weighing paper
Wire loop, platinum or nichrome

What to Do

a. Get 3 grams of borax. Remember the weighing paper.
b. Place the borax in an aluminum foil dish.
c. Place a few crystals of chromium(III) sulfate in another aluminum foil dish.
d. Light and adjust the burner for a hot flame.
e. Heat the wire loop until it is hot.
f. Dip the loop into the mound of borax. Some borax should stick to the wire.

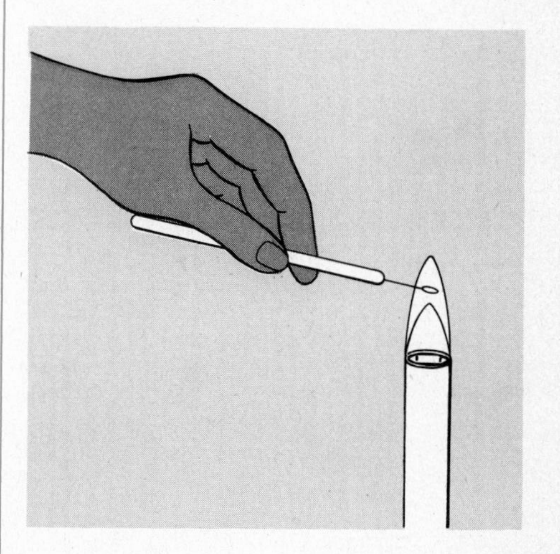

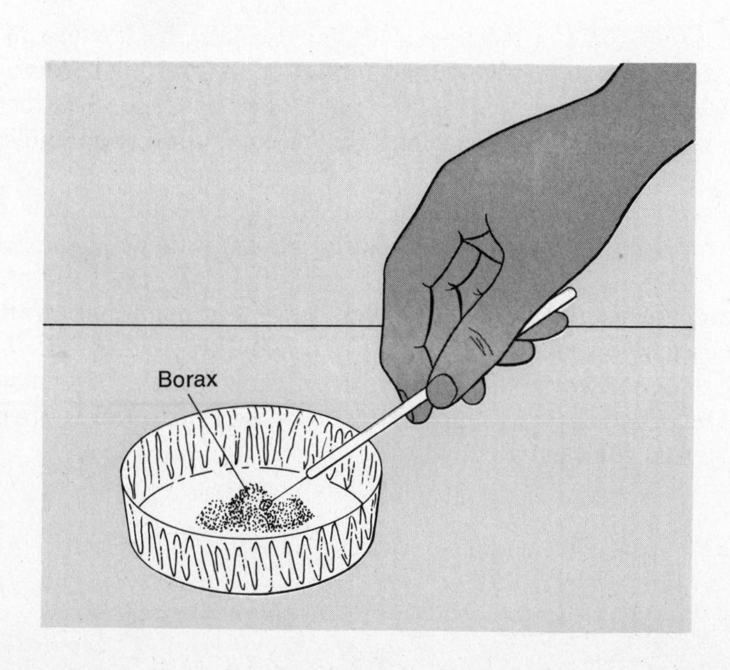

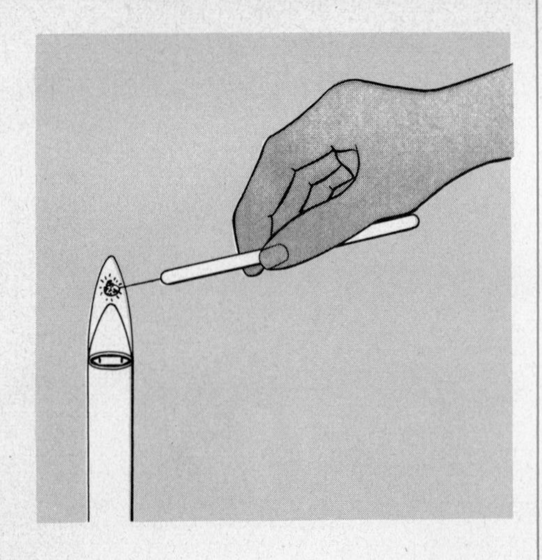

g. Reheat the loop with the borax. Heat until the borax melts into a ball and becomes red hot.
h. Dip the hot borax into the chromium(III) sulfate. Reheat it for two or three minutes.
i. Place the wire on the asbestos pad to cool.
j. Observe the bead.

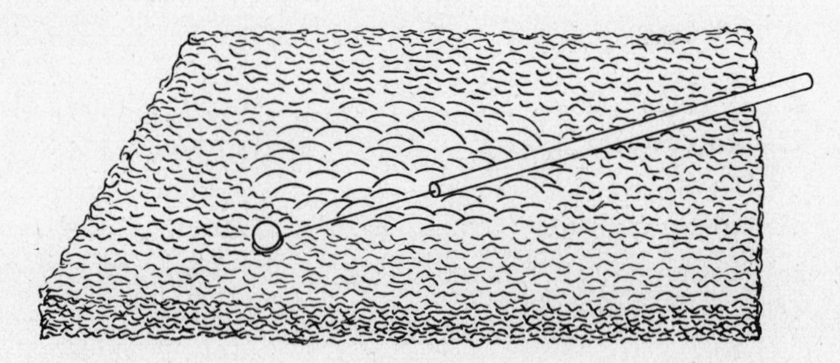

17. Record the color in your notebook in a table like Table 3.

TABLE 3: Results of Borax Bead Test

Compound	*Metal in the compound*	*Color of bead*

The melted borax forms a glass bead. This bead reacts with different elements to give specific colors.

k. Try to remove the bead without breaking it. If you can't, tap the bead gently with a hammer or an iron ring to remove it. Keep it.
l. Repeat steps **e** through **k** for each compound supplied.

18. What compound from Table 3 could be used in making red colored glass? green colored glass?

You have tested many substances. Table 4 is a summary of the elements you have tested.

19. Copy Table 4 into your notebook. Fill in the way you tested for each element and the result of each test.

TABLE 4: Tests for Elements

Element	*Way you tested for the element*	*Result of test*
Chromium (Cr)		
Cobalt (Co)		
Copper (Cu)		
Hydrogen (H)		
Iron (Fe)		
Lithium (Li)		
Manganese (Mn)		
Mercury (Hg)		
Neon (Ne)		
Nickel (Ni)		
Oxygen (O)		
Sodium (Na)		
Strontium (Sr)		
Sulfur (S)		

You have tested for many elements. And you have done a good job. Most of the elements have different properties. Do any of these elements have similar properties? Check out the next investigation.

INVESTIGATION

8

It's All in the Family

A. Family Ties

You have performed some very complicated experiments. Think about all the ways you tested elements. When you ran these tests, you learned some of the innermost secrets of nature. What next?

Do you and your family have some similarities? Think about it. Is your hair texture, height, walking style, eye color, or blood type similar? You and your family are part of a total community. But you are still a family, unique in its own ways.

Do elements have similarities? There are more than 100 different elements. Perhaps some elements have similarities so that they can be grouped together into families of elements.

To find out if elements are similar, their properties have to be tested. Your instructor will test the properties of a few elements. You can try to group the elements by similarities of reactions.

What You Need

Aluminum (Al)
Copper (Cu)
Hydrochloric acid (HCl), 6 *M*
Iron (Fe)
Lead (Pb)
Lithium (Li)
Magnesium (Mg)
Phenolphthalein indicator
Potassium (K)
Sodium (Na)
Zinc (Zn)

Beakers, 250 ml, 9

Asbestos pads, 9
Forceps
Razor blade, single edge
Safety goggles

What to Do

a. Observe the nine elements.

1. Copy Table 1 into your notebook.

TABLE 1: Properties of Nine Elements

Element	*Solid, Liquid, or Gas*	*Color*	*Reaction in water*	*Reaction with phenolphthalein*	*Reaction with 6 M HCl*
Lithium (Li)					
Sodium (Na)					
Potassium (K)					
Magnesium (Mg)					
Zinc (Zn)					
Aluminum (Al)					
Iron (Fe)					
Copper (Cu)					
Lead (Pb)					

2. Are the elements solids, liquids, or gases? Record in your table.

b. Observe the colors of the elements.

3. Record your observations in the table.

You have just recorded some properties of nine elements. Other properties have to be tested before we can group these elements by their similarities. The elements will now be tested in water and acid. A small piece of each element will be added to water and quickly covered with an asbestos pad.

c. Observe the metals in water.

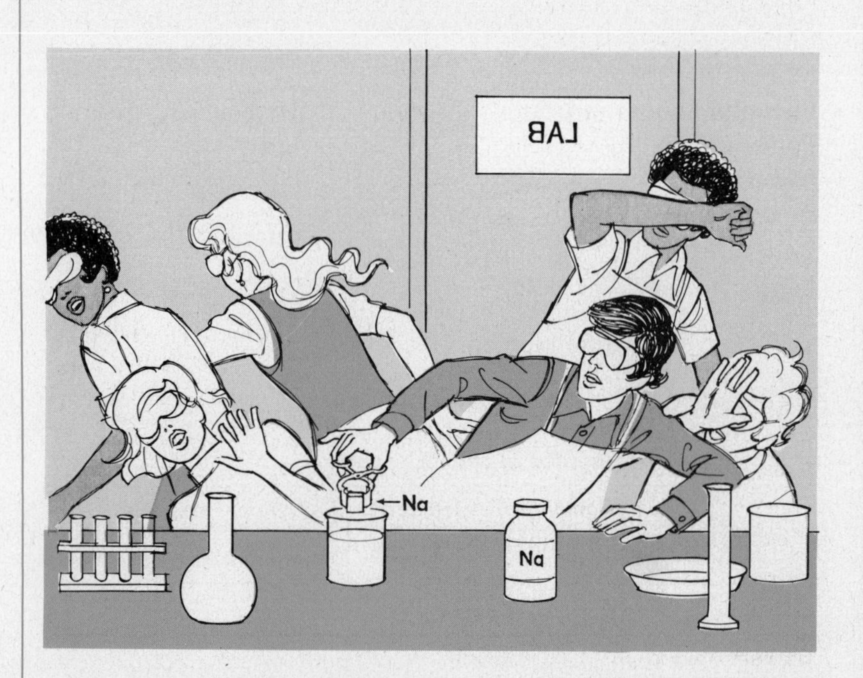

4. Record your observations.

d. Observe what happens when phenolphthalein is added to each beaker of water.

5. Record any changes.

For those metals that didn't react with water, acid will be added to the beaker.

e. Observe the reaction when acid is added.

6. Record any reactions.
7. Which elements sank in water?
8. Which elements didn't sink in water?

f. Look at all the data in Table 1.

9. Did any elements react the same? Which ones?
10. How many different groups can you put the elements in?

If elements react similarly under the same conditions, they are called a *family.*

You now have a good start on understanding the organization of the *periodic table.* The elements are grouped by their similar properties. You have just grouped different elements. Properties like melting point, boiling point, and density might also help in grouping elements.

All elements in a family will not have identical properties. In fact, different families may have some properties that are similar. For example, lead, magnesium, and sodium were probably not grouped together in your experiment. But they were all metals with shiny, silvery colors.

Now try another experiment.

B. Stay in Line, Fella!

In this experiment you will observe the reactions of various compounds. As a result, you will be able to determine which elements react similarly.

What You Need

(in dropper bottles)
Barium chloride ($BaCl_2$), 0.1 *M* solution

Stirring rod
Watch glass

Safety goggles

Barium nitrate
[$Ba(NO_3)_2$], 0.1 *M* solution
Calcium chloride
($CaCl_2$), 0.1 *M* solution
Calcium nitrate
[$Ca(NO_3)_2$], 0.1 *M* solution
Magnesium chloride
($MgCl_2$), 0.1 *M* solution
Magnesium nitrate
[$Mg(NO_3)_2$], 0.1 *M* solution
Sodium carbonate
(Na_2CO_3), 0.1 *M* solution
Strontium chloride
($SrCl_2$), 0.1 *M* solution
Strontium nitrate
[$Sr(NO_3)_2$], 0.1 *M* solution

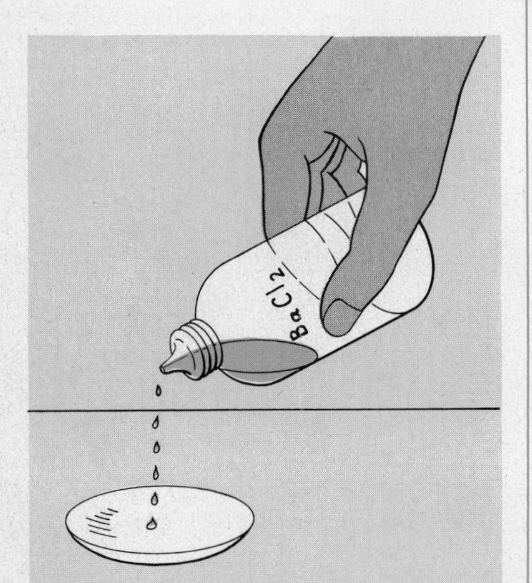

What to Do

a. Place 5 or 6 drops of barium chloride solution in a watch glass.

b. Add 5 or 6 drops of sodium carbonate solution to the watch glass.

c. Stir with a stirring rod.

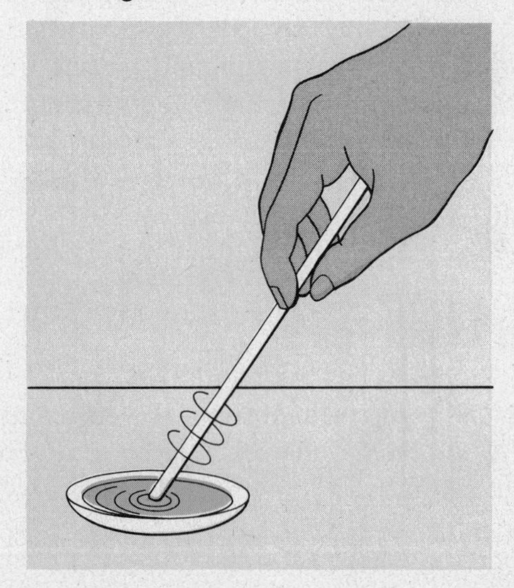

d. Observe any change.

11. Record your results in your notebook in a table like Table 2.

element
mixture
molecule
compound

TABLE 2

Compound mixed with sodium carbonate	Elements in the compound	Observations
Barium chloride	Ba, Cl	

e. Wash and dry your watch glass and stirring rod.
f. Repeat steps **a** through **d** using barium nitrate in step **a.**

12. Were the reactions similar or different?

Look at the elements in the two compounds you mixed with sodium carbonate.

13. What element was in both compounds?

14. What element must have given you the reaction?

g. Repeat steps **a** through **e** using calcium chloride in step **a.**
h. Repeat the same steps using calcium nitrate in step **a.**

15. Were the reactions similar or different?

16. What element was in both compounds?

17. What element must have given you the reaction?

i. Repeat steps **a** through **e** using magnesium chloride in step **a.**
j. Repeat the same steps using magnesium nitrate in step **a.**

18. Were the reactions the same or different?

19. What element was in both compounds?

20. What element must have given you the reaction?

k. Repeat steps **a** through **e** using strontium chloride in step **a.**
l. Repeat the same steps using strontium nitrate in step **a.**

21. Were the reactions the same or different?

22. What element was in both compounds?

23. What element must have given you the reaction?

24. List the answers to questions 14, 17, 20, and 23.

25. Did these elements react with sodium carbonate similarly or differently?

Below is a periodic table.

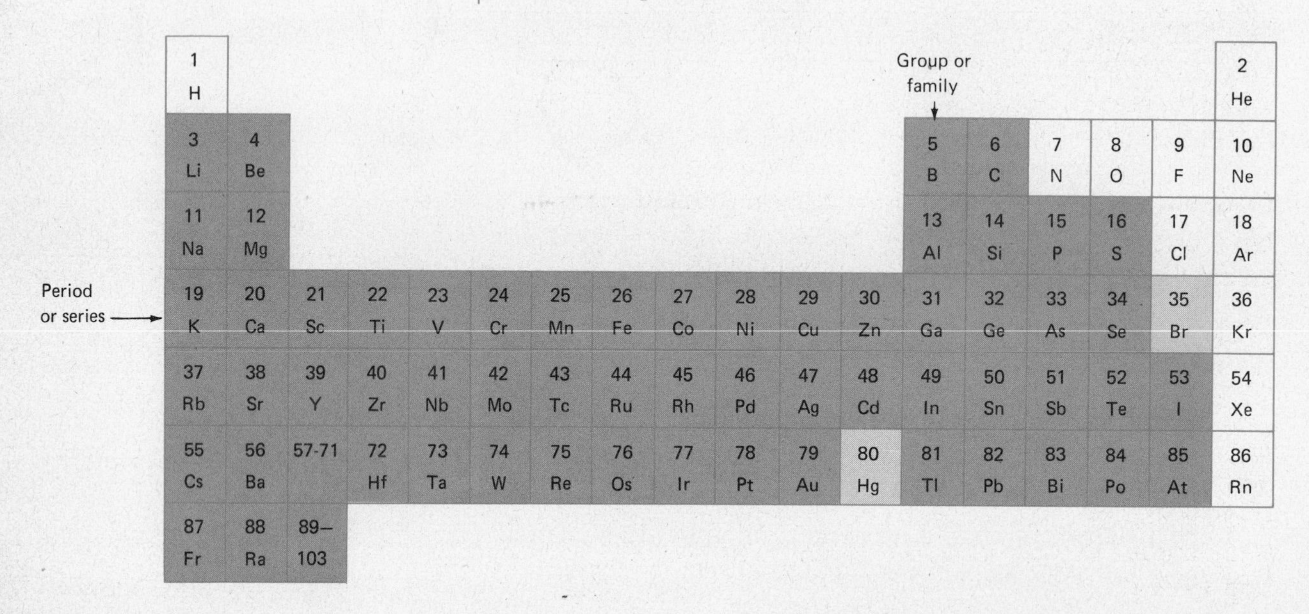

1 H																	2 He
3 Li	4 Be											5 B	6 C	7 N	8 O	9 F	10 Ne
11 Na	12 Mg											13 Al	14 Si	15 P	16 S	17 Cl	18 Ar
19 K	20 Ca	21 Sc	22 Ti	23 V	24 Cr	25 Mn	26 Fe	27 Co	28 Ni	29 Cu	30 Zn	31 Ga	32 Ge	33 As	34 Se	35 Br	36 Kr
37 Rb	38 Sr	39 Y	40 Zr	41 Nb	42 Mo	43 Tc	44 Ru	45 Rh	46 Pd	47 Ag	48 Cd	49 In	50 Sn	51 Sb	52 Te	53 I	54 Xe
55 Cs	56 Ba	57-71	72 Hf	73 Ta	74 W	75 Re	76 Os	77 Ir	78 Pt	79 Au	80 Hg	81 Tl	82 Pb	83 Bi	84 Po	85 At	86 Rn
87 Fr	88 Ra	89–103															

A horizontal row on a periodic table is called a *period* or a *series.* A vertical column is called a *family* or a *group.*

m. Examine the periodic table. Locate the elements in question 24.

26. Do they form part of a period or a family?

You have just discovered another family of elements. The elements in this family are metals.

The most important uses of barium are in paint (barium sulfate) and rat poison (barium carbonate).

Calcium is the fifth most abundant element in the earth's crust. Calcium is an essential part of leaves, bones, teeth, and shells.

You have probably all used magnesium and its compounds. Magnesium is used in flash bulbs and flares, and in airplane construction.

Strontium gives a bright scarlet color with a flame test. It is therefore used in fireworks and flares.

C. Background on the Halogen Family

The halogen family consists of the elements fluorine (F), chlorine (Cl), bromine (Br), iodine (I), and astatine (At). Under normal conditions, chlorine is a greenish gas. Bromine is a dark red liquid, and iodine a shiny purplish-black solid.

Fluorine is a pale yellow gas. It smells something like chlorine, but it is *extremely* dangerous to inhale. It is a very active compound—a little too active for us to study in this course.

Fluorine reacts violently with water. A mixture of hydrogen and fluorine has been studied for possible use as a rocket fuel.

A variety of useful compounds contain fluorine. You have probably seen many ads for different brands of toothpaste which contain fluoride compounds, usually stannous fluoride. Some communities also add fluorides to their water supply to help stop tooth decay.

Compounds of fluorine with carbon and hydrogen, known commercially as Freons and Teflons, have many uses. Freons are used as refrigerants. Teflons are resistant to corrosive chemicals.

27. What uses do you know for Teflon.

Chlorine is a gas of great importance. It has many uses which help us live more safely. Chlorine is added to drinking water to kill bacteria. It is also added to swimming pools for the same reason. (Have you ever smelled that "funny" odor in a swimming pool?) When you bleach your clothing, you use a strong chlorine compound known as sodium hypochlorite. Too much chlorine, however, can kill. It was used as a weapon in World War I.

Bromine was discovered in 1826 by a French chemist named Balard while he was studying the minerals found in sea water. It is one of the two elements which are liquid at room temperature.

An important use of bromine today is in the manufacture of anti-knock gasoline. Another use is in medicine. Compounds containing bromine are used as sedatives or sleeping pills.

The element iodine is found in ocean water and in certain kinds of seaweed. Small amounts of iodine are essential to human life. Iodine is added to common table salt to make sure that people get enough of it. Goiter is a disease caused by lack of iodine. This same element, dissolved in alcohol, forms the familiar "tincture of iodine."

Astatine was not discovered until 1940. It is a very rare element. Astatine is radioactive and very unstable. Therefore compounds containing astatine are not commercially useful.

All of the halogen elements are poisonous and corrosive. Use extreme caution in handling them. Do not breathe the vapors given off by these elements.

Some of the properties of the halogens are shown in Table 3.

TABLE 3: Common Properties of Halogens

Element	*State*	*Color*	*Boiling point*
Fluorine (F_2)	gas	yellow	−187°C
Chlorine (Cl_2)	gas	greenish-yellow	−34.6°C
Bromine (Br_2)	liquid	dark red	58.8°C
Iodine (I_2)	solid	purplish-black	184.4°C

28. What happens to the colors of the halogen elements as you go down the halogen family (from fluorine to iodine)?

29. What happens to the boiling points as you go down the halogen family?

30. In what ways are these elements alike?

31. In what ways are they different?

In studying families of elements, you have found there is an order in the family. Now try to find out the organization of the periodic table.

INVESTIGATION 9

Order in the Court

A. We Will Not Table This Matter!

In the last few investigations you have learned that:

a. Elements are composed of atoms.
b. All materials are composed of elements.
c. Elements can be solids, liquids, or gases.
d. Elements can be identified with specific tests.
e. Elements have characteristic spectra.
f. Elements have properties such as density, color, melting point, magnetic attraction, and solubility.
g. Some elements have similar properties and form a family or group.

You have been busy and doing a good job. You have a wealth of information. You've been stockpiling this information about the elements. Now it's time to get that information organized and put it to work. So, get organized!

To begin, how are you going to get all of those elements into some sort of order? How would you organize them?

The early chemists organized, or classified, the elements according to convenience and practical use. The chemists first grouped elements as metals and non-metals; then as heavy elements, light elements, and gaseous elements; and later, according to their chemical properties.

TABLE 1: **The Periodic Table**

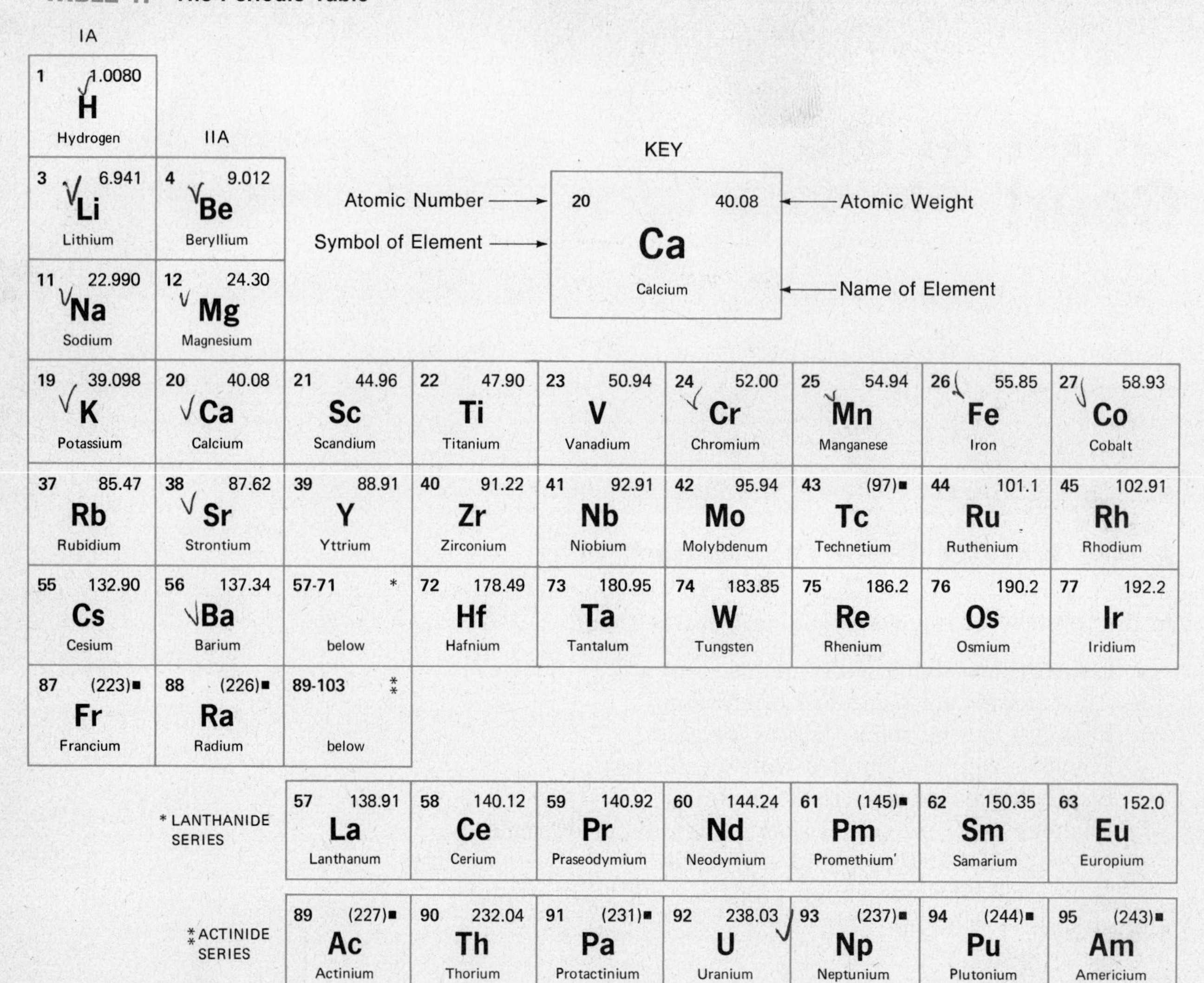

IA	IIA							
1 1.0080 **H** Hydrogen								
3 6.941 **Li** Lithium	4 9.012 **Be** Beryllium							
11 22.990 **Na** Sodium	12 24.30 **Mg** Magnesium							
19 39.098 **K** Potassium	20 40.08 **Ca** Calcium	21 44.96 **Sc** Scandium	22 47.90 **Ti** Titanium	23 50.94 **V** Vanadium	24 52.00 **Cr** Chromium	25 54.94 **Mn** Manganese	26 55.85 **Fe** Iron	27 58.93 **Co** Cobalt
37 85.47 **Rb** Rubidium	38 87.62 **Sr** Strontium	39 88.91 **Y** Yttrium	40 91.22 **Zr** Zirconium	41 92.91 **Nb** Niobium	42 95.94 **Mo** Molybdenum	43 (97)■ **Tc** Technetium	44 101.1 **Ru** Ruthenium	45 102.91 **Rh** Rhodium
55 132.90 **Cs** Cesium	56 137.34 **Ba** Barium	57-71 * below	72 178.49 **Hf** Hafnium	73 180.95 **Ta** Tantalum	74 183.85 **W** Tungsten	75 186.2 **Re** Rhenium	76 190.2 **Os** Osmium	77 192.2 **Ir** Iridium
87 (223)■ **Fr** Francium	88 (226)■ **Ra** Radium	89-103 ** below						

KEY

Atomic Number → 20	40.08 ← Atomic Weight
Symbol of Element → **Ca**	
Calcium ← Name of Element	

* LANTHANIDE SERIES	57 138.91 **La** Lanthanum	58 140.12 **Ce** Cerium	59 140.92 **Pr** Praseodymium	60 144.24 **Nd** Neodymium	61 (145)■ **Pm** Promethium	62 150.35 **Sm** Samarium	63 152.0 **Eu** Europium
** ACTINIDE SERIES	89 (227)■ **Ac** Actinium	90 232.04 **Th** Thorium	91 (231)■ **Pa** Protactinium	92 238.03 **U** Uranium	93 (237)■ **Np** Neptunium	94 (244)■ **Pu** Plutonium	95 (243)■ **Am** Americium

As the number of known elements increased, and more was learned about them, things got pretty confusing.

Many clever and interesting plans for grouping the elements were proposed from time to time.

			IIIA	IVA	VA	VIA	VIIA	VIIIA
								2 4.003 **He** Helium
			5 10.81 **B** Boron	6 12.011 **C** Carbon	7 14.007 **N** Nitrogen	8 15.9994 **O** Oxygen	9 19.00 **F** Fluorine	10 20.179 **Ne** Neon
			13 26.98 **Al** Aluminum	14 28.09 **Si** Silicon	15 30.974 **P** Phosphorus	16 32.064 **S** Sulfur	17 35.453 **Cl** Chlorine	18 39.948 **Ar** Argon
28 58.70 **Ni** Nickel	29 63.55 **Cu** Copper	30 65.38 **Zn** Zinc	31 69.72 **Ga** Gallium	32 72.59 **Ge** Germanium	33 74.92 **As** Arsenic	34 78.96 **Se** Selenium	35 79.904 **Br** Bromine	36 83.80 **Kr** Krypton
46 106.4 **Pd** Palladium	47 107.868 **Ag** Silver	48 112.40 **Cd** Cadmium	49 114.82 **In** Indium	50 118.69 **Sn** Tin	51 121.75 **Sb** Antimony	52 127.60 **Te** Tellurium	53 126.90 **I** Iodine	54 131.30 **Xe** Xenon
78 195.09 **Pt** Platinum	79 197.0 **Au** Gold	80 200.59 **Hg** Mercury	81 204.37 **Tl** Thallium	82 207.2 **Pb** Lead	83 208.98 **Bi** Bismuth	84 (209)■ **Po** Polonium	85 (210)■ **At** Astatine	86 (222)■ **Rn** Radon

64 157.25 **Gd** Gadolinium	65 158.93 **Tb** Terbium	66 162.50 **Dy** Dysprosium	67 164.93 **Ho** Holmium	68 167.26 **Er** Erbium	69 168.93 **Tm** Thulium	70 173.04 **Yb** Ytterbium	71 174.97 **Lu** Lutetium
96 (247)■ **Cm** Curium	97 (247)■ **Bk** Berkelium	98 (251)■ **Cf** Californium	99 (254)■ **Es** Einsteinium	100 (257)■ **Fm** Fermium	101 (258)■ **Md** Mendelevium	102 (259)■ **No** Nobelium	103 (260)■ **Lr** Lawrencium

Chemistry is basically an organized way of studying the elements and their properties.

Above is a periodic table. Scientists have organized the table this way for a reason. Maybe we can discover the order and the reason. You have already experimented with some families. What about other families? Why is the table organized into series? Try to find out.

Atomic number and *atomic weight* will be defined later, when you understand the order in the periodic table.

TABLE 2

1 1.0080 **H** Hydrogen							2 4.003 **He** Helium
3 6.941 **Li** Lithium	4 9.012 **Be** Beryllium	5 10.81 **B** Boron	6 12.011 **C** Carbon	7 14.007 **N** Nitrogen	8 15.9994 **O** Oxygen	9 19.00 **F** Fluorine	10 20.179 **Ne** Neon
11 22.990 **Na** Sodium	12 24.30 **Mg** Magnesium	13 26.98 **Al** Aluminum	14 28.09 **Si** Silicon	15 30.974 **P** Phosphorus	16 32.064 **S** Sulfur	17 35.453 **Cl** Chlorine	18 39.948 **Ar** Argon
19 39.098 **K** Potassium	20 40.08 **Ca** Calcium						

B. We've Got Your Number

Above is a periodic table of the first twenty elements. Studying just a few elements makes it easier to understand the organization of the table. Look at the atomic numbers and answer the following questions.

1. As you go down the first group (family), do the atomic numbers increase or decrease?

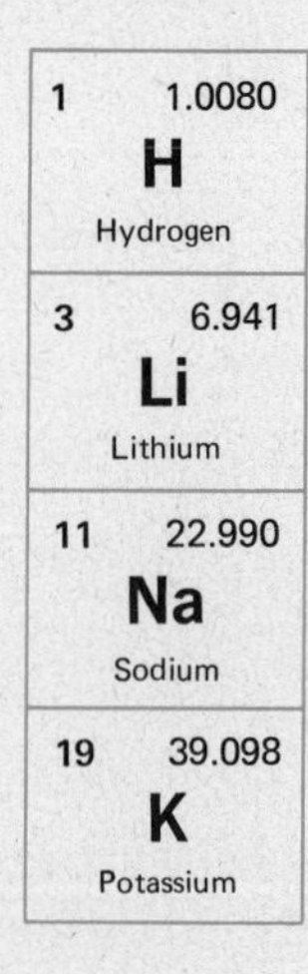

2. Look at the other families you have studied. Do the atomic numbers increase or decrease as you go down a family?

3. Do atomic numbers increase or decrease as you go across a series?

11 22.990 Na Sodium	12 24.30 Mg Magnesium

13 26.98 Al Aluminum	14 28.09 Si Silicon	15 30.974 P Phosphorus	16 32.064 S Sulfur	17 35.453 Cl Chlorine	18 39.948 Ar Argon

4. Do the atomic numbers increase by one as you go down a family or as you go across a series?

That's right! The atomic number increases by one as you go across a series. So the chart above could be made into a row of twenty elements.

1 H	2 He	3 Li	4 Be	5 B	6 C	7 N	8 O	9 F	10 Ne	11 Na	12 Mg	13 Al	14 Si	15 P	16 S	17 Cl	18 Ar	19 K	20 Ca

But if the chart were one long row, it wouldn't show families of elements. And you know that elements with similar properties can be grouped into families. So the chart has both rows and columns. The elements are put in rows because of the atomic number. And they are put in columns because of similar properties.

3 Li	4 Be	5 B	6 C	7 N	8 O	9 F	10 Ne
11 Na	12 Mg	13 Al	14 Si	15 P	16 S	17 Cl	18 Ar

Look at the atomic weights to answer the following questions.

5. Do the atomic weights increase or decrease as you go down a family of elements?

6. Do atomic weights increase or decrease as you go across a series of elements?

C. Charge!

All of the atoms of an element have the same properties. What makes them have the same properties? What makes them different from other atoms? To understand the answers to these questions, a study of the smaller particles of an atom will help. There are three basic parts to an atom.

$$\text{ATOM}\begin{cases}\text{neutrons, n}\\ \text{protons, } p^+\\ \text{electrons, } e^-\end{cases}$$

Protons have a positive charge, and are written as $\mathbf{p^+}$.
Electrons have a negative charge, and are written as $\mathbf{e^-}$.
Neutrons are neutral, with no charge, and are written as **n.**

Atoms are made up of these particles. How are these particles important to the periodic chart's organization? Look at what has been found and try to understand the periodic chart.

Below is part of a periodic chart. After answering questions 7–20, you can complete the information for the rest of the first twenty elements.

TABLE 3

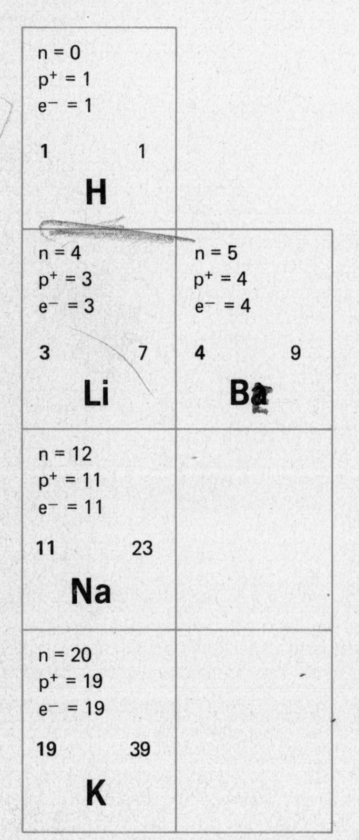

n = 6
p^+ = 5
e^- = 5
5 11
B

Atomic weights are rounded off.

7. How many positive charges (protons, p^+) does hydrogen have?

8. How many negative charges (electrons, e^-) does hydrogen have?

9. What is the atomic number of hydrogen?

10. How many positive charges does lithium have?

11. How many negative charges does lithium have?

12. What is the atomic number of lithium?

Look at the other elements and compare the number of protons and electrons.

13. Is the number of protons equal or unequal to the number of electrons?

14. Does the number of protons equal the atomic number or atomic weight?

15. Does the number of electrons equal the atomic number or the atomic weight?

The answers to the questions above can be summarized in the following way:

$$\text{Atomic number} = \underset{(p^+)}{\text{number of protons}} = \underset{(e^-)}{\text{number of electrons}}$$

16. Uranium has an atomic number of 92. How many protons and how many electrons are in an atom of uranium?

Now examine the periodic table and see if you can account for the atomic weight. What numbers give the atomic weight? Almost all the weight of an atom is due to protons and neutrons. The weight of an electron is very small compared to that of a proton or a neutron.

17. Does the number of protons plus the number of neutrons equal the atomic number or atomic weight for:
   a. lithium (Li)?
   b. boron (B)?

This can be summarized in the following way:

$$\text{Atomic weight} = \underset{(p^+)}{\text{number of protons}} + \underset{(n)}{\text{number of neutrons}}$$

Therefore, you can summarize atomic number and atomic weight as follows:

Either one equals the atomic number — {Protons, p^+; Electrons, e^-}

{Neutrons, n; Protons, p^+} — added together give the atomic weight

18. Which particle of an atom helps you find both the atomic number and atomic weight?

19. Given the information below, determine the number of p^+, e^-, and n.

26
Fe (iron)
56

20. Copy the table below into your notebook and fill in the blanks.

TABLE 4

Element	*Symbol*	*n*	p^+	e^-	*Atomic number*	*Atomic weight*
Cobalt	Co	32			27	
Uranium	U			92		238
Arsenic	As	42	33			
Mercury	Hg				80	201
Chromium	Cr	28		24		

Now go back to Table 3 on page 138.

21. Copy Table 3 into your notebook. Complete the information for the other elements in the table.

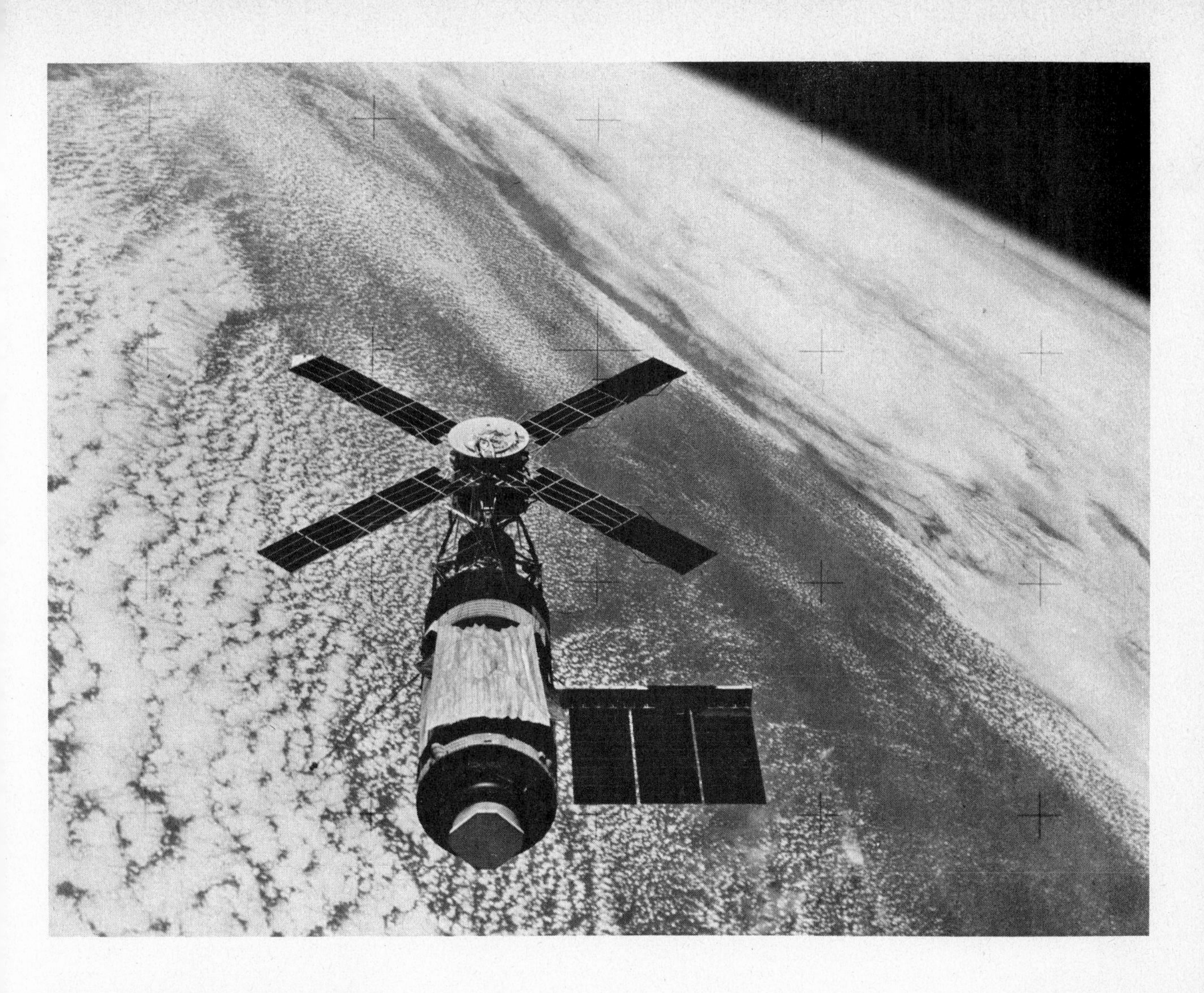

D. Let's Go Into Orbit

In your previous studies, you used the idea of a model. Atoms and molecules were too small to see. You used nuts and bolts to represent atoms and molecules.

The parts of atoms are even smaller. Therefore, you will again adopt a model. Protons, electrons, and neutrons don't look like the models you will draw. But it helps to draw them to understand them.

Every atom has a nucleus. Neutrons and protons are located in the nucleus. In addition, all atoms have electrons.

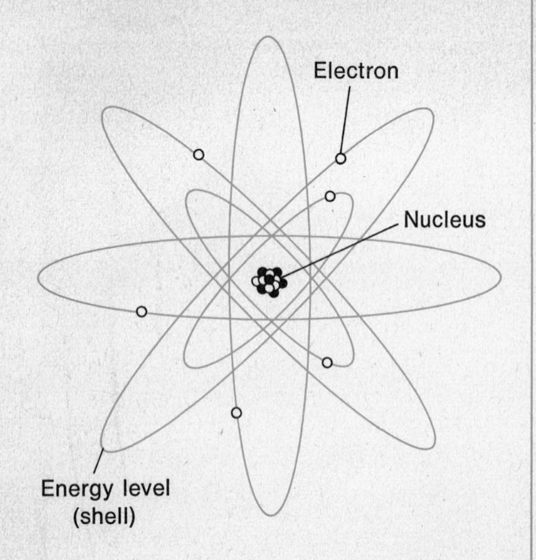

The electrons are very light particles which orbit about the nucleus of the atom. An electron has a negative electrical charge. Because they have similar charges, which act to push each other away, electrons don't bump into each other. The distance from the electrons to the nucleus is more than 10,000 times the diameter of the nucleus.

The proton is one of the two kinds of particles found in the nucleus. It is about 2,000 times heavier than the electron. It has a positive electrical charge.

The neutron is the other kind of particle found in the nucleus. It has the same weight as the proton, but is neutral. This means that it has no electrical charge.

If you were to imagine a nucleus the size of a small marble, the whole atom would be as big as a giant balloon, with a diameter greater than the length of a football field.

In the atom, the protons and neutrons are packed together to form a tight nucleus. Whirling around the nucleus at great distances are the electrons. If you think about it, you'll see that the atom is mostly space.

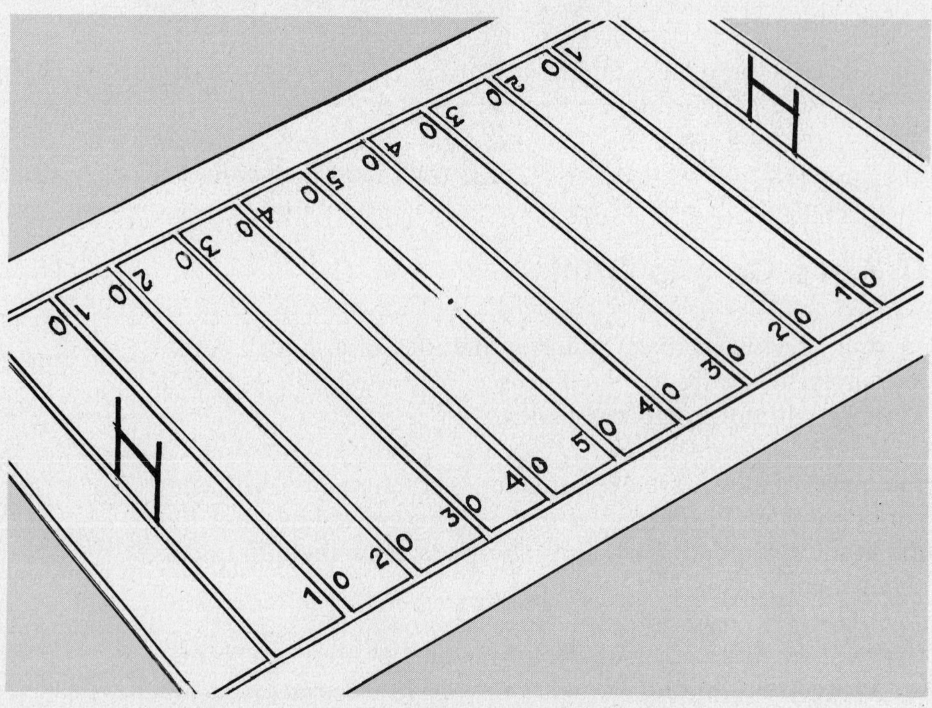

If the atom were the diameter of a football field, this would be the nucleus.

At the right is a more useful model of the atom. It is easier to work with. The energy levels (shells) represent the paths of the electrons. The electrons move around the nucleus, but not in circles. For convenience, a circle is used as a model to represent the path of an electron.

Energy levels (shells)

Nucleus (protons, neutrons)

1 2 3

Electrons

Only a part of the above model will be used in discussing the periodic table. Instead of a complete circle, only part of the circle will be used.

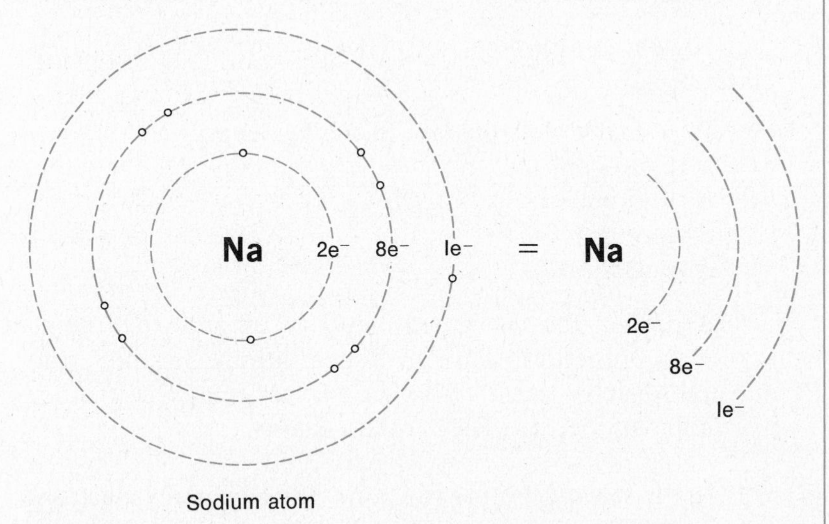

Sodium atom

Below is a chart with some of the elements drawn. The questions which follow will help you to see the organization of the elements in the periodic chart.

TABLE 5

Energy level 1, H, $1e^-$							
1, 2, Li, $2e^-$, $1e^-$	1, 2, Be, $2e^-$, $2e^-$	1, 2, B, $2e^-$, $3e^-$					
1, 2, 3, Na, $2e^-$, $8e^-$, $1e^-$	1, 2, 3, Mg, $2e^-$, $8e^-$, $2e^-$						
1, 2, 3, 4, K, $2e^-$, $8e^-$, $8e^-$, $1e^-$							

22. What number represents the energy level closest to the nucleus?

23. What number represents the energy level second from the nucleus?

24. How many energy levels do the following atoms have:
    a. hydrogen?
    b. lithium?
    c. sodium?
    d. potassium?

25. As you go down a family, the number of energy levels increases by how much?

26. How many electrons are in the last energy level for:
    a. hydrogen?
    b. lithium?
    c. sodium?
    d. potassium?

The electrons in the last energy level of an atom determine many of its properties. Since hydrogen, lithium, sodium, and potassium all have one electron in the last energy level, they have similar properties. They form a family.

27. In the second column (Be, Mg, Ca), how many electrons are in the last energy level?

28. How many electrons would you expect aluminum (Al) to have in its last energy level?

Fantastic! You now know why elements are grouped into families. And why they have similar properties. Nice work!

What about the rows (series) in the periodic table? How do the energy levels and electrons change as you go across a row?

29. Look at the second row (Li, Be, B, . . .). How many energy levels does each of these elements have?

30. How many electrons are in the last energy levels of:
    a. lithium?
    b. beryllium?
    c. boron?
    d. carbon?

Go back to Table 5 on page 143.

31. Copy the table into your notebook. Draw in the electrons and energy levels for the rest of the elements.

E. The Element Series

Your instructor will do a demonstration with three elements in a series. The elements are sodium, magnesium, and aluminum. In each reaction hydrogen will be produced. The quantity of hydrogen produced by each element can be compared. This should help you understand a series of elements. Equal numbers of atoms of each element will be reacted.

What You Need

Aluminum (Al), 0.27 g
Hydrochloric acid (HCl), 6 *M*
Magnesium (Mg), 0.24 g
Methanol (CH_3OH)
Sodium (Na), 0.23 g

Flask, 250 ml
Glass bend, right angle
Graduated cylinder, 100 ml
Graduated cylinder, 500 ml

Ring or clamp
Ring stand
Rubber tubing
Stopper, 1-hole
Trough, water

What to Do

a. Set up the apparatus as shown below, filling the graduate with water.

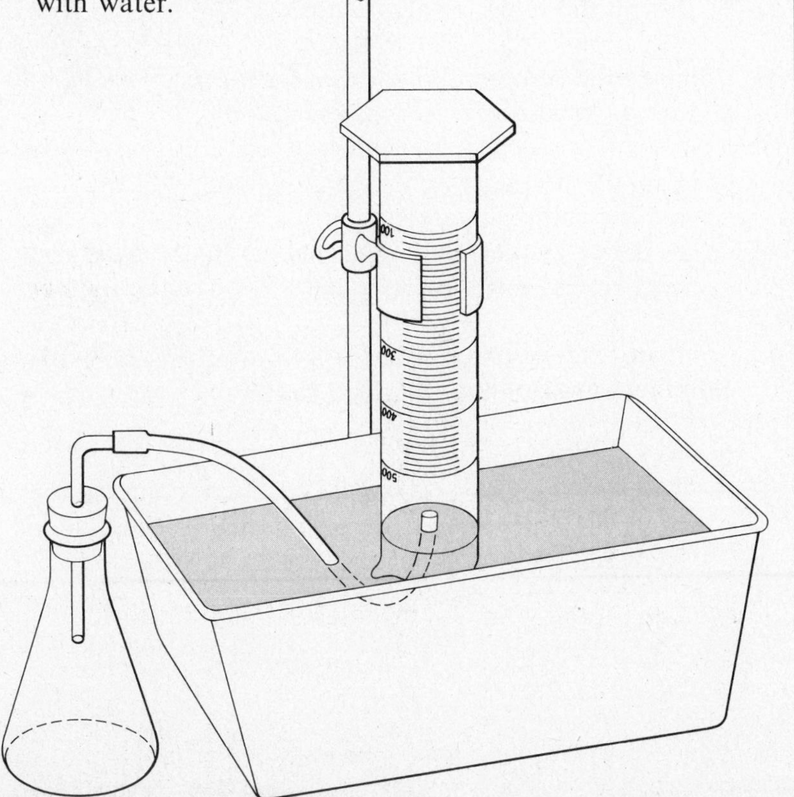

b. Place 100 ml of methanol in the flask.
c. Add the sodium to the flask and stopper quickly.
d. Observe the volume of hydrogen collected.

32. Copy Table 6 into your notebook. Fill it in for sodium.

e. Clean the flask and refill the graduate with water.
f. Place 150 ml of hydrochloric acid in the flask.
g. Add the magnesium and stopper quickly.
h. Observe the volume of hydrogen collected.

33. Fill in the table for magnesium.

TABLE 6

Element	*Volume of hydrogen collected*
Sodium (Na)	
Magnesium (Mg)	
Aluminum (Al)	

i. Repeat steps **e** through **h** using aluminum in step **g.**

34. Fill in the table for aluminum.

j. Compare the volumes of hydrogen for the reactions of magnesium and sodium.

35. Pick the best answer:
    a. The volumes were equal.
    b. Twice as much was collected for magnesium.
    c. Twice as much as collected for sodium.

k. Compare the volumes of hydrogen for the reactions of aluminum and sodium.

36. Pick the best answer:
    a. The volumes were equal.
    b. Three times as much was collected for aluminum.
    c. Three times as much was collected for sodium.

37. Copy Table 7 into your notebook and fill it in.

TABLE 7

Element	*Volume of hydrogen collected*	*Number of electrons in the last energy level*
Sodium (Na)	100 ml	
Magnesium (Mg)	200 ml	
Aluminum (Al)	300 ml	
Silicon (Si)		
Carbon (C)		

You have just completed a large task. And you know a lot. What you have learned will help you to write formulas and equations, and to predict other information.

You discovered that calcium, barium, strontium, and magnesium form a family. They have similar properties.

38. Calcium oxide has the formula CaO. What do you think will be the formulas of barium oxide, strontium oxide, and magnesium oxide? Write them in your notebook.

To find out why these formulas are correct, tune in to the next Unit.

Home Activities

Investigation 1

1. Identify home products that are solids, liquids, and gases. List in a table what they are made of. Then estimate whether they would have high or low melting and boiling points.

Brand name	Solid, liquid, or gas	Melting point	Boiling point

2. Why do different foods have to be cooked at different temperatures?
3. Why does ice float on water?
4. Completely fill a glass with room temperature water. Carefully place an ice cube in the water. What do you expect to happen to the level of water in the glass as the ice cube melts? What does happen? Why?
5. Do all kinds of wood float? Try several kinds of wood and list their relative density in water. The more wood there is above water, the less dense the wood. Which wood is most dense? least dense?

Investigation 2

1. What home products are pure substances? mixtures?
2. Place drops of some liquid foods on the surface of the water in a glass. Do they mix, float, or sink? Make a data table of your results.

Investigation 3

1. What foods in the kitchen dissolve in water?
2. What paints and oils dissolve in water?

Investigation 4

1. Put an empty soda bottle in the refrigerator until it is cold. When it is cold, wet the rim and place a dime on it. Make sure there is water between the dime and rim. Wait a while. What happens? Why?
2. Visit a water purification plant and report how it cleans water.
3. Look up in a book various ways of purifying water. Make charts and diagrams to help you report your findings to the class.
4. Keep a chart of how much water you use in one day or one week.
   a) How much would you use in one month?
   b) How much would you use in one year?
   c) How much would your town use in one year?

Investigation 5

1. From the labels of foods, cut out the parts listing the ingredients. Bring several to class. Read a list of ingredients to a classmate and ask him or her to guess what the food is.

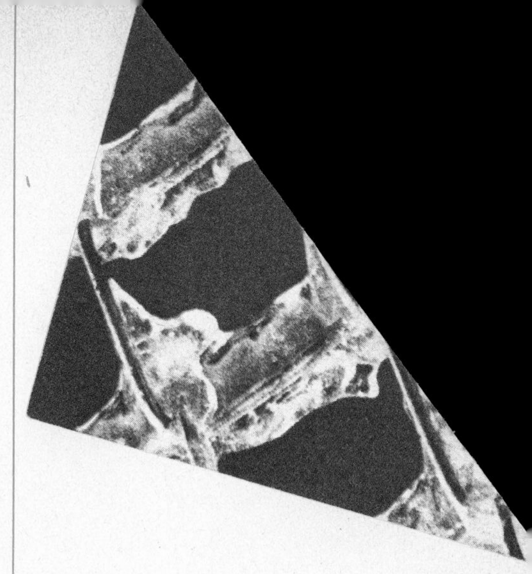

Investigation 6

1. Examine several products found in the kitchen and workshop area. What elements are in each? What are the properties of these elements? Use a table:

Product	Element in the product	Properties of the element

2. Pick an element from the periodic table and look up its properties and uses.

Investigation 7

1. Bring to class small samples of toothpaste, baking soda, etc. Identify some of the elements in them using the flame test.
2. Make some logs (possibly by rolling up magazines or newspapers) that burn colored flames.

Investigation 8

1. Look up two elements in a family and compare their boiling points, melting points, and uses. List some similarities and differences in the elements.

	Element	Boiling point	Melting point	Uses	Similarities	Differences
Smaller atomic number						
Larger atomic number						

Investigation 9

1. To see what kind of substances give up a charge, do the following. Blow up a balloon and tie the end. Rub the balloon on a substance (nylon, rayon, wool, carpet, polystyrene) and see if it will stick to your clothing or to the wall.
2. When your hair is clean and dry, run a comb through it several times. Slowly bring the teeth of the comb towards a stream of water running from a faucet. The smaller the stream the better. Explain what happens.

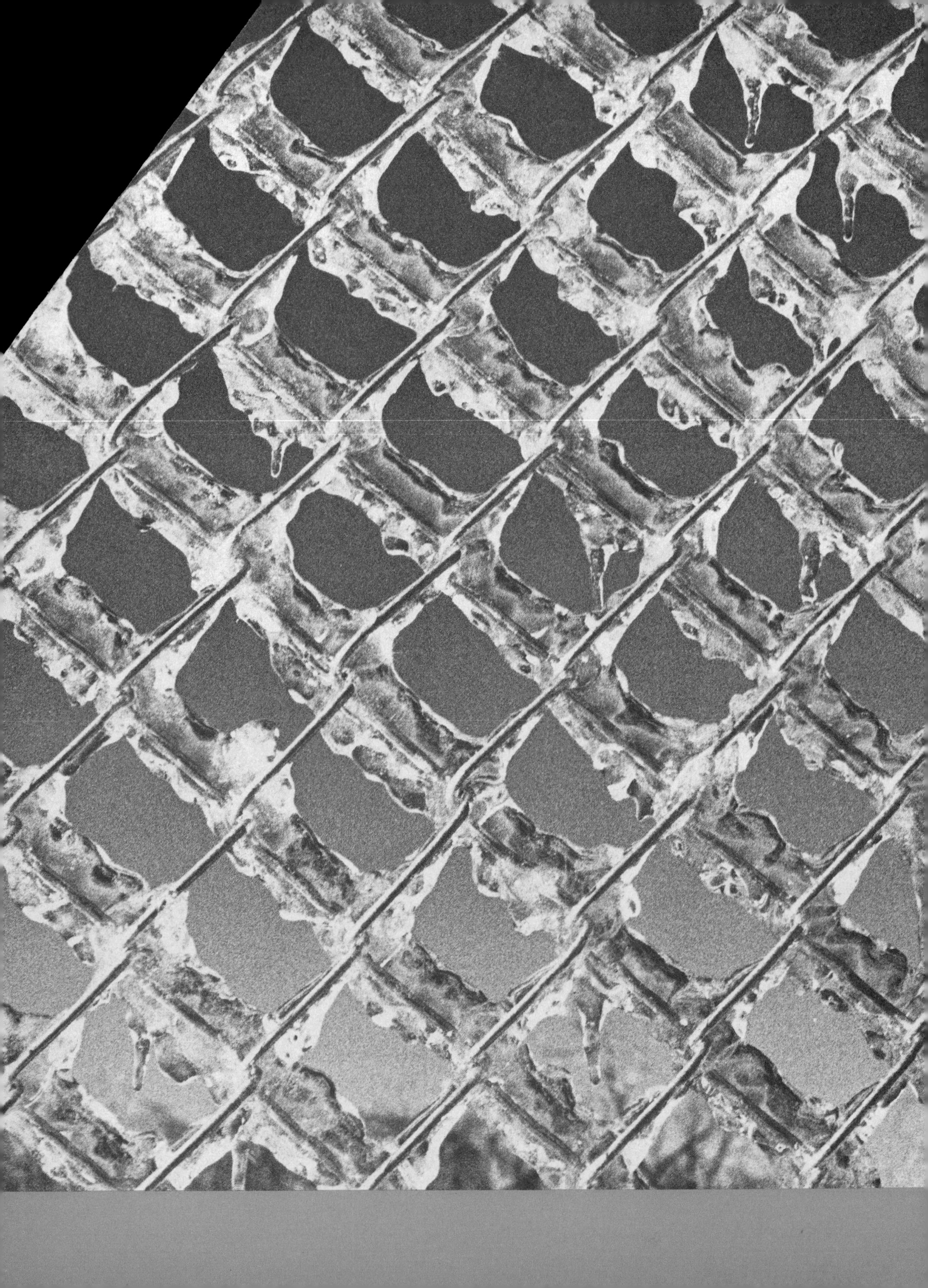

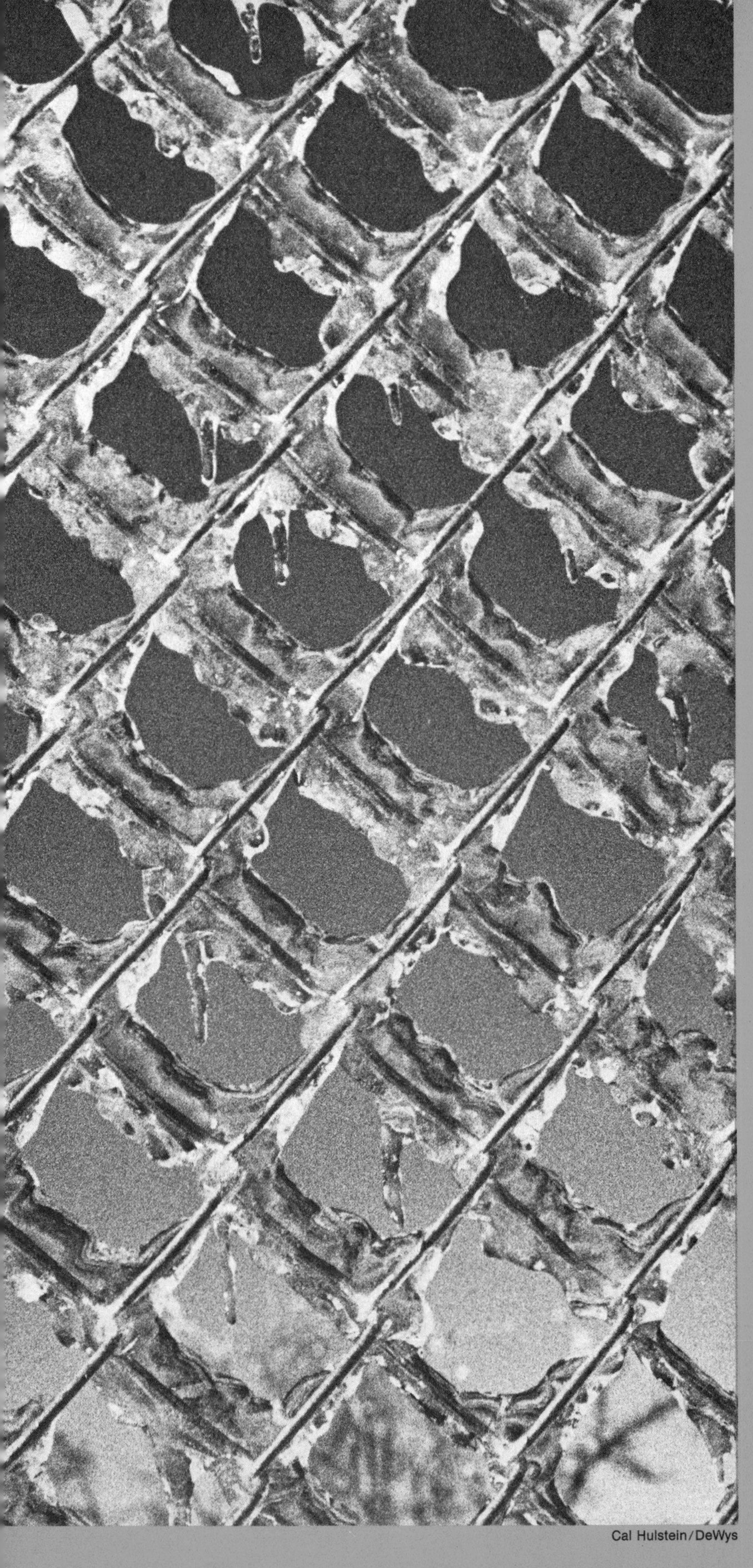

Cal Hulstein/DeWys

UNIT 3

INVESTIGATION 1

According to Hoyle

Learn the rules!

As you have worked through the first two Units, you have undoubtedly noticed the chemical formulas. You may have wondered why water is always H_2O, and never H_5O or HO or HO_2; or why hydrochloric acid is always HCl, and never H_2Cl or HCl_2. The formulas are a shorthand in the language of chemistry. This shorthand is used to describe chemical properties and reactions.

A Scholastic Magazine cartoon

"Any questions?"

A. Keep Your "Ion" This!

The inert elements are a group (family) of elements that seldom react. Helium, neon, argon, and the other inert elements generally don't form compounds. All other elements form compounds. So the inert elements must somehow be special.

Sodium (Na) Chlorine (Cl) Magnesium (Mg) Oxygen (O)	form compounds, for example, Sodium chloride (NaCl) Magnesium oxide (MgO)
Neon (Ne) Argon (Ar)	will not form compounds

Let's try to find out why the inert elements don't react. The inert elements are all gases. Because they don't react, they are called *chemically stable.* Since electrons are important in chemical reactions, examine how many electrons the inert gases have.

1. Copy Table 1 in your notebook. Complete the table.

TABLE 1: Inert Elements

Element	*Helium He*	*Neon Ne*	*Argon Ar*	*Krypton Kr*	*Xenon Xe*	*Radon Rn*
Number of electrons the element has						

An element becomes stable if it has the same number of electrons as one of the inert gases.

Elements that form compounds can become chemically stable. Therefore, the elements must end up with the same number of electrons as one of the inert elements. An element can form different stable compounds. For example,

Copper (Cu) will react to form copper nitrate, $Cu(NO_3)_2$. Copper nitrate will react to form copper hydroxide, $Cu(OH)_2$.

Copper hydroxide will react to form copper oxide, CuO. Copper oxide will react to form copper sulfate, $CuSO_4$. Copper sulfate will react to form copper, Cu.

The compounds $Cu(NO_3)_2$, $Cu(OH)_2$, CuO and $CuSO_4$ are all stable. Each compound will still react to form other stable compounds.

Copper, copper, who has the copper?

Try an example of how many electrons atoms will lose or gain. Refer back to the table on page 138.

2. How many electrons does a sodium atom have?

3. What element in Table 1 has the closest number of electrons to sodium?

4. Must a sodium atom lose or gain electrons to have the same number as neon (10 electrons)?

5. How many electrons?

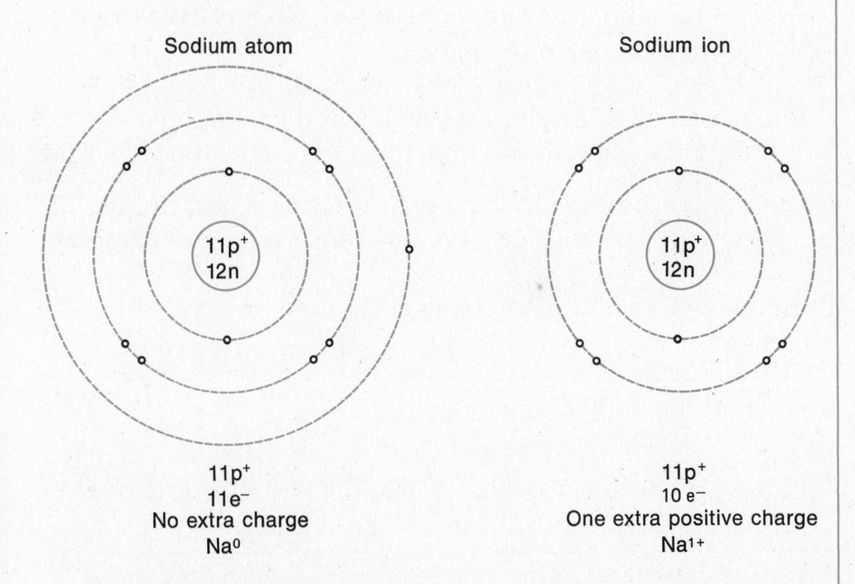

6. How many electrons does a chlorine atom have?

7. What element in Table 1 has the closest number of electrons to chlorine?

8. Must a chlorine atom lose or gain electrons to have the same number as argon?

9. How many electrons?

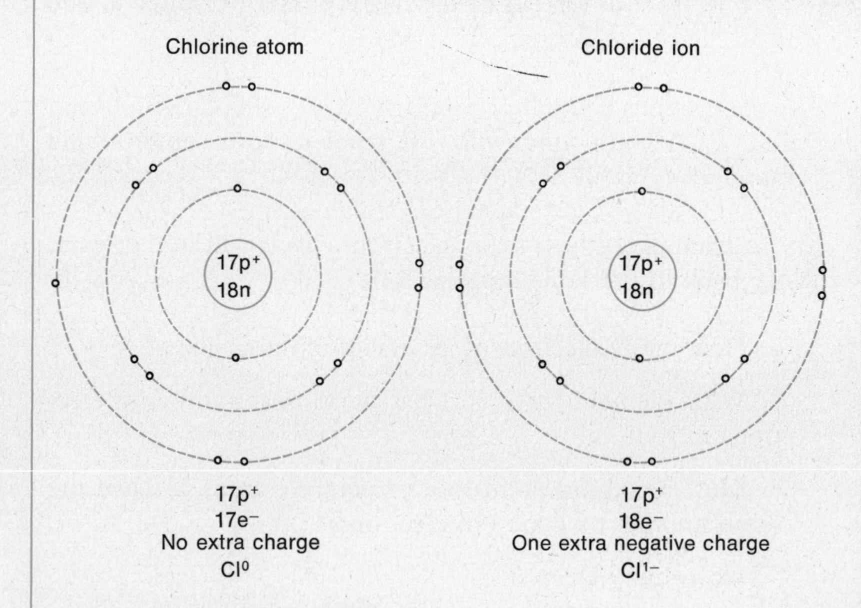

An *ion* is a substance with a charge. An ion has extra negative charges or extra positive charges.

If a substance gains electrons (e^-), it forms a negative ion.
If chlorine gains one electron, it becomes a chloride ion, Cl^{1-}.

If a substance loses electrons (e^-), it forms a positive ion.
If sodium loses one electron, it becomes a sodium ion, Na^{1+}.

Atoms will lose or gain enough electrons to have the same number of electrons as the inert gases.

10. Copy Table 2 into your notebook.

TABLE 2: Non-Inert Elements

Element	*Sodium Na*	*Calcium Ca*	*Aluminum Al*	*Carbon C*		*Nitrogen N*	*Oxygen O*	*Chlorine Cl*
Number of electrons the atom has	11							
Number of electrons lost or gained to become stable	lose 1							
Charge on the ion	Na^{1+}			C^{4+}	C^{4-}			

11. Fill in Table 2. Use Table 1 to find out how many electrons these atoms need to lose or gain.

12. Copy Table 3 into your notebook.

TABLE 3

H^{1+}							He^{0}
			$C^{4\pm}$	N^{3-}	O^{2-}		
Na^{1+}		Al^{3+}				Cl^{1-}	
	Ca^{2+}						

13. Fill in the rest of Table 3, giving the symbol and charge on each ion.

The charge on an ion is called its *combining number.* The combining numbers of elements are very useful in writing formulas.

B. Kem-I-Know

To help you learn the symbols, and charges on ions, you can play the Kem-I-Know Game. It is similar to dominoes. Here are the rules:

Rules for Kem-I-Know (for 2, 3, or 4 players)

1. Dealer shuffles the cards and deals six cards, face down, to each player.

2. Dealer places the rest of the cards, face down, in a pile. This is called the draw pile.

3. Dealer turns over one card from his or her hand and places it, face up, in the middle of the table.

4. Player to dealer's left tries to play on this card. If a play cannot be made, the player draws a card from the draw pile. If a play still cannot be made, player continues to draw until he or she can play.

5. Cards are played as follows:

   Each card has two halves. Each half has the symbol of an element and its combining number. (Some halves are blank.)

   Cards are placed end to end, with equal but opposite combining numbers touching. For example, if the first card was:

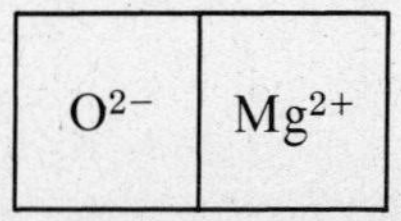

   a 2+ could be played on the left half or a 2− on the right half:

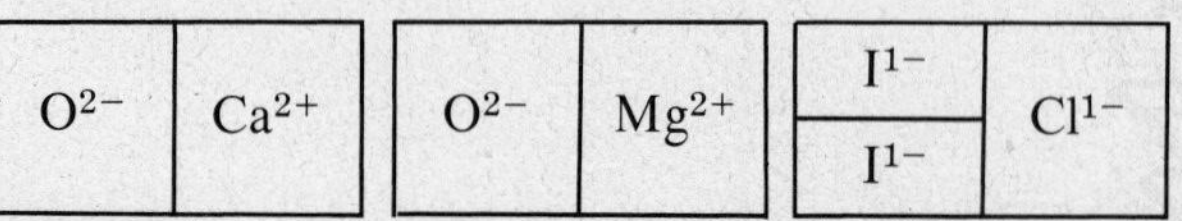

   A blank can be called whatever the player who plays it wants.

6. When a play is made, the player must state the name of the compound formed.

7. Play continues in a clockwise direction.

8. The winner is the player who uses up all his or her cards first.

9. Play continues until only one player has cards left. He or she then becomes the dealer for the next round.

INVESTIGATION

2

What Secret Formula?

DOONESBURY by Garry Trudeau

Formulas tell a chemist many things. They tell what elements are present in a compound. For example, in the formula H_2O, the letters H and O are the symbols for the elements hydrogen and oxygen.

A. One Molecule of Atoms, Please

Formulas also tell how many atoms of each element are present in one molecule of the compound.

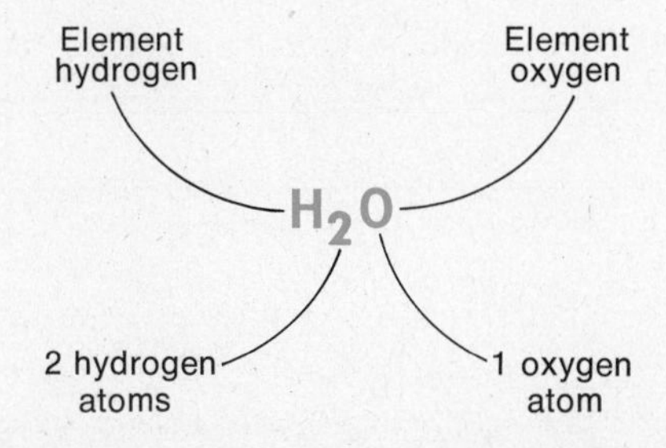

The number after the H above is called a *subscript.* If there is no subscript after a symbol, there is only one atom of that element in the molecule.

Look at the formula for propane, C_3H_8.

1. How many elements are in a molecule of propane?
2. What are these elements?
3. How many atoms of each element are in one molecule of propane?

Look at the formula for sulfuric acid, H_2SO_4.

4. How many elements are in a molecule of sulfuric acid?
5. What are these elements?
6. How many atoms of each element are in one molecule of sulfuric acid?

If there is a number in front of a formula, it tells how many molecules of the substance are present. For example, a formula that reads $2\ H_2O$ means that two water molecules are present.

$$2\ H_2O = \begin{cases} H_2O \\ H_2O \end{cases}$$

$$3\ H_2O = \begin{cases} H_2O \\ H_2O \\ H_2O \end{cases}$$

7. Copy Table 1 into your notebook and fill it in.

TABLE 1

	1 Molecule (HNO_3)	*2 Molecules ($2\ HNO_3$)*	*6 Molecules ($6\ HNO_3$)*
How many hydrogen atoms?			
How many nitrogen atoms?			
How many oxygen atoms?			

Chemical formulas can simplify your study of chemistry. For one thing, they save you a lot of writing. They also help you to understand what happens when elements combine. It's time to learn how to write and to use chemical formulas.

If you know the combining number of each element in a compound, it is easy to write the formula of the compound. The total number of electrons lost by elements in a compound must equal the total number of electrons gained by other elements in the compound.

In a molecule, the positive charges = the negative charges

Why not work through an example? Try magnesium chloride.

a. First write the symbols of the elements which are going to combine. These are:

$$Mg \quad Cl$$

b. Then write the combining number of each element above and to the right of its symbol.

$$Mg^{2+}Cl^{1-}$$

c. Look at the charges on the ions. The total charge on the formula would be zero if the charges were 2+ and 2−.
d. You have *one* Mg^{2+}, which gives a total positive charge of 2+.
e. You need *two* Cl^{1-} to get a total negative charge of 2−.
f. Therefore, the formula can be written as:

one magnesium ion	*two chloride ions*	
Mg^{2+}	$\begin{cases} Cl^{1-} \\ Cl^{1-} \end{cases}$	
2+	2−	⟵ total charge equals zero

g. This can be written as:

$$Mg_1Cl_2$$

h. When there is a subscript of 1, the number is not included in the formula. Therefore, the formula is written:

$$MgCl_2$$

Now, try another formula for yourself. Try aluminum oxide, a compound of aluminum and oxygen.

8. What are the symbols for the two elements in this compound?

9. What are the combining numbers of these two elements?

10. Write the combining number above and to the right of each symbol.

Aluminum is 3+ and oxygen is 2−. The numbers 3 and 2 will both divide into 6.

i. You need two Al^{3+} to get a total positive charge of 6+.
j. You need three O^{2-} to get a total charge of 6−.
k. Therefore, the formula can be written as:

two aluminum ions	*three oxide ions*	
$\left.\begin{matrix} Al^{3+} \\ Al^{3+} \end{matrix}\right\}$	$\left\{\begin{matrix} O^{2-} \\ O^{2-} \\ O^{2-} \end{matrix}\right.$	
6+	6−	⟵ total charge equals zero

l. Which is written more simply as:

$$Al_2O_3$$

Using Table 3 in the last investigation (page 157), write the formulas for the following:

11. Sodium fluoride.

12. Calcium chloride.

13. Aluminum chloride.

14. Calcium oxide.

15. Boron oxide.

Not all ions are elements. Some are groups of atoms with a charge. Examples are:

Phosphate	PO_4^{3-}
Sulfate	SO_4^{2-}
Ammonium	NH_4^{1+}
Hydroxide	OH^{1-}
Carbonate	CO_3^{2-}

How are formulas written for these ions? The same way! If you need two hydroxides, you write $(OH^{1-})_2$. For example, work through the formula for calcium hydroxide.

m. First write the symbols with their combining numbers:

$$Ca^{2+} \qquad OH^{1-}$$

n. You can see that two hydroxides are needed. Therefore,

$$Ca^{2+}\left\{\begin{matrix} OH^{1-} \\ OH^{1-} \end{matrix}\right. = Ca^{2+}(OH^{1-})_2 = Ca(OH)_2$$

A hydroxide ion contains two different kinds of atoms. When an ion has two or more atoms and you have more than one of them, use parentheses. For example,

$$(OH)_3 \text{ means three hydroxides } \left\{\begin{matrix} OH^{1-} \\ OH^{1-} \\ OH^{1-} \end{matrix}\right.$$

$$(SO_4)_2 \text{ means two sulfates } \left\{\begin{matrix} SO_4^{2-} \\ SO_4^{2-} \end{matrix}\right.$$

o. Here's another example. Aluminum hydroxide starts with:

$$Al^{3+} \qquad OH^{1-}$$

p. Three hydroxides are needed.

$$Al^{3+}\left\{\begin{matrix} OH^{1-} \\ OH^{1-} \\ OH^{1-} \end{matrix}\right. = Al(OH)_3$$

Write the formulas for the following:

16. Calcium sulfate.
17. Sodium carbonate.
18. Aluminum sulfate.
19. Aluminum phosphate.
20. Ammonium sulfate.

These formulas are a little more difficult to write. If you did them correctly, you have done a good job, and you should be proud of how much you have learned.

Wide World Photos

B. Let's Keep in Shape

In this Part, you will be making models of molecules. You will be using a set of molecular models for this. There are several kinds of molecular models. Yours may be made of foam or plastic, or possibly some other material. Colored spheres or other geometric shapes are used to represent the atoms. Different colors represent different elements. The atom models have small holes in them. The number of holes may or may not equal the combining number of the element the model represents. Your model kit will also have small pieces of plastic, pipe cleaner, wood, or possibly springs. These represent the forces which hold the atoms together in the molecule.

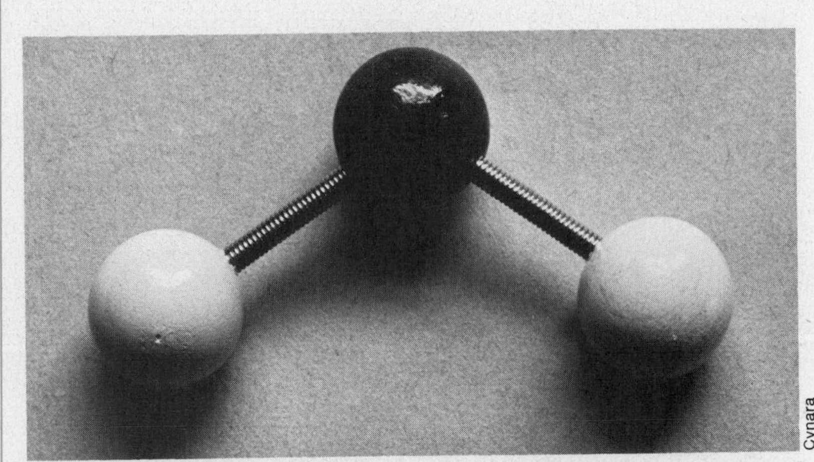

Cynara

Ball and stick model of water.

Space filling model of water.

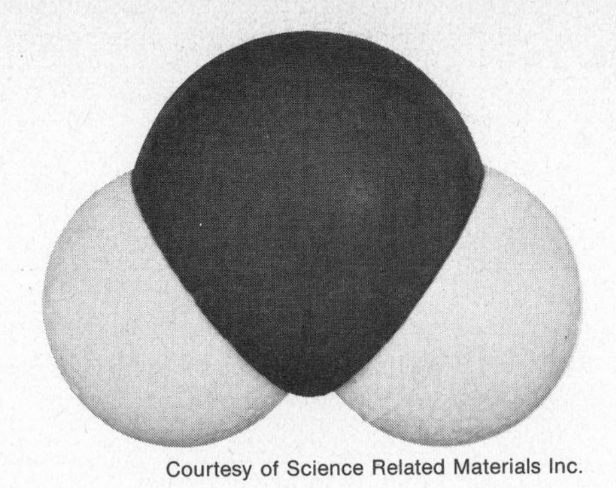

Courtesy of Science Related Materials Inc.

What You Need

Molecular model set

What to Do

a. Examine your set of molecular models.

21. Copy Table 2 into your notebook.

TABLE 2

Element	*Color*	*Combining number*	*Sketch*

22. In your table, record the elements which are represented in your model set. Record the color of each element, and enter its correct combining number. Sketch each model.

b. Now you are ready to use your models. First, make a model of a hydrogen chloride molecule.

23. What is the combining number of hydrogen?

24. What is the combining number of chlorine?

25. How many atoms of hydrogen should you combine with an atom of chlorine to make one molecule of hydrogen chloride?

c. Put the model of hydrogen chloride together. Your model should look something like the illustration.

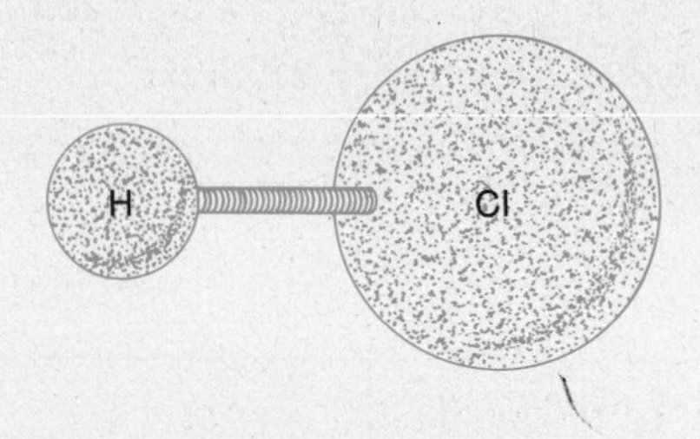

26. What is the chemical formula of hydrogen chloride?

Try another compound—your old friend, water.

27. What two elements are in water?

28. What is the combining number of hydrogen?

29. What is the combining number of oxygen?

30. How many atoms of hydrogen combine with one atom of oxygen in a water molecule?

d. Put together a model of a water molecule.

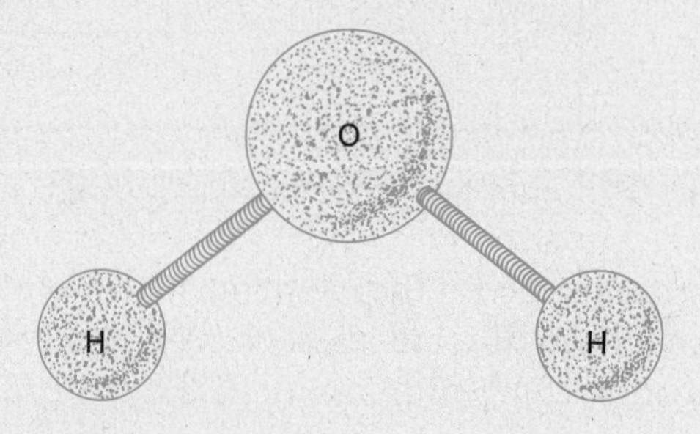

31. Copy Table 3 into your notebook.

TABLE 3

Name of compound	Formula	Sketch
Ammonia	NH_3	
Methane	CH_4	
Carbon tetrachloride	CCl_4	
Methanol (Methyl alcohol)	CH_3OH	

e. Now that you have the idea, make models of each of the compounds listed in Table 3.

32. Sketch each model in Table 3. Label each atom.

f. Your teacher may suggest other compounds to make models of. Make these models and draw sketches of them in your notebook.

You have come a long way in your study of chemistry. You've learned a great deal about elements, the building blocks of the world. You can separate, identify, and describe them, as well as predict their behavior. You can even combine them into useful compounds. Now it is time to put the elements and compounds to work for you.

INVESTIGATION 3

B Is for Bond

In the previous investigation, you learned that the atoms of most elements can combine with the atoms of other elements to form compounds. Now you are ready to investigate how atoms combine. What holds the atoms together?

A *bond* holds the atoms together. A bond is really just a force of attraction between atoms. This attraction is in the form of energy.

Bonds are classified into two main types. The type of bond in any particular compound depends primarily on the number of electrons in the outer energy levels of the elements involved. The two main types of bonds are *ionic* and *covalent.*

An example of ionic bonding.

Lithium chloride (LiCl)

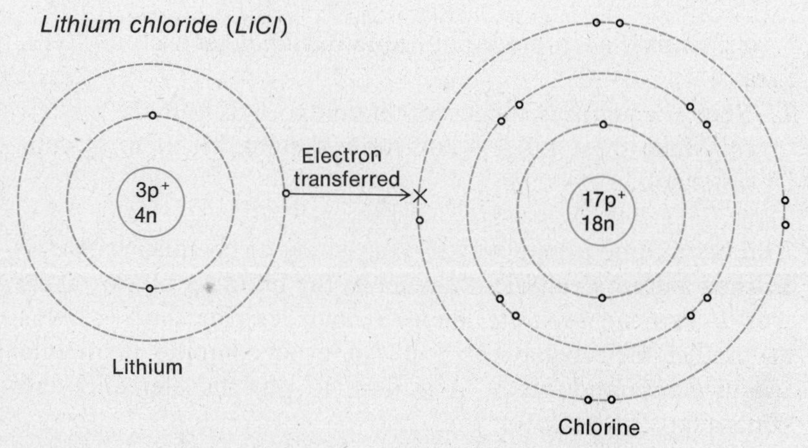

In an ionic bond, one or more electrons from the outer energy level of one atom are *transferred* to the outer energy level of another atom.

Lithium loses one electron. Lithium ends up with the same number of electrons as the inert gas helium, He ($2e^-$). By losing one electron, lithium becomes stable. Chlorine gains one electron from lithium. Chlorine ends up with the same number of electrons as the inert gas argon, Ar ($18e^-$). By gaining one electron, chlorine becomes stable.

In covalent bonding, no electrons are transferred. Instead, the electrons are *shared.*

The shaded areas show electrons which are shared. In each shaded pair, one electron comes from the carbon atom and one electron comes from the hydrogen atom. Shared electrons move around both atoms. They are counted for both atoms when counting the electrons. Each hydrogen has two electrons. The carbon has ten electrons.

Compounds with ionic bonding have different properties from compounds with covalent bonding. Let's look at these different properties in the laboratory, and then discuss what is happening.

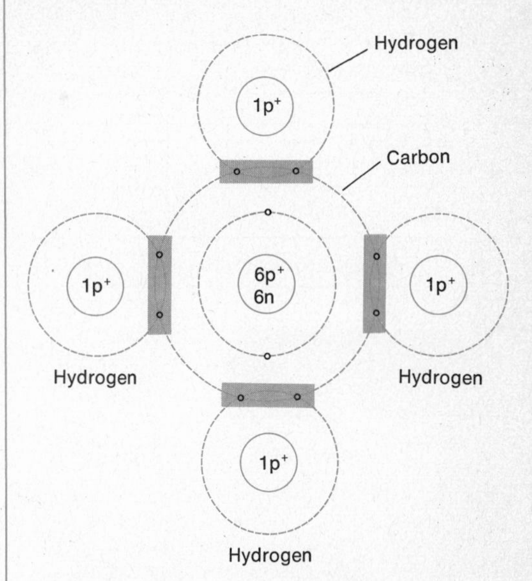

An example of covalent bonding.

Methane (CH_4)

A. Have a Covalent Ion

What You Need

Calcium chloride ($CaCl_2$), an ionic compound
Para-dichlorobenzene ($C_6H_4Cl_2$), a covalent compound
Silicon dioxide (SiO_2), a covalent compound
Sodium chloride (NaCl), an ionic compound
Starch [$(C_6H_{10}O_5)_n$], a covalent compound
Water (H_2O), distilled

Beaker, 250 ml
Evaporating dish
Graduated cylinder, 100 ml
Stirring rod
Test tubes, 5

Asbestos pad
Balance
Burner
Electrical conductivity apparatus
Matches
Ring
Ring stand
Safety goggles
Spatula
Test tube rack
Wax pencil
Weighing paper

TABLE 1

Compound	Formula	*Ionic or Covalent*	*Volatility*	*Solubility*	*Electrical conductivity*	*Melting time*
Calcium chloride	$CaCl_2$					
Sodium chloride	$NaCl$					
Silicon dioxide	SiO_2					
Starch	$(C_6H_{10}O_5)_n$					
Para-dichlorobenzene	$C_6H_4Cl_2$					

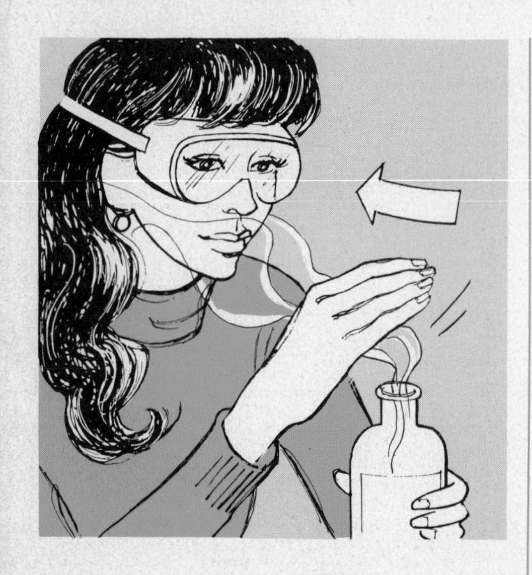

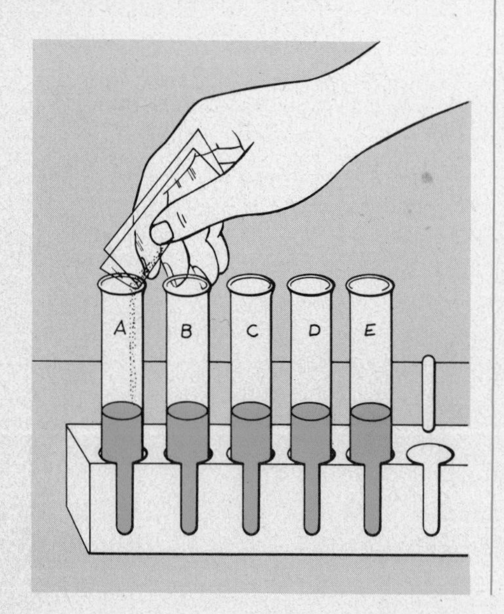

What to Do

1. Copy Table 1 into your laboratory notebook.

Volatility refers to the ability of a substance to evaporate easily—to form a gas. The easiest way to tell if a compound is volatile is to smell it. Remember how that is done correctly?

If you can detect an odor, you can assume that the compound is volatile. If there is no odor, then it is non-volatile.

a. Sniff away—**carefully!**

2. Enter your results in Table 1 in your notebook.

3. Generally, are the ionic compounds volatile or non-volatile?

4. Are the covalent compounds volatile or non-volatile?

Next, test the compounds for solubility.

b. Set up 5 test tubes in a rack. Label them **A** through **E**.
c. Add 10 ml of water to each test tube.
d. Then add samples of the compounds as follows:
To tube **A**, add 0.5 grams of $CaCl_2$.
To tube **B**, add 0.5 grams of NaCl.
To tube **C**, add 0.5 grams of SiO_2.
To tube **D**, add 0.5 grams of starch.
To tube **E**, add 0.5 grams of para-dichlorobenzene.
e. Use a stirring rod to mix the contents of each tube. Be sure to rinse the rod between uses.

5. Record the results of your solubility tests in Table 1.

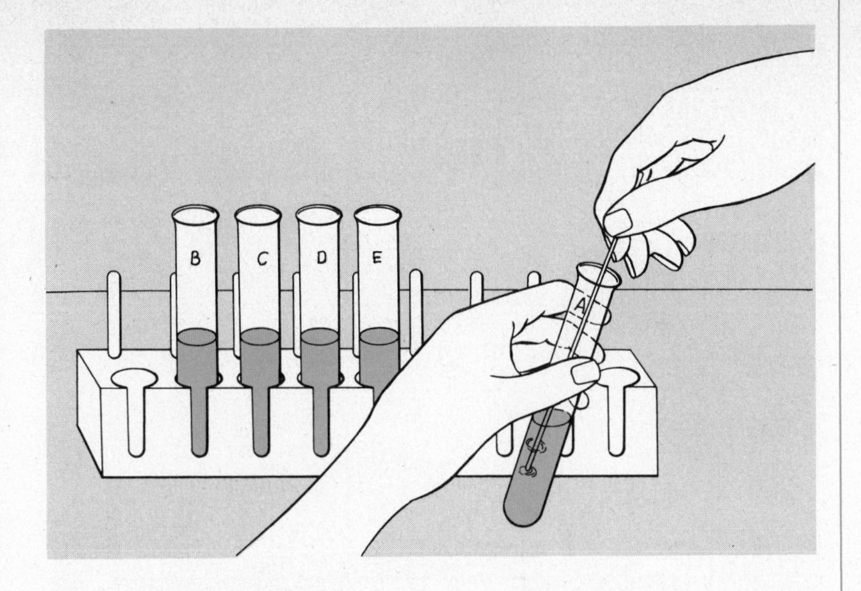

6. Are the ionic compounds soluble in water?

7. Are the covalent compounds soluble in water?

Next, test the compounds for electrical conductivity.

f. Label the beaker with the name of the compound you are testing. Dissolve 5 grams of the compound in 100 ml of water in the beaker.
g. Test the contents of the beaker with the electrical conductivity apparatus. You should rest it on the top of the beaker so the prongs are in the liquid. **Then** plug it in. If the bulb lights, the compound conducts electricity. (With pure water, the bulb may glow dimly.)
h. **Unplug the apparatus before removing it from the beaker.** Rinse the prongs carefully and then dry them. The apparatus should be plugged in **only** when it is resting on one of the beakers.
i. Rinse the beaker thoroughly. Repeat steps **f** through **h** for the other compounds. (If you know that the compound does not dissolve in water, test a small amount of the solid for electrical conductivity.)

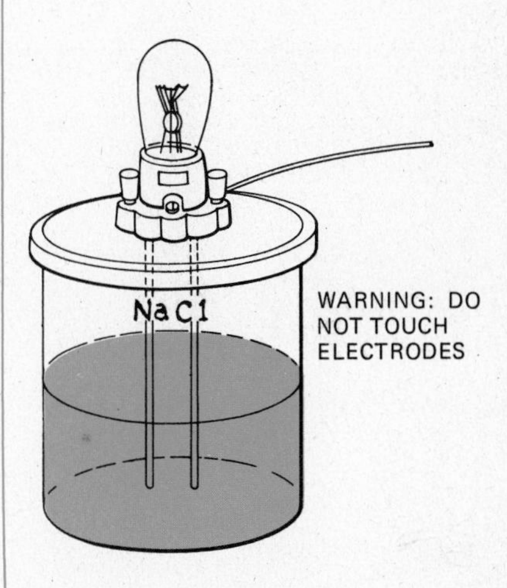

8. Record the results of the conductivity tests in Table 1.

9. Do the solutions of ionic compounds conduct electricity?

10. Do the covalent compounds conduct electricity?

Now for the melting time!

j. Get 1 gram of NaCl.
k. Place the sample in an evaporating dish.
l. Heat it with a hot flame. **Caution: do not get too close.**

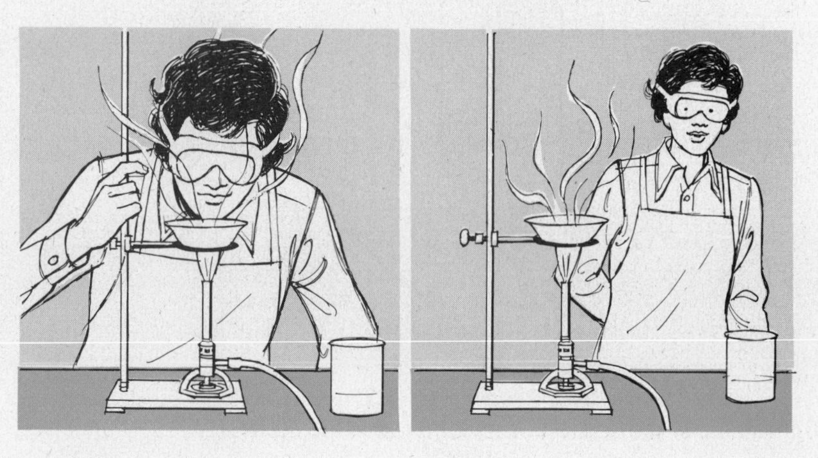

Incorrect *Correct*

m. Measure the time it takes for the sample to melt.

11. Record the time in Table 1. (If the sample has not melted in 5 minutes, write "did not melt" in your table.)

n. Allow the evaporating dish to cool. Clean it thoroughly.
o. Test the other compounds in the same way. Between tests, allow the dish to cool before cleaning it.

12. How do the melting times of covalent and ionic compounds compare?

13. How do covalent compounds and ionic compounds differ?

14. Copy Table 2 into your notebook. Using the results of your experiments, predict the properties of the two compounds listed.

TABLE 2

Compound	*Formula*	*Ionic or Covalent*	*Volatility*	*Solubility*	*Electrical conductivity*	*Melting time*
Magnesium sulfate	$MgSO_4$	ionic				
Benzoic acid	C_6H_5COOH	covalent				

p. If these two compounds are available, test your predictions.

15. Record your results in your laboratory notebook.

B. Time for Talk

At the beginning of this investigation, a covalent bond was described as electrons that are shared. Let's see how that happens. For example, a single atom of chlorine has seven electrons in its outer shell.

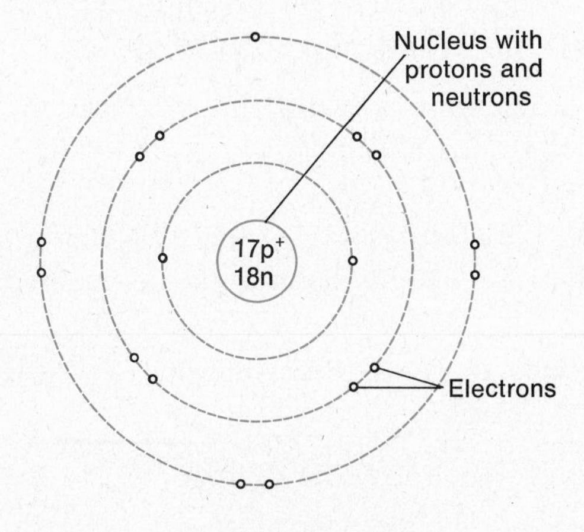

A chlorine atom

When two chlorine atoms join up to form a chlorine molecule, they can share electrons.

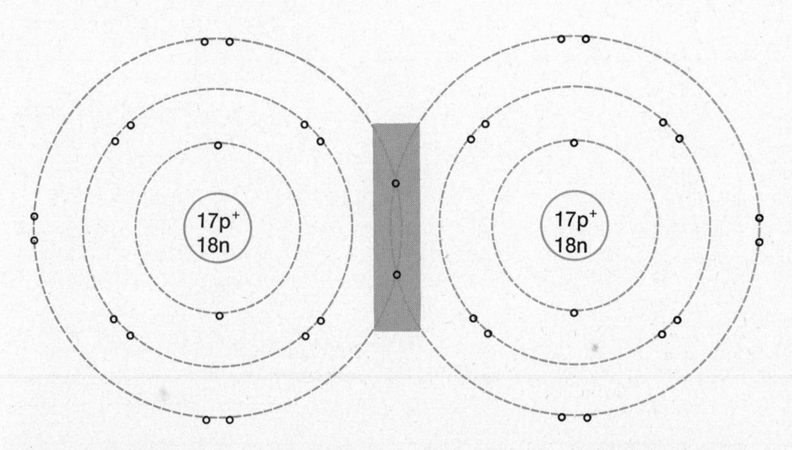

A chlorine molecule. The shared electrons are indicated by the shaded area.

The last drawing shows a chlorine molecule. The shared electrons are those within the shaded area. These electrons count in filling up both atoms' outer energy levels. Both atoms have 18 electrons. The element argon (inert) has 18 electrons also.

Covalent bonds can also be formed between atoms of different elements. Hydrogen has one electron in its energy level. Chlorine has seven electrons in its outer energy level. Hydrogen and chlorine can join together in a covalent bond. The electrons are shared unequally. Hydrogen and chlorine are *unlike* atoms.

16. Draw a diagram of a hydrogen chloride (HCl) molecule, showing how the electrons are shared.

17. Draw a diagram of a fluorine (F_2) and of a carbon tetrachloride (CCl_4) molecule.

In ionic bonds, the electrons are actually transferred from one atom to another. Let's look at this with sodium fluoride.

18. How many electrons does fluorine (F) need to fill its outer energy level?

19. How many electrons does sodium (Na) have in its outer energy level?

When a sodium atom combines with a fluorine atom to form sodium fluoride (NaF), the sodium atom loses an electron. At the same time, the fluorine atom gains that one electron.

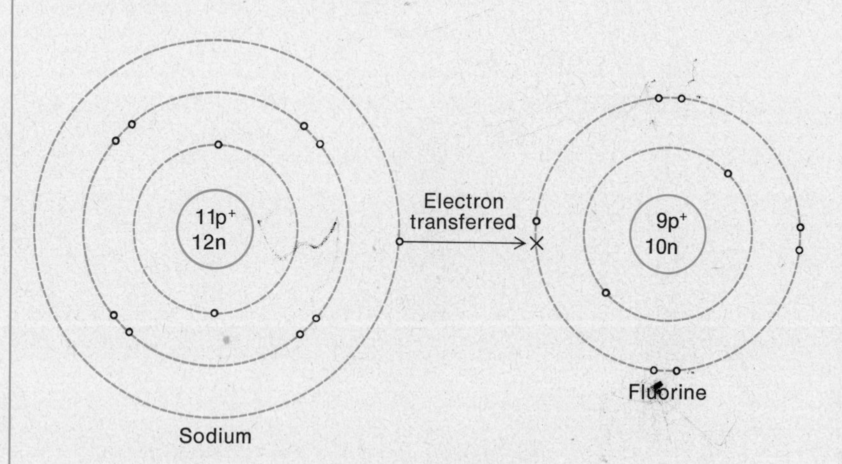

Sodium fluoride (NaF)

20. Draw diagrams for the following ionic bonded molecules:

   a. sodium iodide (NaI);
   b. calcium chloride ($CaCl_2$).

The following statements are a summary of these two types of bonds.

a. When atoms are held together by shared electrons, the attraction between them is called a *covalent bond.*
b. Compounds that are held together by covalent bonds are called *covalent compounds.*
c. When atoms are held together by the transfer of electrons from one atom to another atom, the attraction between them is called an *ionic bond.*
d. Compounds that are held together by ionic bonds are called *ionic compounds.*

You have already learned about the combining numbers of elements. Now you understand how atoms are put together in different ways to form chemical compounds. There are hundreds of thousands of chemical compounds.

In fact, in the next few investigations, you will find out how these compounds react. And you will be able to know what new substances are formed.

INVESTIGATION

4

More Than One Way to Skin a Cat

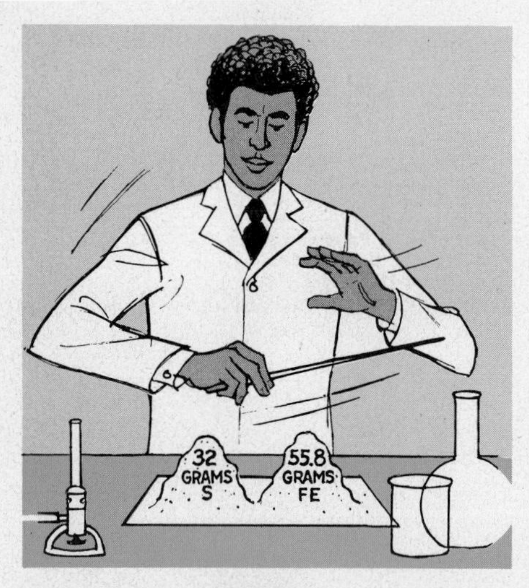

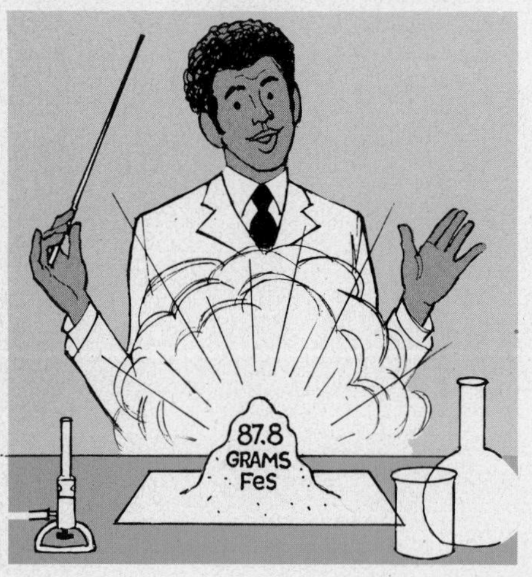

By this time—whether or not you realize it—you have solved many problems. But solving problems creates more problems. How are you supposed to predict what happens when elements or compounds react together to form new products?

Believe it or not, it is possible for a chemist to predict accurately what will happen in a chemical reaction. A chemist can predict not only *what* will happen, but what *products* will be formed. And, to top it off, it is possible to predict *how much* of the products will form.

It is impossible for anyone to memorize all of the chemical reactions known. Fortunately, reactions can be classified into different types. If you learn to write equations for each type, you can easily write hundreds of equations. An equation is a chemical shorthand for a word equation. You used the word equation for the decomposition of water earlier. The word equation was:

$$\text{Water} \longrightarrow \text{Hydrogen} + \text{Oxygen}$$

The chemical shorthand equation would be:

$$2\,H_2O \longrightarrow 2\,H_2 + O_2$$

First, you will investigate how some reactions occur in the lab; then you will look at them on paper.

A. You Two Make a Beautiful Combo

What You Need

Magnesium (Mg) ribbon

Burner
Matches
Safety goggles
Tongs

What to Do

a. Put on your safety goggles.
b. Light the burner.
c. Hold a 6-cm piece of magnesium ribbon with the tongs.
d. Ignite the magnesium in the burner flame. **Do not look directly at the burning magnesium.**

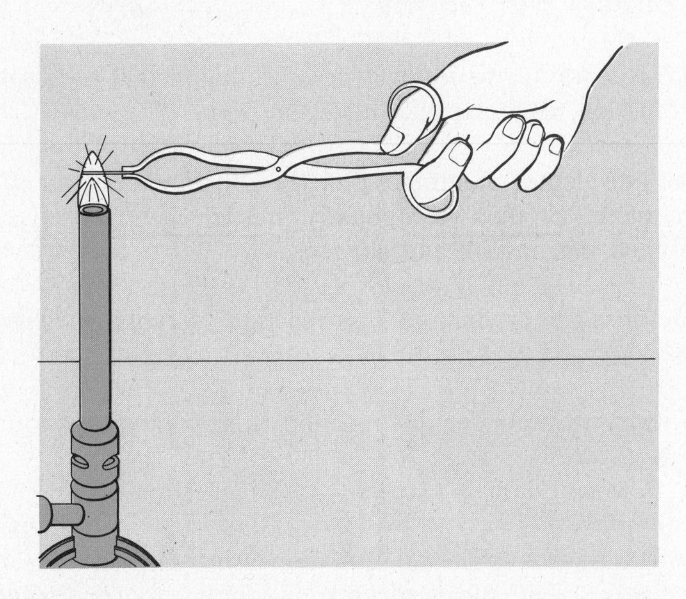

1. What happens when you ignite the magnesium? Record your observation.

The magnesium has combined with an element in the air.

2. With what element in the air did the magnesium combine?

The reaction you've just observed is an example of a *combination reaction.* A combination reaction occurs when two or more elements join to form a compound.

The following equations are examples of combination reactions.

Carbon + Oxygen ⟶
Carbon dioxide (*di*oxide means *two* oxygens)

$$C + O_2 \longrightarrow CO_2$$

Sulfur + Oxygen ⟶
Sulfur trioxide (*tri*oxide means *three* oxygens)

$$2\,S + 3\,O_2 \longrightarrow 2\,SO_3$$

Water + Sulfur trioxide ⟶ Sulfuric acid

$$H_2O + SO_3 \longrightarrow H_2SO_4$$

B. Slowly But Surely

There is a way to write the type of equation that goes with a combination reaction. Take it step by step.

When one element combines directly with another element, the name of the product (compound) ends in *-ide.* You have just combined magnesium and oxygen.

Therefore, the product of this reaction is called *magnesium oxide.*

The word equation for this reaction is easy to write.

Magnesium + Oxygen ⟶ Magnesium oxide

Now you have to write the equation using the formulas. The first step is to write the correct formula for one molecule of each of the reacting substances. An oxygen molecule is diatomic, O_2. *Diatomic* means two atoms.

So, the formulas for the reacting substances are written like this:

$$Mg + O_2 \longrightarrow$$

The next step is to write the correct formula for one molecule of the product which is formed. To do this, you need to check the combining numbers of the elements involved. (See Unit 3, Investigation 1.) The combining number of magnesium is 2+;

the combining number of oxygen is 2−. So the correct formula for magnesium oxide is MgO. The equation becomes

$$Mg + O_2 \longrightarrow MgO$$

Now here's the kicker—the equation has to be *balanced.* This means that the number of atoms of each element on both sides of the arrow has to be the same.

3. How many atoms of magnesium (Mg) are on the left side of the equation?
4. How many atoms of magnesium are on the right side of the equation?
5. Are they equal?

Okay; leave the magnesium alone for a while. What about oxygen?

6. How many atoms of oxygen (O) are on the left side of the equation?
7. How many atoms of oxygen are on the right side of the equation?
8. Are they equal?

The same number of oxygen atoms is needed on both sides.

9. How many molecules of magnesium oxide are needed to do this?

Right! If you have two molecules of magnesium oxide, the oxygen atoms will be equal (balanced).

$$Mg + O_2 \longrightarrow \left\{\begin{matrix} MgO \\ MgO \end{matrix}\right.$$

You can simplify the right hand side like this:

$$Mg + O_2 \longrightarrow 2\ MgO$$

The number "2" goes in front of the formula for magnesium oxide. This shows that there are two molecules of magnesium oxide.

10. Does the number of oxygen atoms on one side of the equation now equal the number of oxygen atoms on the other side of the equation?

11. Does the number of magnesium atoms on one side of the equation now equal the number of magnesium atoms on the other side of the equation?

You've got a problem. There are two magnesium atoms on the right side and only one on the left.

The number of magnesium atoms on both sides must be equal.

12. What number must you put in front of the magnesium atom on the left side of the equation?

If you said "2," you're right. Congratulations!

By now, your equation looks like this:

$$2\,Mg + O_2 \longrightarrow 2\,MgO$$

or

$$\left.\begin{matrix} Mg \\ Mg \end{matrix}\right\} + O_2 \longrightarrow \left\{\begin{matrix} MgO \\ MgO \end{matrix}\right.$$

This equation is now *balanced.*

13. Why?

There are four steps in writing a correct equation.

a. Write the correct formulas for the reacting substances.
b. Decide what will be produced.
c. Write the correct formulas for the products.
d. Balance the equation.

Look at this word equation.

$$\text{Carbon} + \text{Oxygen} \longrightarrow \text{Carbon dioxide}$$

Using symbols, the above equation can be written as:

$$C + O_2 \longrightarrow CO_2$$

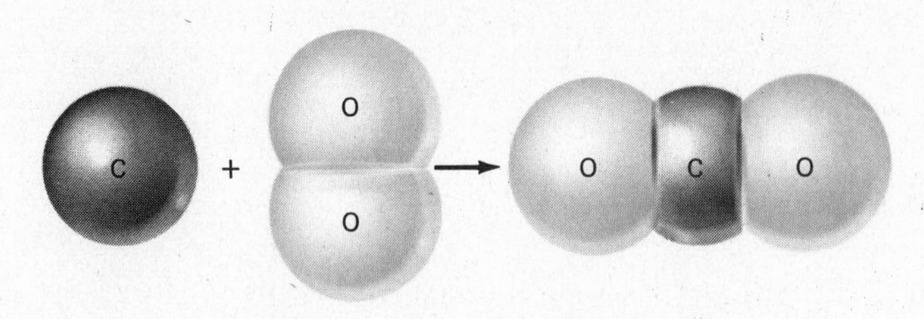

14. Is this equation balanced? Why?

Here's another equation.

$$\text{Nitrogen} + \text{Hydrogen} \longrightarrow \text{Ammonia}$$

$$N_2 + H_2 \longrightarrow NH_3$$

15. Is this equation balanced? Why?

16. Balance the equation, following the same steps as before.

Try one on your own.

17. Write a balanced equation for the reaction of aluminum and oxygen to give aluminum oxide. Show all of your work in your notebook.

So, no one ever said chemistry was a snap; but you can figure it out.

In this section, you have looked at *combination reactions.*

18. What is a combination reaction?

C. The Decomposition Waltz

A *decomposition reaction* occurs when a compound breaks down into two or more simpler substances. Decomposition is the reverse of combination. In Unit 2, you studied the decomposition of water into hydrogen and oxygen. Let's look at some other decomposition reactions now.

What You Need

Lead(IV) oxide (PbO_2)

Test tube, Pyrex

Burner
Matches
Ring stand
Safety goggles
Spatula
Test tube clamp
Wood splint

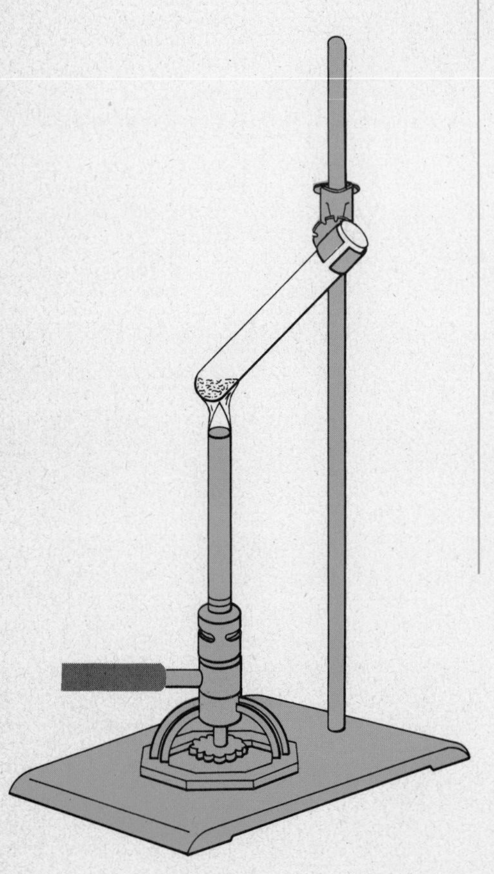

What to Do

a. Put a small amount of PbO_2 in a dry Pyrex tube. There should be just enough to fill the rounded bottom of the tube.
b. Clamp the tube at a slant on the ring stand. The tube should be pointing away from you.
c. Heat the powder over the burner. **Be sure to wear your safety goggles.**
d. After heating for 15 to 20 seconds, insert a *glowing* splint into the tube.

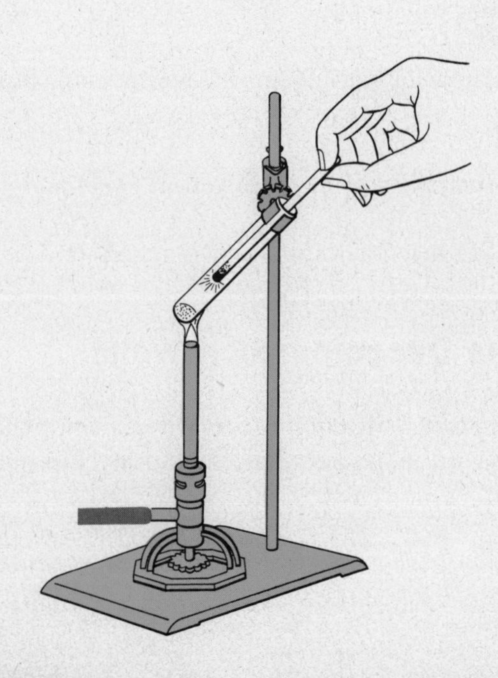

19. What gas is being given off?

e. Turn off the burner.
f. Wait a few minutes. Then use a *glowing* splint to test if gas is still being given off.

20. Does this chemical reaction continue when the flame is removed?

21. What condition was needed to start this decomposition reaction?

Now, let's write a balanced equation for the reaction you just did in the lab.

The word equation for the reaction is

$$\text{Lead(IV) oxide} \longrightarrow \text{Lead} + \text{Oxygen}$$

The formula for the reacting substance is PbO_2.

22. What are the formulas for the products?

23. What does your equation look like at this point?

Examine each substance to see if there are the same number of atoms on both sides of the equation.

$$PbO_2 \longrightarrow Pb + O_2$$

As you can see, the lead atoms are balanced. There is one lead atom on each side of the equation. There are two oxygen atoms on the left side and two on the right side of the equation.

24. Is the equation balanced now? Why?

You've done so well; don't stop here!

25. Write a balanced equation for the decomposition of mercury(II) oxide (HgO).

Here are equations for two more decomposition reactions. You already know that water decomposes to give hydrogen and oxygen.

$$\text{Water} \longrightarrow \text{Hydrogen} + \text{Oxygen}$$

$$2\,H_2O \longrightarrow 2\,H_2 + O_2$$

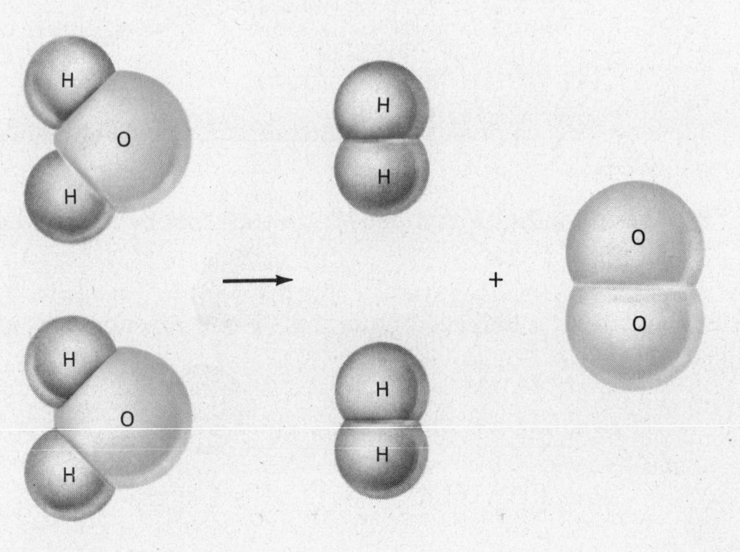

You can also decompose sulfurous acid to get water and sulfur dioxide.

$$\text{Sulfurous acid} \longrightarrow \text{Water} + \text{Sulfur dioxide}$$

$$H_2SO_3 \longrightarrow H_2O + SO_2$$

26. Are these equations balanced? Why?

D. Don't Give Me Any Rough Stuff, or Out You Go

Here is another type of reaction that works well in the lab.

What You Need

Copper (Cu) strip
Copper(II) sulfate ($CuSO_4$), 0.1 *M* solution
Iron (Fe) nail
Silver nitrate ($AgNO_3$), 0.1 *M* solution
Beakers, 50 ml, 2
Safety goggles

What to Do

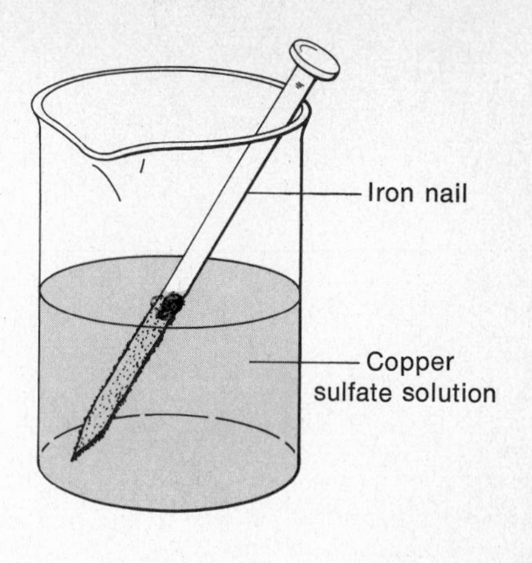

a. Half-fill a beaker with $CuSO_4$ solution.
b. Place a clean iron (Fe) nail in the solution.
c. Leave the nail in the solution for a few minutes.

27. What happens to the nail?

28. What element is coating the nail?

29. How could you test the element coating the nail to identify it definitely?

d. Leave the nail in the beaker until the next class period. Then examine the solution in the beaker.

30. What happened to the color of the solution in a day's time?

31. What do you think caused this change?

32. What is being formed in the beaker?

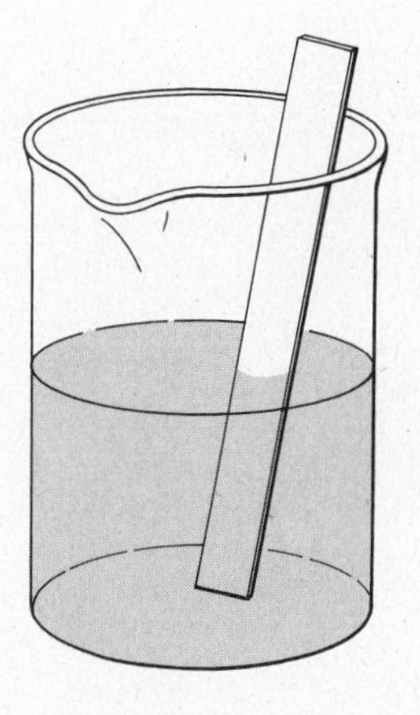

e. Half-fill another beaker with $AgNO_3$ solution.
f. Place a copper (Cu) strip in this solution.
g. Observe the copper strip for a few minutes.

33. What happens to the copper?

h. Leave the copper strip in the beaker until the next class period. Then observe it again.

34. Describe the reaction that occurred.

These two reactions are examples of *single displacement reactions.* They are sometimes called *single replacement reactions.* Actually, you've already done a great deal of work with displacement reactions.

Single displacement reactions occur when one element in a compound is replaced by another element. An example is the reaction between zinc and copper(II) sulfate.

$$\text{Zinc} + \text{Copper(II) sulfate} \longrightarrow \text{Zinc sulfate} + \text{Copper}$$

$$Zn + CuSO_4 \longrightarrow ZnSO_4 + Cu$$

The zinc replaces (or displaces) the copper in the copper(II) sulfate. Zinc sulfate is formed, leaving the element copper behind. One element (zinc) has taken the place of another element (copper).

Another example is the reaction between magnesium and hydrochloric acid.

Magnesium + Hydrochloric acid $\longrightarrow$
Magnesium chloride + Hydrogen

$$Mg + 2\,HCl \longrightarrow MgCl_2 + H_2$$

35. Which element has been replaced in HCl?

36. What element combined with the Cl?

37. What are the three types of reactions you have investigated so far?

Now it's time for more practice on equation writing.

Actually the same steps are used in balancing all these equations. Try the above reaction.

First, write the word equation.

Magnesium + Hydrochloric acid $\longrightarrow$
Magnesium chloride + Hydrogen

Then write the correct formulas for the chemicals that react.

$$Mg + HCl \longrightarrow$$

Then write the correct formulas for the products.

$$Mg + HCl \longrightarrow MgCl_2 + H_2$$

Next, examine the number of atoms of each element on each side of the equation. The magnesium atoms balance, but the hydrogen and chlorine atoms do not. There are two hydrogens and two chlorines on the right side of the equation, but only one of each on the left side. Putting a "2" in front of HCl gives:

$$Mg + 2\,HCl \longrightarrow MgCl_2 + H_2$$

38. Is the equation balanced now? Why?

Table 1 below is a list of ions which will be useful. Their formulas and combining numbers are also listed.

TABLE 1: Formulas and Combining Numbers of Ions

Ion	*Formula*	*Combining number*
Sulfate	SO_4^{2-}	2−
Nitrate	NO_3^{1-}	1−
Acetate	CH_3COO^{1-}	1−
Carbonate	CO_3^{2-}	2−
Sulfite	SO_3^{2-}	2−
Nitrite	NO_2^{1-}	1−
Hydroxide	OH^{1-}	1−
Chromate	CrO_4^{2-}	2−

Now, Ms. or Mr. Expert, take the reaction of the iron nail in the copper sulfate solution.

39. First write the word equation.

40. Next write the correct chemical formulas for the reacting substances.

41. Write the correct formulas for the products.

42. Balance the equation.

E. Hang in There; You're Almost Home Free

A *double displacement reaction* occurs when substances in two different compounds replace each other. These reactions are also called *double replacement reactions.* An example is the reaction of zinc sulfate and barium chloride.

Zinc sulfate + Barium chloride $\longrightarrow$ Zinc chloride + Barium sulfate

$$ZnSO_4 + BaCl_2 \longrightarrow ZnCl_2 + BaSO_4\downarrow$$

The zinc from the $ZnSO_4$ has taken the place of the barium in the $BaCl_2$. At the same time, the barium from the $BaCl_2$ has taken the place of the zinc in the $ZnSO_4$. The arrow beside the $BaSO_4$ indicates that it is a precipitate. This means that it settles out of the solution. Try it in the lab.

What You Need

Barium chloride ($BaCl_2$), 0.1 *M* solution
Lead(II) acetate [$Pb(C_2H_3O_2)_2$], 0.1 *M* solution
Potassium chromate (K_2CrO_4), 0.1 *M* solution
Silver nitrate ($AgNO_3$), 0.1 *M* solution
Sodium chloride (NaCl), 0.1 *M* solution
Zinc sulfate ($ZnSO_4$), 0.1 *M* solution
Dropper
Graduated cylinder, 10 ml
Test tube
Safety goggles
Test tube rack

What to Do

a. Add 5 ml of $ZnSO_4$ solution to a test tube.
b. Add a few drops of $BaCl_2$ solution.

43. What happens?

44. Write the names and formulas of the compounds formed by the exchange of ions between zinc sulfate and barium chloride.

45. Write a balanced equation for this reaction.

c. Rinse out the test tube and the graduated cylinder thoroughly.
d. Add 5 ml of NaCl solution to the test tube.
e. Add a few drops of $AgNO_3$ solution.

46. What happens?

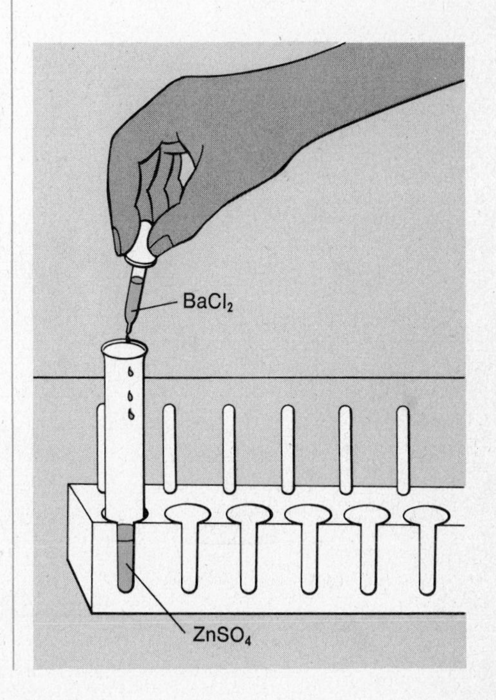

47. Write the names and formulas of the compounds formed by the exchange of ions between sodium chloride and silver nitrate.

48. Write a balanced equation for this reaction.

f. Rinse out the test tube and the graduated cylinder thoroughly.
g. Add 5 ml of $Pb(C_2H_3O_2)_2$ solution to the test tube.
h. Add a few drops of K_2CrO_4 solution.

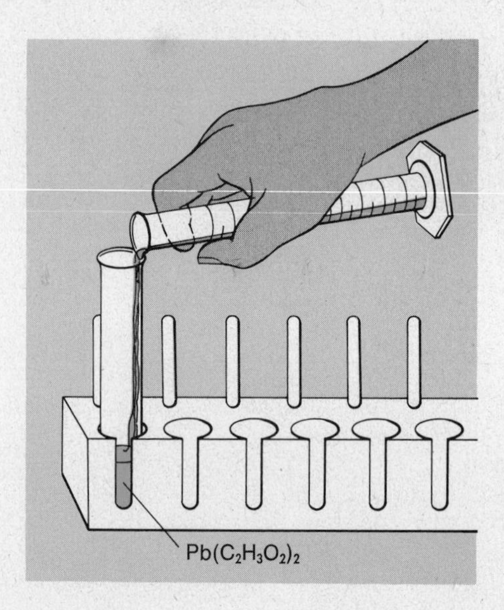

49. What happens?

50. Write a balanced equation for this reaction.

Don't be fooled by these results. Not all double displacement reactions give such obvious results as a precipitate or gas bubbles. Sometimes you won't be able to *see* what has happened; but you should be able to figure it out from the equation.

You have now investigated the four major types of reactions.

51. What is a combination reaction?

52. What is a decomposition reaction?

53. What is a single displacement reaction?

54. What is a double displacement reaction?

You have figured out how elements and compounds react. It's time to investigate how much has reacted and how much was produced.

INVESTIGATION

5

A Mountain Out of a Molehill

What now? How many atoms or molecules does it take to have enough to work with?

A. One Mole Time

Let's look at some of the equations from the past few investigations. They will help answer these questions.

$$1\,C \quad + \quad 1\,O_2 \quad \longrightarrow \quad 1\,CO_2$$

one atom of carbon	one molecule of oxygen	$\longrightarrow$	one molecule of carbon dioxide

Suppose twice as much of each substance reacted. Each *coefficient* (number in front) would be multiplied by two (2).

$$2\,C \quad + \quad 2\,O_2 \quad \longrightarrow \quad 2\,CO_2$$

two atoms of carbon + two molecules of oxygen ⟶ two molecules of carbon dioxide

There are many ways of balancing equations.

1. Copy Table 1 into your notebook. Fill in the missing numbers. For each example one quantity will be given.

TABLE 1

$1\,C$ + atoms of carbon	$1\,O_2$ ⟶ molecules of carbon	$1\,CO_2$ molecules of carbon dioxide
3 C	___ O_2	___ CO_2
___ C	___ O_2	3 CO_2
___ C	6 O_2	___ CO_2
600 C	___ O_2	___ CO_2
6×10^2 C	___ O_2	___ CO_2
6×10^{23} C	___ O_2	___ CO_2

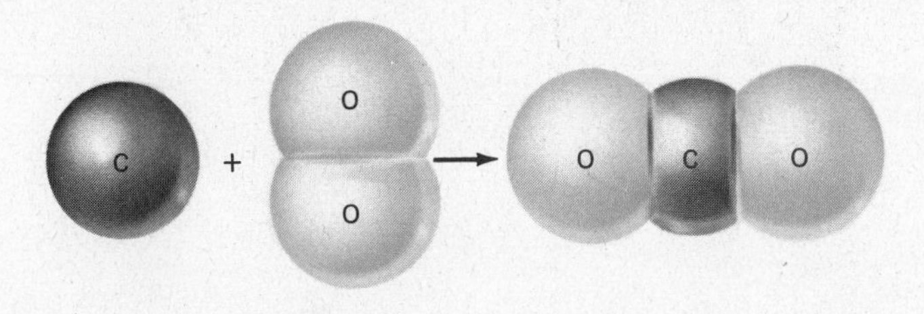

Atoms and molecules are small—too small to work with. But the quantity of 6×10^{23} atoms or 6×10^{23} molecules is large enough to work with. This number has a special name, a *mole.*

1 mole $= 6 \times 10^{23}$ of anything

1 mole of bicycles $= 6 \times 10^{23}$ bicycles

1 mole of carbon $= 6 \times 10^{23}$ carbon atoms

2 moles of carbon $= 2(6 \times 10^{23})$ carbon atoms

$\frac{1}{2}$ mole of carbon $= \frac{1}{2}(6 \times 10^{23})$ carbon atoms

Since $6 \times 10^{23} = 1$ *mole,* the equation can be written

$$6 \times 10^{23} \text{ atoms C} + 6 \times 10^{23} \text{ molecules } O_2 \longrightarrow 6 \times 10^{23} \text{ molecules } CO_2$$

or

$$1 \text{ mole C} + 1 \text{ mole } O_2 \longrightarrow 1 \text{ mole } CO_2$$

Try another equation that you just balanced in the last investigation.

2. Copy Table 2 into your notebook. Fill in the blanks to balance the equation.

TABLE 2

2 HgO *2 molecules of HgO* $\longrightarrow$	*2 Hg* *2 molecules of Hg* +	*1* O_2 *1 molecule of* O_2
4 HgO	_____ Hg	_____ O_2
6 HgO	_____ Hg	_____ O_2
6×10^{23} HgO	_____ Hg	_____ O_2
1 mole HgO	_____ Hg	_____ O_2
12×10^{23} HgO	_____ Hg	_____ O_2
2 moles HgO	_____ Hg	_____ O_2

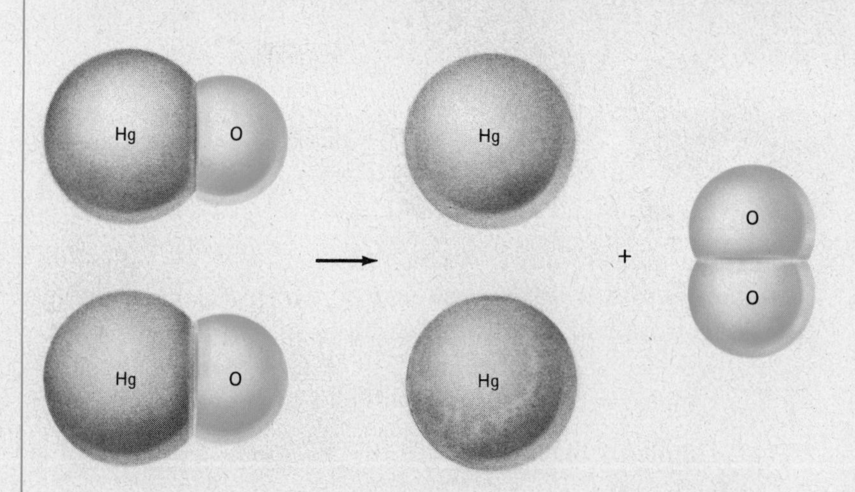

You have done very well. The concept of a mole is difficult. Good work!

B. That's Heavy

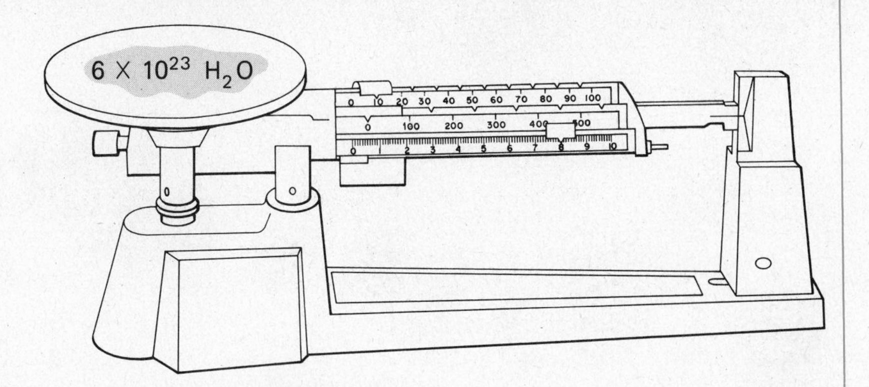

In Unit 2, Investigation 9, you worked with the atomic weight of elements. Protons and neutrons give the atoms their weight. Different elements have different numbers of protons and neutrons. Therefore, atoms of different elements have different weights. How do you find the weight of one molecule of a compound?

Try an example. Water is H_2O. Water contains two hydrogens and one oxygen.

$$H_2O = \begin{matrix} H \\ H \end{matrix} > O$$

The atomic weights of hydrogen and oxygen are:

Hydrogen = 1

Oxygen = 16

Since water is H_2O,

	Atomic weight
2 Hydrogens would =	2
1 Oxygen would =	16
Therefore H_2O =	18

The sum of the atomic weights is 18, which is called the *molecular weight* of water. To find the mass of a mole of a compound, you need to know the molecular weight. The mass of one mole of a substance equals the sum of the atomic weights of the elements, in grams. Therefore, one mole of water (H_2O) has a mass of 18 grams.

$$1 \text{ mole of water} = 18 \text{ grams}$$

Try one more example, carbon dioxide—CO_2.

$$\text{atomic weight of C} = 12$$
$$\text{atomic weight of O} = 16$$

Carbon dioxide contains two oxygens.

	Atomic weight
1 Carbon =	12
2 Oxygens =	32
CO_2 =	44
1 mole of CO_2 =	44 grams

If you have one mole of a substance, you add the atomic weights of the elements to find the mass in grams of that one mole.

3. What is the mass of one mole of HgO?
4. What is the mass of two moles of HgO?
5. What is the mass of one mole of sulfuric acid (H_2SO_4)?
6. What is the mass of $\frac{1}{2}$ mole of sulfuric acid?

This can all be summarized as follows:

$$\begin{array}{c} 6 \times 10^{23} \text{ atoms} \\ \text{or molecules} \end{array} = 1 \text{ mole} = \begin{array}{c} \text{sum of the atomic} \\ \text{weights in grams} \end{array}$$

Wow! That is a lot of information. Let's see if all this information can be put together.

$$6 \times 10^{23} \text{ water molecules} = 1 \text{ mole} = 18 \text{ g of water}$$

Every time you take a swallow of water (18 grams of water = 18 ml = 1 swallow), you swallow 6×10^{23} water molecules. That's a lot. You can see how really small the molecules are. Every time you take a swallow of water, you drink a mole!

7. What is the mass of one mole of NaOH in grams?

Atomic weight of sodium = 23
Atomic weight of oxygen = 16
Atomic weight of hydrogen = 1

8. Copy Table 3 into your notebook. Fill in the blanks using the information given in the table.

TABLE 3

6×10^{23} molecules of NaOH	=	1 mole of NaOH	=	40 g of NaOH
12×10^{23} molecules of NaOH	=	____ moles of NaOH	=	____ g of NaOH
____ molecules of NaOH	=	3 moles of NaOH	=	____ g of NaOH
____ molecules of NaOH	=	____ moles of NaOH	=	20 g of NaOH

You have just had your introduction to a very difficult concept. Congratulations are in order. This information will be useful in the rest of your activities in this class.

INVESTIGATION 6

Go-Go-Go!

Do you think we dare take a lunch break?

In Unit 3, Investigation 4, you looked at a number of chemical reactions. How does a chemical reaction start? When does it stop? Can you make it go faster or slower?

These are some of the questions you will try to answer in this investigation.

A. The Heat's On

In any experiment, scientists change only one thing at a time. In this activity, you will *change* the *temperature* of the reactions. Everything else will stay the same. Then you can see what effect different temperatures have on the reaction.

What You Need

Hydrochloric acid (HCl), 1 *M* solution
Ice (H_2O)
Magnesium (Mg) ribbon, 11 cm
Beaker, 250 ml
Graduated cylinder, 25 ml
Stirring rod
Test tubes, 4
Thermometer
Asbestos pad
Burner
Matches
Ring
Ring stand
Ruler, metric
Safety goggles
Sandpaper or steel wool
Scissors
Test tube rack
Wax pencil

What to Do

1. Copy Table 1 into your notebook.

TABLE 1

Tube	Temperature in °C	Reaction time in seconds
A		
B		
C		
D		

a. Place four test tubes in a test tube rack. Label them **A, B, C,** and **D.**

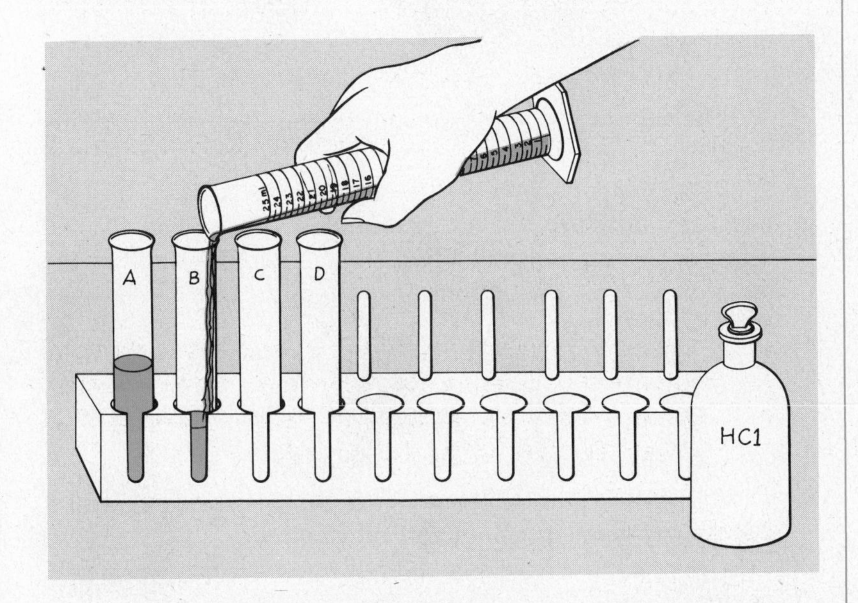

b. Add 5 ml of 1 M HCl to each tube.

c. Carefully clean an 11-cm strip of magnesium ribbon with sandpaper or steel wool.

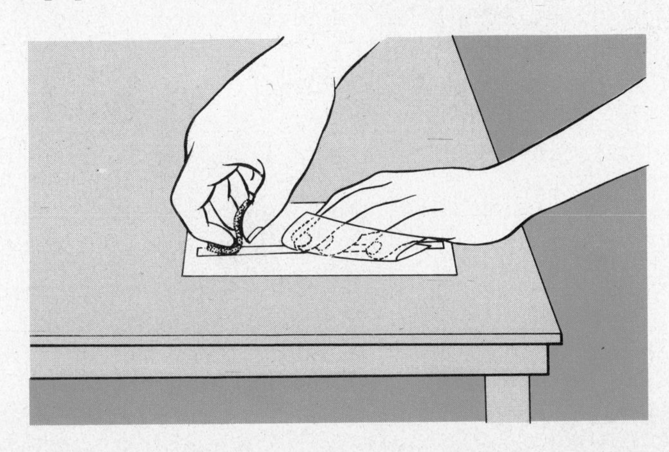

d. Cut the ribbon into 1-cm strips. This is enough for all parts of this investigation.

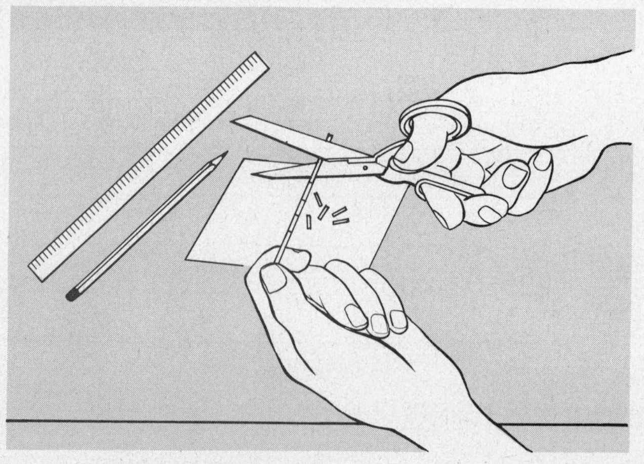

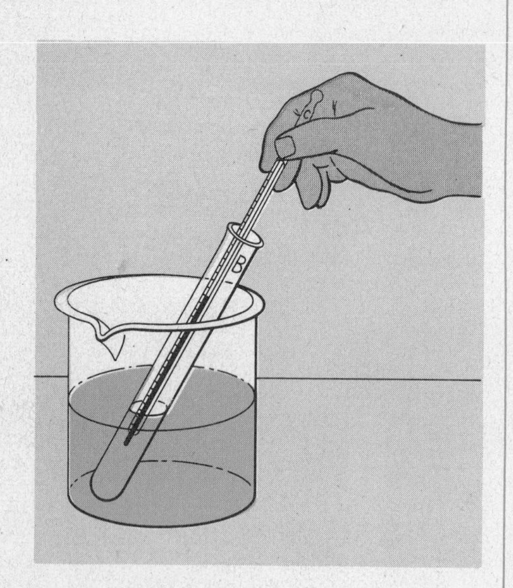

e. Half-fill a 250 ml beaker with tap water.
f. Place tube **B** in the beaker.

2. Why should you wait two or three minutes before taking the temperature?

g. Carefully measure the temperature of the acid in tube **B** using a Celsius thermometer. **Do not let the bulb of the thermometer hit the bottom of the test tube.**

3. Record the temperature of the acid in Table 1 in your notebook.

h. Remove the thermometer from the tube.
i. Carefully rinse the thermometer in water. **Then put it in a safe place where it cannot roll off the desk.**

Read steps **j** and **k** before going on.

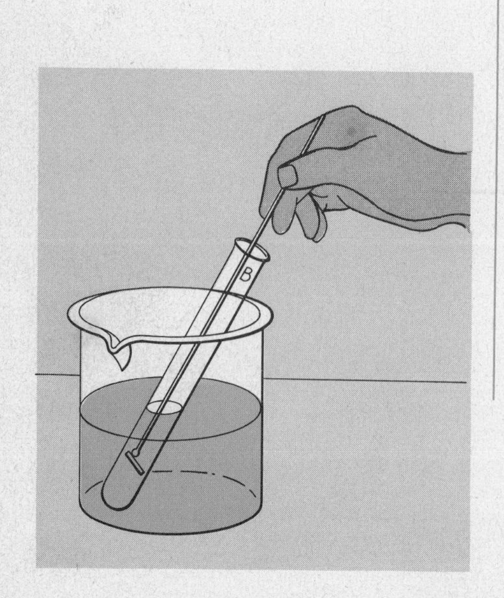

j. Place one of the strips of magnesium ribbon in tube **B**. Push it below the surface of the acid with a stirring rod and stir.
k. Measure the reaction time in seconds. Begin timing as soon as the metal comes in contact with the acid. Stop timing when all of the magnesium has reacted.

4. Record the reaction time in Table 1 in your notebook.

5. Write a balanced chemical equation for the reaction of the magnesium strip and hydrochloric acid.

l. Put tube **B** back in the test tube rack.
m. Pour out about half of the water in the beaker.
n. Then add crushed ice to the beaker until it is about $\frac{1}{2}$-full with ice water.
o. Repeat steps **f** through **l** using tube **A** and another strip of magnesium.

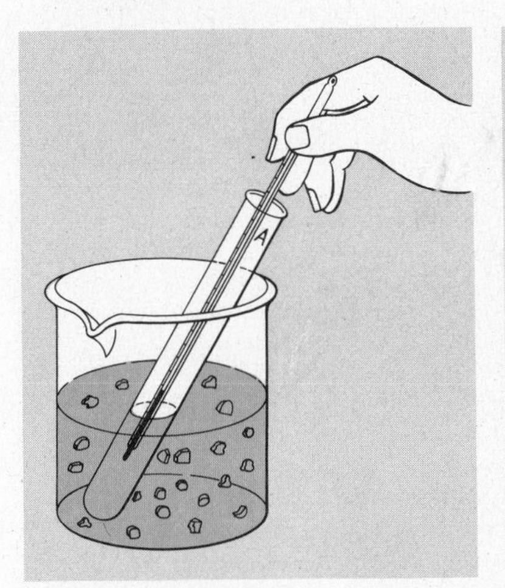

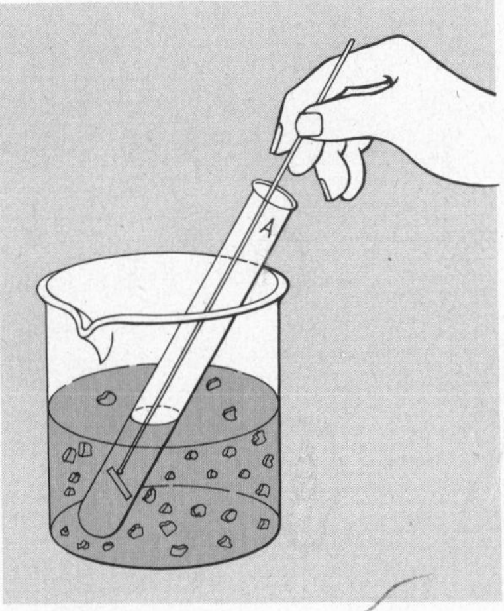

6. Record the results in Table 1.

p. Heat the water in the beaker to about 50°C.
q. Then repeat steps **f** through **l** using tube **C** and another strip of magnesium.

7. Record your results.

r. Heat the water in the beaker to boiling.
s. Then repeat the experiment using tube **D** and another strip of magnesium.

8. Make a graph showing the relationship between the temperature and reaction time. Use your data from Table 1. The axes of your graph should be like those shown in the illustration on the next page.

t. Look at your graph.

9. What effect does increasing the temperature have on the reaction time?

10. What effect does decreasing the temperature have on the reaction time?

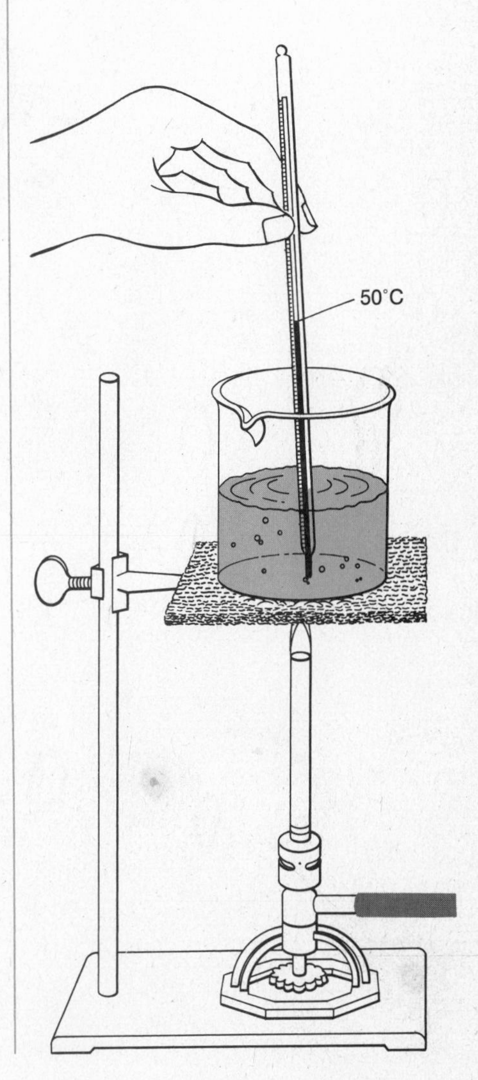

GRAPH 1

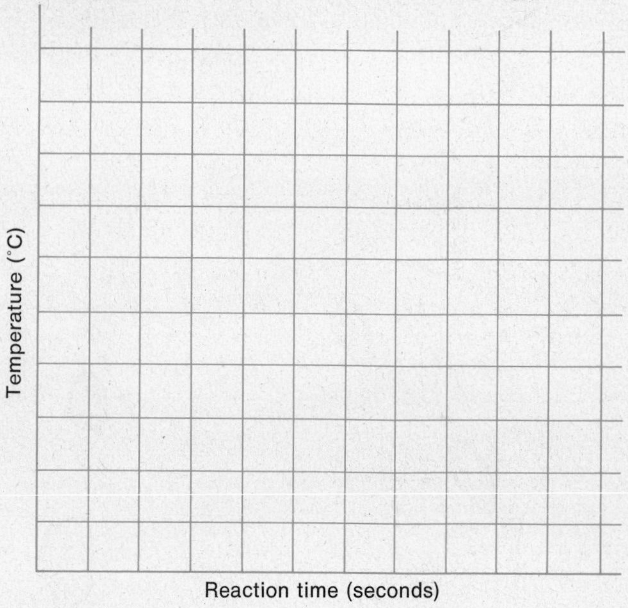

Landre/dpi

11. What would be the reaction time at 60°C? 35°C?

Most chemical reactions occur when atoms, molecules, or ions collide and interact. Atoms are rearranged. Chemical bonds are broken and new ones are formed to give new substances. You can tell that something has happened when there is a change in color, a change in temperature, or the formation of a precipitate or a gas.

Do all chemical reactions proceed at the same speed? Think for a minute. How long does it take iron to rust? How long does it take wood to burn? How long does it take a stick of dynamite to explode?

What determines the speed of a chemical reaction? The *rate of reaction* means the amount of substance undergoing some change in a unit of time. The unit of time might be a second, a minute, or an hour. You could measure the number of tons of coal mined per day. You could also measure the amount of acid formed or gas generated or precipitate produced in a minute or an hour.

Your work in the lab has shown that temperature can change the rate of a chemical reaction. Now do a little more digging to find out what else affects the rate of reaction.

B. My, What a Tower of Strength

You will now do the same experiment, but with a different twist. Last time you changed the temperature of the reaction and kept everything else the same. Now you'll keep the temperature, volume of acid, and amount of magnesium the same. But you will use acids of different *concentrations.* They will all be solutions of acid and water. Some have more acid and less water; they are more concentrated. Some of the solutions are less concentrated (contain less acid). You will look at your results to see if different concentrations make reactions go faster or slower.

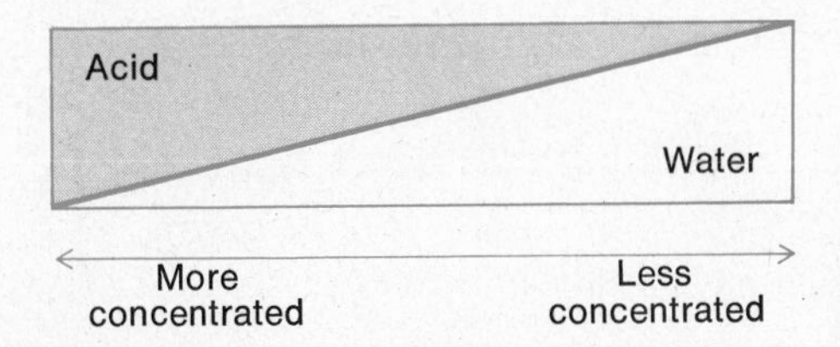

What You Need

Magnesium (Mg) ribbon
Solution **A** (6 *M* HCl)
Solution **B** (3 *M* HCl)
Solution **C** (1 *M* HCl)
Solution **D** (0.5 *M* HCl)
Beaker, 250 ml
Graduated cylinder, 25 ml
Stirring rod
Test tubes, 4
Safety goggles
Test tube rack
Wax pencil

What to Do

12. Copy Table 2 into your laboratory notebook.

TABLE 2

Tube	Concentration	Reaction time in seconds
A	6 *M*	
B		
C		
D		

a. Label four clean test tubes **A, B, C,** and **D**. Place them in a test tube rack.

b. Add 10 ml of solution **A** to tube **A**.
c. Add 10 ml of solution **B** to tube **B**.
d. Add 10 ml of solution **C** to tube **C**.
e. Add 10 ml of solution **D** to tube **D**.

13. What is the difference between solutions **A, B, C,** and **D**?

f. Place tube **A** in a 250 ml beaker $\frac{1}{2}$-full with room temperature water. (This is a safety precaution. Why?)
g. Insert a strip of magnesium ribbon into the test tube. Push it below the surface of the acid with a stirring rod and stir.

14. Record the reaction time in Table 2.

h. Repeat steps **f** and **g** with tubes **B, C,** and **D.**

15. Record the reaction times.

Remember that solutions **A, B, C,** and **D** are all different in concentration.

16. Which of the four solutions is the most concentrated? Why?

17. Which of the four solutions is the least concentrated? Why?

18. Make a graph showing the relationship between concentration and reaction time. Use your data from Table 2.

19. What effect does increasing the concentration have on the reaction time?

20. What effect does decreasing the concentration have on the reaction time?

21. If the concentration were 4.5 *M*, what would be the reaction time?

C. You're in Fantastic Shape Now

So far you have done the same experiment twice. First you changed, or varied, the temperature. Then you varied the concentration. Temperature and concentration were *variables* in the experiment. This means they are things that can be changed, as you have been doing. Only one variable is changed during each experiment. This allows you to see if it affects the reaction.

In this Part, you will do the experiment once again. This time you will vary the shape of the reacting magnesium. The volume and concentration of the acid will be kept constant.

What You Need

- Hydrochloric acid (HCl), 1 *M*
- Magnesium (Mg) ribbon
- Beaker, 250 ml
- Graduated cylinder, 25 ml
- Stirring rod
- Test tubes, 3
- Safety goggles
- Scissors
- Test tube rack
- Wax pencil
- Weighing paper

What to Do

22. Copy Table 3 into your notebook.

TABLE 3

Tube	Shape of strip	Reaction time in seconds
A		
B		
C		

a. Label three test tubes **A, B,** and **C.** Place them in a test tube rack.
b. Add 5 ml of 1 *M* HCl to each tube.
c. Place tube **A** in a 250 ml beaker $\frac{1}{2}$-full with room temperature water.
d. Roll one of the magnesium strips into a tight ball.
e. Drop it into tube **A.**

23. Record the reaction time in Table 3.

f. Repeat the experiment using tube **B** and a flat strip of magnesium ribbon.

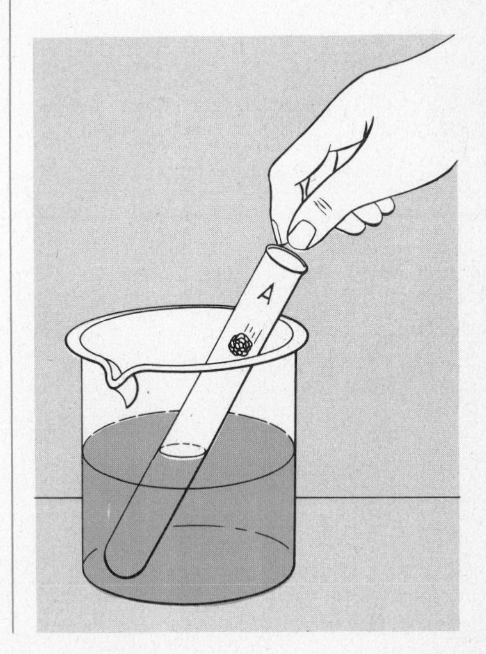

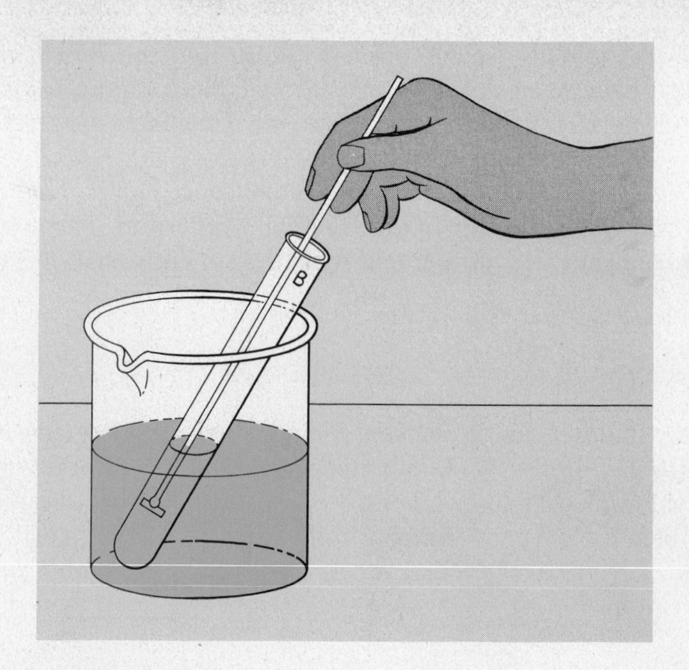

24. Record the reaction time in your table.

g. Cut a strip of magnesium into as many slivers as possible. Do this over a piece of weighing paper.
h. Repeat the experiment using tube **C** and the slivers.

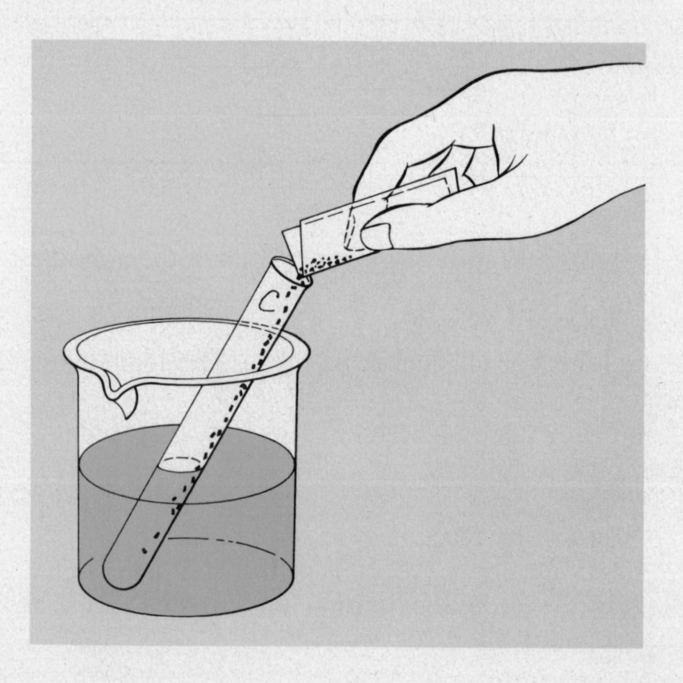

25. Record the reaction time.

26. Is the volume of the acid the same in all three tubes?

27. Is the strength of the acid the same in all three tubes?

28. Was the temperature about the same in all three tubes?

29. What is the *variable* that you changed in the three tubes?

30. How does the shape of the magnesium strip affect the reaction rate?

The *surface area* of a material is the amount of its surface that is exposed. Chemical reactions take place at the surface. Examine the surface area of the three samples of magnesium in the illustration.

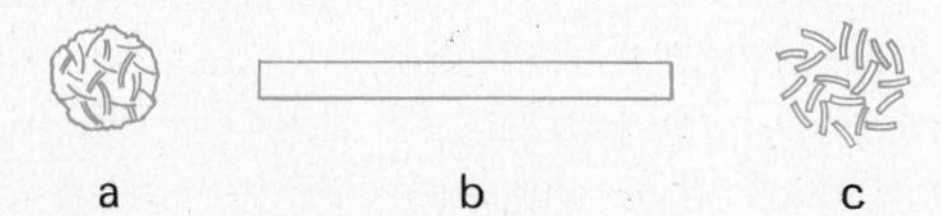

31. Which magnesium sample has the greatest surface area?

32. Examine your results in Table 3. How does the surface area affect the rate of reaction?

In the next investigation you will apply what you have learned so far this year.

INVESTIGATION 7

Cu Later!

This investigation is a summary of all your work so far this year. You will be using all the knowledge and most of the laboratory techniques you have acquired. In this investigation, copper metal will be reacted, forming many copper compounds. Then you will recover the copper.

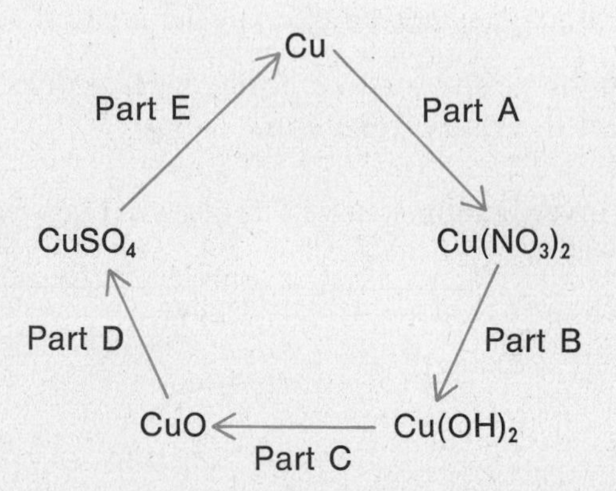

You can compare the amount of copper at the beginning of the investigation with the amount at the end.

1. Would you expect the copper at the end to be equal to, more than, or less than the amount you started with? Why?

A. The Disappearing Act

The copper (Cu) will react and form copper(II) nitrate [$Cu(NO_3)_2$]. The "II" in "copper(II) nitrate" means the copper ion has a positive charge of 2; it is written Cu^{2+}.

What You Need

Copper (Cu) wire
Nitric acid (HNO_3), concentrated
Beaker, 250 ml
Graduated cylinder, 25 ml
Balance
Safety goggles
Scissors or wirecutters
Weighing paper

What to Do

2. Copy Table 1 into your notebook.

TABLE 1

	Mass in grams
Mass of copper wire	______
Mass of evaporating dish and copper	______
Mass of evaporating dish	______
Mass of copper at end of investigation	______

a. Cut off about 1 gram of copper wire.
b. Find the exact mass of the wire.

3. In the first line of Table 1, record the mass of the copper wire.

c. Place the wire flat on the bottom of a 250 ml beaker.
d. Place the beaker in the hood.

e. Add 10 ml of concentrated HNO_3. **Be extremely careful in handling nitric acid. Do not breathe the fumes.**
f. Allow all the copper to react. Leave the beaker in the hood until the reaction is complete.

The color in the beaker is due to copper(II) nitrate. The brown fumes are nitrogen dioxide, a poisonous gas. The equation for the reaction is:

$$Cu + 4\,HNO_3 \longrightarrow Cu(NO_3)_2 + 2\,NO_2 + 2\,H_2O$$

4. What is the formula of nitrogen dioxide?
5. Copy Table 2 into your notebook.
6. In Table 2, fill in the information for Part A of your experiment.

TABLE 2

	Substance produced	*Formula of substance produced*	*Description of substance produced*	*Equation for the reaction*
Part A	Copper(II) nitrate			
Part B	Copper(II) hydroxide			
Part C	Copper(II) oxide			
Part D	Copper(II) sulfate			
Part E	Copper			

B. Cooling It

Copper(II) nitrate [$Cu(NO_3)_2$] will form copper(II) hydroxide [$Cu(OH)_2$]. This reaction needs to be slowed down, so you can just "cool it."

What You Need

Copper(II) nitrate [$Cu(NO_3)_2$] solution from Part A
Ice, crushed
Litmus paper, red
Sodium hydroxide (NaOH), 6 *M* solution
Beaker, 400 ml
Glass plate
Graduated cylinder, 25 ml
Stirring rod
Safety goggles

What to Do

a. Fill a 400 ml beaker $\frac{1}{4}$-full with crushed ice.
b. Add just enough water to cover the ice.

c. Place your 250 ml beaker from Part A inside the 400 ml beaker.
d. While stirring, allow the $Cu(NO_3)_2$ to cool for 2–3 minutes.
e. Get 15 ml of 6 *M* NaOH. **Handle with care.**
f. While stirring, add 2–3 ml of NaOH at a time.

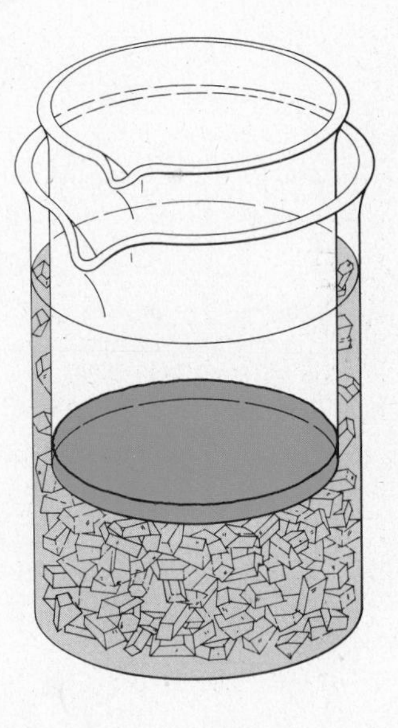

After all the sodium hydroxide has been added, you will test the liquid with litmus paper.

g. Dip the stirring rod in the liquid and put a drop on your litmus paper. If the paper turns blue, the reaction is complete.
h. If the reaction is not complete, repeat steps **f** and **g.**

7. In Table 2, fill in the information for Part B of the investigation. The reaction between $Cu(NO_3)_2$ and NaOH is a double displacement reaction.

C. The Heat's on Again

In this reaction, copper(II) hydroxide [$Cu(OH)_2$] will react to form copper(II) oxide (CuO). Copper hydroxide has to be heated to form copper oxide.

What You Need

Copper(II) hydroxide [$Cu(OH)_2$] solution from Part B
Water (H_2O), distilled
Graduated cylinder, 100 ml
Stirring rod
Asbestos pad
Burner
Matches
Ring
Ring stand
Safety goggles

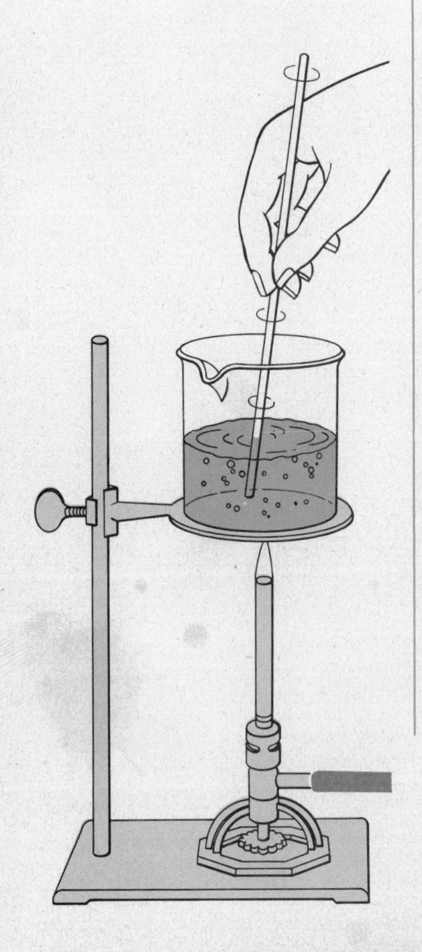

What to Do

a. Add 100 ml of distilled water to your 250 ml beaker, which contains the $Cu(OH)_2$ from Part B.
b. While stirring the $Cu(OH)_2$, gently boil it for about 5 minutes.

8. In Table 2, fill in the information for Part C of your experiment. The reaction of $Cu(OH)_2$ to produce CuO and H_2O is a decomposition reaction.

D. Don't Be Blue

In this Part, copper(II) oxide (CuO) will react with sulfuric acid (H_2SO_4) to form copper(II) sulfate ($CuSO_4$).

What You Need

- Copper(II) oxide (CuO) from Part C
- Sulfuric acid (H_2SO_4), 3 *M*
- Graduated cylinder, 25 ml
- Stirring rod
- Safety goggles

What to Do

a. Make sure the CuO from Part C has settled.

b. Carefully *decant* and discard the clear liquid. (To decant means to pour off the liquid and leave the solid behind.)

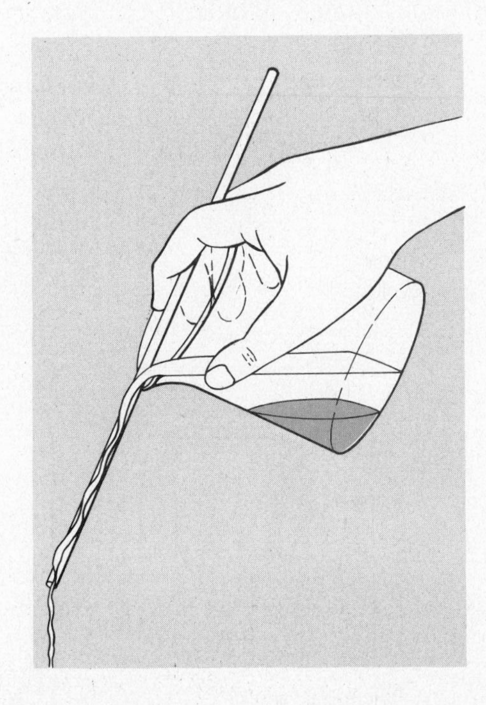

c. Add 25 ml of H_2SO_4 to the beaker. Stir. The black CuO should react, and the black color should disappear.

d. If the black color isn't gone, add H_2SO_4, 2–3 ml at a time, until the solution clears.

9. In Table 2, fill in the information for Part D of your experiment. The reaction of CuO and H_2SO_4 is a double displacement reaction.

E. The Reappearing Act

In this reaction the copper(II) sulfate ($CuSO_4$) will react with zinc (Zn), producing pure copper (Cu).

10. How much copper do you expect at the end of this investigation?

11. Will small or large pieces of zinc react faster with the copper(II) sulfate?

What You Need

- Copper(II) sulfate ($CuSO_4$) from Part D
- Hydrochloric acid (HCl), 3 *M*
- Water (H_2O), distilled
- Zinc (Zn)
- Evaporating dish
- Graduated cylinder, 25 ml
- Stirring rod
- Asbestos pad
- Balance
- Burner
- Matches
- Ring
- Ring stand
- Rubber policeman
- Safety goggles
- Weighing paper

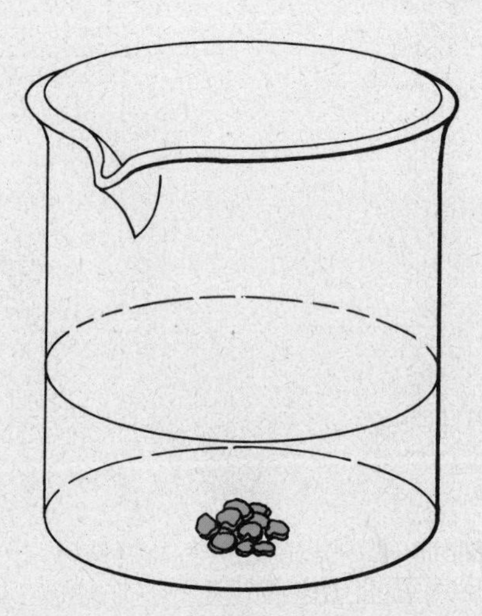

What to Do

a. Add 5 grams of Zn to the beaker of $CuSO_4$.

b. Warm the contents while stirring.

12. Copper(II) sulfate is blue. How do you know when all the copper sulfate has reacted?

This reaction may take a period or two of warming and stirring. Once the blue color is gone, you need to react the unused zinc. This will leave only solid copper in the bottom of the beaker.

13. In Table 2, fill in the information for Part E of your investigation. The reaction of Zn with $CuSO_4$ is a single replacement reaction.

c. To react the remaining Zn, add 20 ml of 3 *M* HCl.

14. How can you speed up this reaction? Try it.

When the zinc has all reacted, you need to clean and dry the copper.

d. Decant the liquid from the Cu.
e. Rinse the Cu with 50 ml of distilled water.
f. Decant the liquid.
g. Repeat steps **e** and **f** again.
h. Find the mass of a clean, dry evaporating dish.

15. Record the mass in Table 1.

i. Transfer the Cu to the evaporating dish using your rubber policeman. Use a little distilled water if necessary.

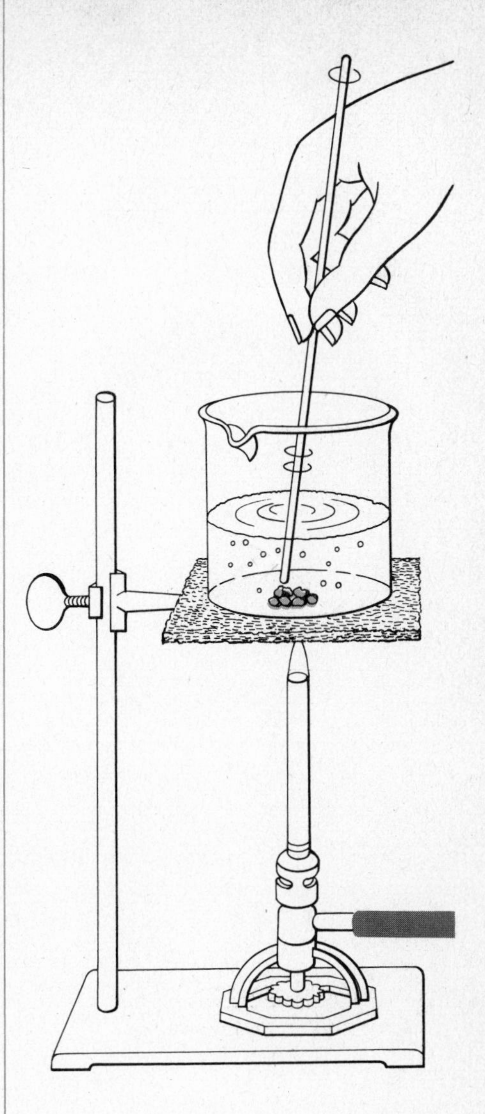

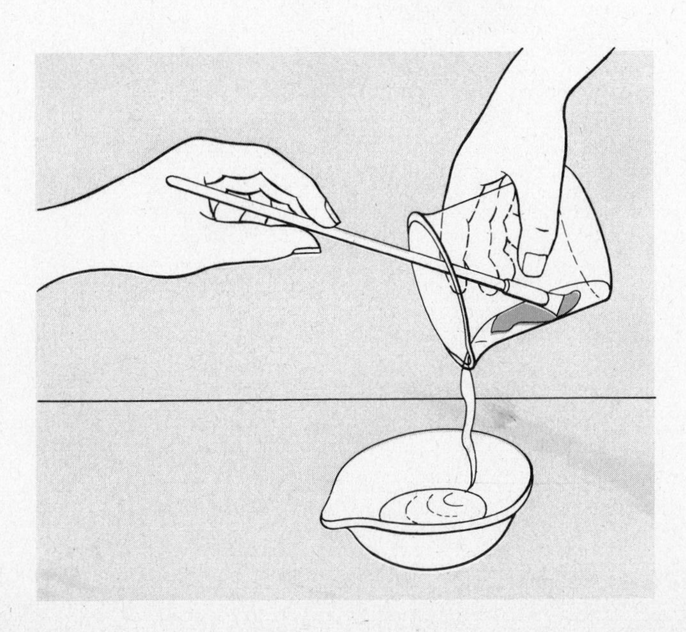

j. **Gently** warm the evaporating dish until the Cu is dry.
k. Find the mass of the dry Cu plus the evaporating dish.

16. Record the mass in Table 1.

17. Record in Table 1 the mass of copper at the end of the investigation.

18. If your copper is not pure, examine its color and look at Table 2. What substance could be the impurity?

19. Would impurities make your final copper sample heavier or lighter?

20. Why did you wash the sample in Part E with distilled water?

21. If you didn't decant carefully, what would be your final mass of copper? (same, less, or more)

When you have completed this activity, you know that you have learned a lot of chemistry. Your activities in the last few days have been a summary of all you've studied so far. Good work!

What's Your Identity?

INVESTIGATION 8

So far this year you have used many different substances. Many of them had different properties. But some had similar properties. You can't be expected to learn much about all these substances unless there is some way to put them into groups. Don't worry—there is. One way is to classify (group) compounds as acids, bases, salts, and oxides.

An understanding of acids, bases, salts, and oxides will come gradually. No one expects you to learn all there is to know about them. You will investigate the properties of these substances in the lab so you can understand them and have them work for you.

You will start on acids and bases. Why should we bother with acids and bases? Without acids, we would have no steel prod-

United States Steel

© 1972 Paul Ginsberg/dpi

George Malone/dpi

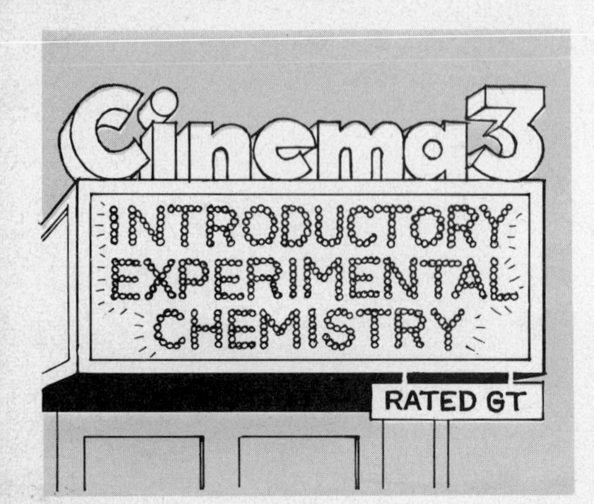

ucts, no explosives, little fertilizer, and no plastics. Without bases, there would be no soap and no synthetic fabrics. Acids and bases have a wide range of uses.

A. Another Opening, Another Show

This section is rated GT: General admission, with Teacher guidance suggested.

What You Need

Acetic acid (CH_3COOH), 0.1 *M*
Hydrochloric acid (HCl), 0.1 *M*
Litmus paper, blue
Litmus paper, red
Phenolphthalein indicator
Sulfuric acid (H_2SO_4), 0.1 *M*
Dropper
Glass plate
Graduated cylinder, 10 ml
Stirring rod
Test tubes, 3
Safety goggles
Test tube rack
Wax pencil

What to Do

1. Copy Table 1 into your notebook.

TABLE 1: Properties of Acids

Acid	Formula	Ions present	Feel	Litmus paper test		Phenolphthalein test	Reaction with $CaCO_3$	Reaction with Mg
				Blue	Red			
A Hydrochloric	HCl							
B Sulfuric	H_2SO_4							
C Acetic	CH_3COOH							

a. Place three test tubes in a test tube rack. Label them **A, B,** and **C.**

b. Add 5 ml of HCl to tube **A.** Add 5 ml of H_2SO_4 to tube **B.** Add 5 ml of CH_3COOH to tube **C.** Rinse out the graduated cylinder after using each acid.

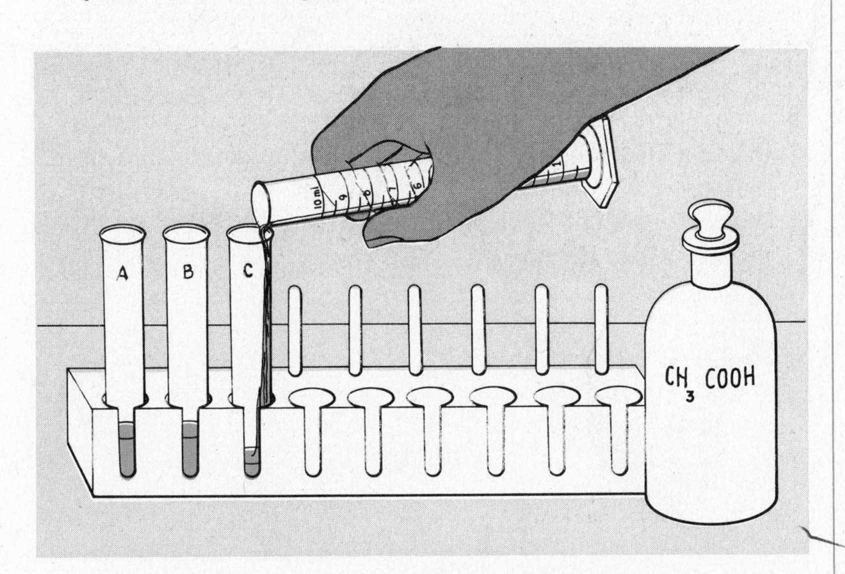

c. Use your stirring rod to place a drop of HCl from tube **A** between your fingers.

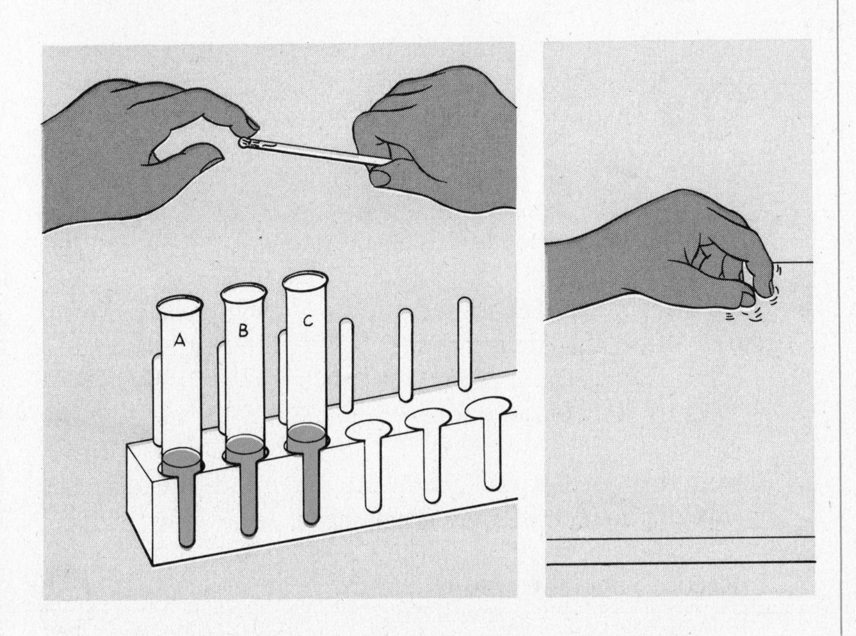

d. Rinse your fingers and the stirring rod thoroughly.

2. In Table 1, record any unusual feeling.

Don't ever feel chemicals without specific instructions. And if you do spill something on your skin, rinse it off immediately.

e. Repeat steps **c** and **d** with the other two acids.

3. Record the feeling of the acids.

f. Place a strip of blue litmus paper and a strip of red litmus paper on a glass plate.

g. Place a drop of HCl on one end of each piece of litmus paper. Rinse the stirring rod.

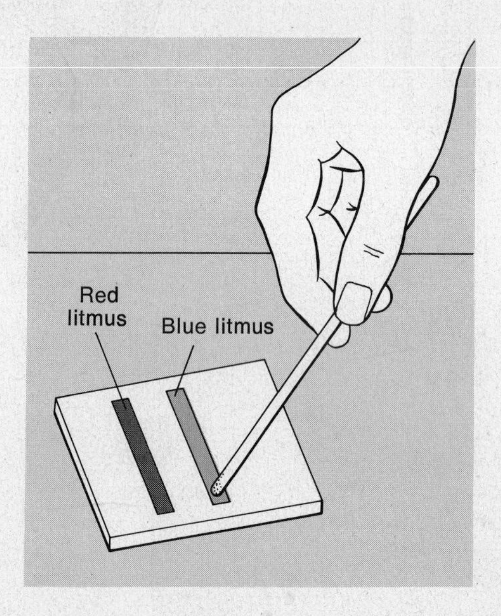

4. Record your observations in the table.

h. Place a drop of H_2SO_4 in the middle of each piece of litmus paper. Rinse the stirring rod.

5. Record your observations.

i. Place a drop of CH_3COOH on the other end of each piece of litmus paper. Rinse the stirring rod.

6. Record your observations.

j. Add a drop of phenolphthalein indicator to each test tube.

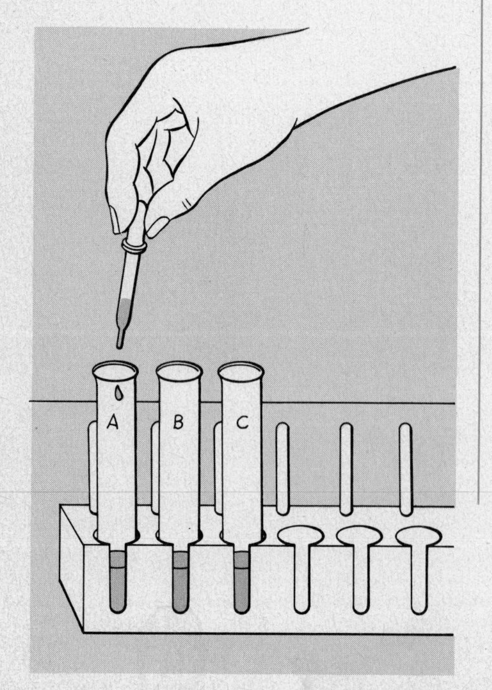

7. Record your observations.

k. Carefully empty the test tubes of acid into the sink. Rinse the sink and test tubes thoroughly with water.

B. Acids, Acids Everywhere, But Not a Drop to Drink

Poor Willy was a chemist.
Now Willy is no more;
For what he thought was H_2O
Was H_2SO_4.

In the next few activities, the reactions will produce gases. You have learned to test for some gases this year. Let's review the tests before completing this investigation.

TABLE 2: Tests for Gases

Gas	*Test*	*Result of test*	*Investigation where studied*
Carbon dioxide, CO_2	gas bubbled through limewater	limewater turns milky (cloudy)	1.4
Hydrogen, H_2	burning splint	a pop or bang	2.4
Oxygen, O_2	glowing splint	glowing splint bursts into flame	2.4

You can react different acids with the same substance. Will these reactions produce the same gas?

What You Need

Acetic acid (CH_3COOH), 3 *M*
Calcium carbonate ($CaCO_3$)
Hydrochloric acid (HCl), 3 *M*
Limewater [$Ca(OH)_2$]
Magnesium (Mg)
Sulfuric acid (H_2SO_4), 3 *M*
Glass bend, right angle
Graduated cylinder, 10 ml
Test tubes, 2
Matches
Safety goggles
Spatula
Stopper, 1-hole
Test tube rack
Wax pencil
Wood splints

What to Do

a. Carefully insert the glass bend into the rubber stopper. Remember to moisten the glass tube and rubber stopper. Grip the glass tube near the end that is being inserted.
b. Fill a test tube $\frac{1}{2}$-full with limewater.
c. Add 5 ml of HCl to another test tube.
d. With a spatula, add a few pieces of $CaCO_3$ to the tube containing acid.

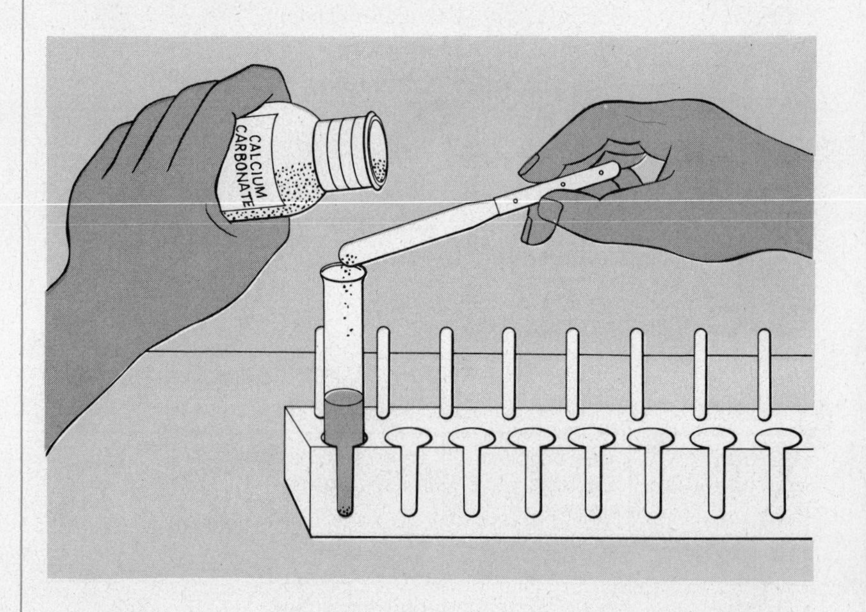

e. Quickly insert the rubber stopper into the tube containing acid.
f. Test the gas by bubbling it through the limewater.

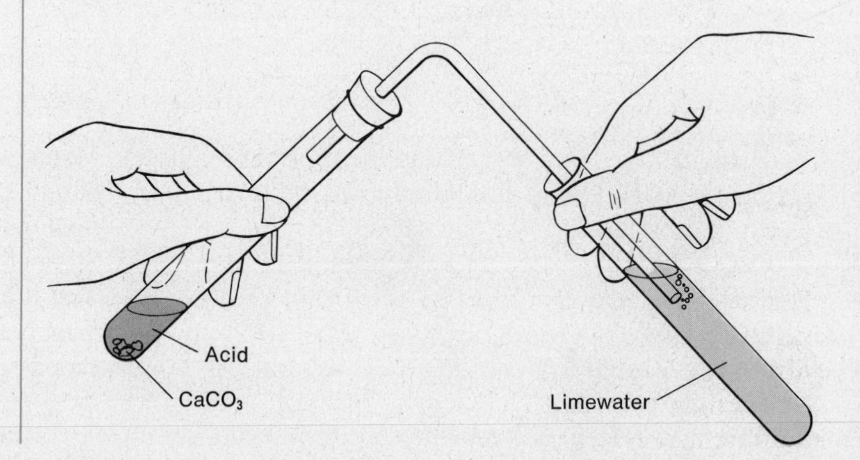

8. Record your results in Table 1.

g. Rinse all glassware.
h. Repeat steps **b** through **g** with the other two acids.

9. Record your results.

Another test for acids is to react them with metals.

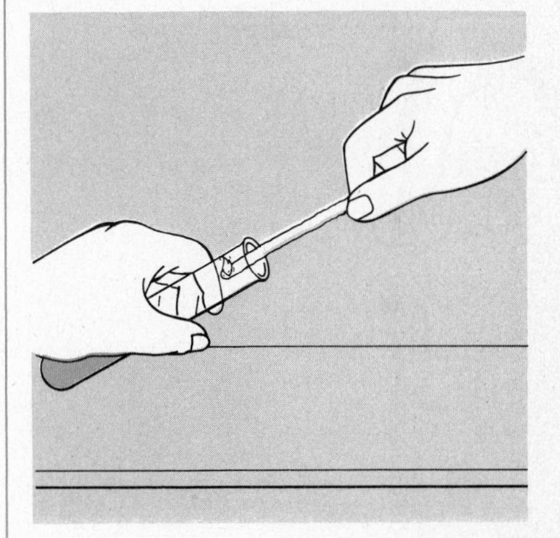

i. Add 5 ml of HCl to a test tube.
j. Place a small piece of Mg in the tube.
k. Test the gas given off with a burning splint.

10. Record your results in Table 1.

l. Rinse your glassware.
m. Repeat steps **i** through **l** with the other acids.

11. Record your results.

12. What gas is given off when magnesium reacts with these acids?

You have tested several properties of acids; and you have found that these acids have similar properties. What might cause these acids to react similarly?

13. Record in Table 1 what ions are present in each acid. (Refer to the table on page 187.)

14. What ion is present in each acid?

15. What ion in each acid causes the acids to have similar properties?

Acids, such as hydrochloric and sulfuric, are used in large amounts in industry and in laboratory work. Hydrochloric acid is used in the preparation and purification of metals and other substances. Sulfuric acid is important in the production of fertilizers, gasoline, and steel.

Several acids are also used in medicine. Hydrochloric acid, normally found in the gastric juices, is necessary for the proper digestion of proteins in the stomach. Patients who have a lower concentration of HCl than normal in their stomachs (hypoacidity) are given *very dilute* HCl to drink.

Nitric acid is used in the laboratory to test for the presence of proteins. Boric acid is widely used as an antiseptic and a germicide. Acetylsalicylic acid (aspirin) is used as a pain killer. Ascorbic acid (vitamin C) is normally found in citrus fruits. It is used in the prevention and treatment of scurvy.

You will experiment with many of these acids later on. But acids are only part of the story. Now it's time for bases.

C. Bases Loaded

What You Need

Limewater [$Ca(OH)_2$]
Litmus paper, blue
Litmus paper, red
Phenolphthalein indicator
Potassium hydroxide (KOH), 0.1 *M* solution
Sodium hydroxide (NaOH), 0.1 *M* solution
Water (H_2O), distilled

Glass plate
Graduated cylinder, 10 ml
Stirring rod
Test tubes, 3

Safety goggles
Spatula
Test tube rack
Wax pencil

What to Do

16. Copy Table 3 into your notebook.

TABLE 3: Properties of Bases

Base	*Formula*	*Ions Present*	*Feel*	*Litmus paper test*		*Phenol-phthalein test*
				Blue	*Red*	
Sodium hydroxide	NaOH					
Potassium hydroxide	KOH					
Calcium hydroxide	$Ca(OH)_2$					

a. Place three clean test tubes in a rack. Label them **A, B,** and **C.**

b. Add 5 ml of NaOH to tube **A.** Add 5 ml of KOH to tube **B.** Add 5 ml of $Ca(OH)_2$ to tube **C.** Rinse out the graduated cylinder after using each base.

c. Place a drop of NaOH from test tube **A** between your fingers.

d. Rinse your fingers and the stirring rod thoroughly.

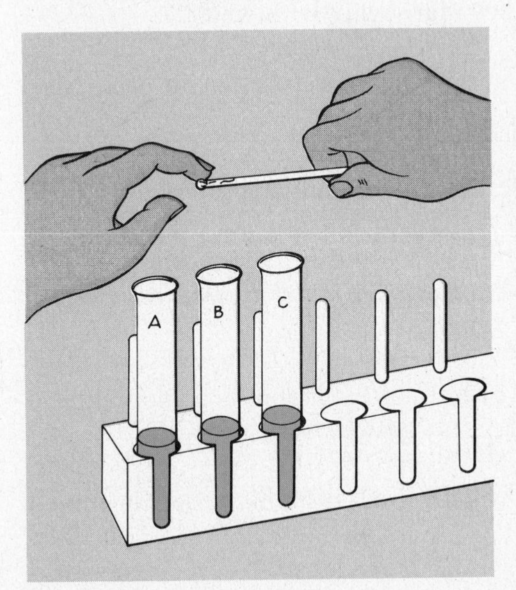

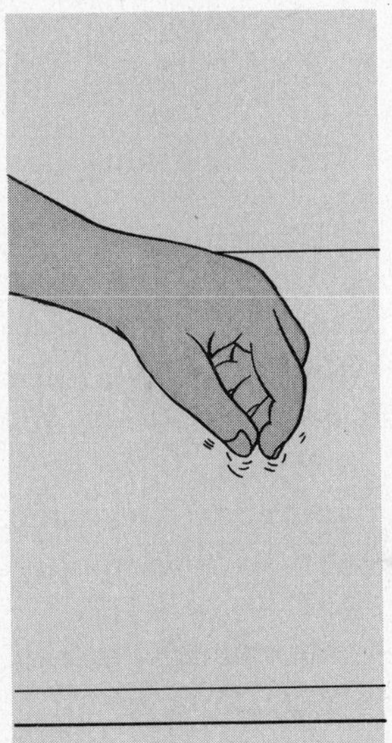

17. Record any unusual feeling in Table 3.

e. Repeat steps **c** and **d** with the other two bases.

18. Record the feeling of the bases.

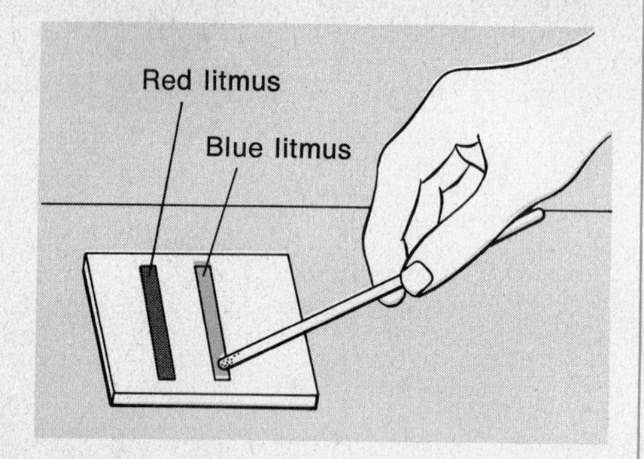

f. Place a strip of blue litmus paper and a strip of red litmus paper on a glass plate.

g. Place a drop of NaOH on one end of each piece of litmus paper. Rinse the stirring rod.

19. Record your observations in the table.

h. Place a drop of KOH on the middle of each piece of litmus paper. Rinse the stirring rod.

20. Record your observations.

i. Place a drop of $Ca(OH)_2$ on the other end of each piece of litmus paper. Rinse the stirring rod.

21. Record your observations.

j. Add a drop of phenolphthalein indicator to each test tube.

22. Record your observations in Table 3.

k. Carefully empty the test tubes of base into the sink. Rinse the sink and test tubes thoroughly with water.

23. In Table 3, fill in what ions are present in each base.

24. What ions are the same in each base?

25. What gives these bases their similar properties?

Many bases are used in industry and in laboratories. Sodium hydroxide, or lye, is used to remove fats and grease from clogged drains and also to make soap.

Magnesium hydroxide, or milk of magnesia, is a good antacid and is used as an antidote for acid poisoning.

Ammonium hydroxide is known as ammonia water. It is used as a cleaning agent.

Let's summarize some of the acid and base properties before the next activity.

Acids contain hydrogen ions, H^{1+}. Acids turn blue litmus paper red.

Acid
Red both end in **d**

Bases contain hydroxide ions, OH^{1-}. Bases turn red litmus paper blue.

Base
Blue both end in **e**

26. Copy Table 4 into your notebook. Fill in the "Acid" and "Base" rows.

TABLE 4

	Feel	Litmus paper test		Phenolphthalein test	Ion that makes the substance an acid, base, or neutral
		Blue	Red		
Acid					
Base					
Distilled water					

Water is listed in Table 4. Water is H_2O, often written HOH. Water contains hydrogen ions, H^{1+}, and hydroxide ions, OH^{1-}. Water contains equal numbers of H^{1+} and OH^{1-}. Therefore, water is *neutral,* neither acid nor base, since it contains equal numbers of hydrogen ions and hydroxide ions.

l. Test distilled water for the properties in Table 4.

27. Fill in the rest of Table 4.

D. Choose Two From Column A and Two From Column B

Some of the most common, everyday substances are acids or bases. You can use the tests you have just performed to see whether these substances are acids or bases.

Cynara

What You Need

Dishwashing liquid
Lemon juice
Litmus paper, blue
Litmus paper, red
Phenolphthalein indicator
Salt (NaCl)
Vinegar (CH_3COOH)
Washing soda (Na_2CO_3)
Water (H_2O), distilled

Glass plate
Graduated cylinder, 10 ml
Stirring rod
Test tubes

Safety goggles
Spatula
Test tube rack
Wax pencil

What to Do

28. Set up a table in your notebook for recording your results. You will need the following columns: *Substance; Feel; Litmus paper test; Phenolphthalein test.*

a. Label test tubes for the substances you will test.
b. Add 1 ml of each liquid substance to the proper test tube.
c. Add a pinch of each solid substance to the proper test tube. Add 1 ml of distilled water, and stir to dissolve.
d. Do steps **e** through **g** for each substance.

29. Record your results in the table.

e. Test to see how a small amount of the substance feels.
f. Test the substance for its reaction with litmus paper.
g. Test the substance with phenolphthalein indicator.

After finishing the tests, study your results.

30. Which of the common substances you tested are acids?

31. Which are bases?

You are now a great deal more familiar with the properties and reactions of acids and bases. Their uses in our society are almost endless. You will keep coming back to the properties and uses of acids and bases as you continue your study of chemistry.

INVESTIGATION

9

Time for a Commercial

You just finished some activities with commercial products. Commercial products that we eat must state what they have in them. Frequently they also state the quantity of the substances in them. Many technicians work on making sure advertisements are honest so that consumers will not be deceived. Some commercial products are both fun and valuable to test.

A. The Fizz That Is

In Unit 2, Investigation 3, you warmed a test tube of soda pop. You observed the gas bubbles that formed. You observed more bubbles as the temperature increased. The gas was carbon dioxide. What effect does the dissolved gas have on the soda pop? You will perform a few tests to see what happens. These tests will be conducted on cold soda pop, room temperature soda pop, and day-old soda pop.

You will add a base to the soda pop to *neutralize* it. To neutralize a substance is to change it so that it is neither acid nor base. A *neutral* substance has equal numbers of hydrogen ions (acid) and hydroxide ions (base).

Neutral

$$H^{1+} = OH^{1-}$$

$$\text{acid} = \text{base}$$

If you add a known amount of base to a substance to neutralize it, you can tell how much acid is present in it. The only trick is to know when the substance is neutral. To show this, you will add an indicator. The indicator will change color when the solution is neutral. This is known as the *end point.*

What You Need

Litmus paper, blue
Litmus paper, red
Phenolphthalein indicator
Soda pop, colorless
Sodium hydroxide ($NaOH$), 0.1 *M* solution
Beakers, 100 ml, 2
Buret
Dropper
Flask, 250 ml
Funnel
Glass plate
Graduated cylinder, 25 ml
Stirring rod
Buret clamp
Paper, white
Ring stand
Safety goggles

What to Do

1. Copy Table 1 into your notebook.

TABLE 1

		Cold soda pop	*Room temperature soda pop*	*Day-old room temperature soda pop*
Litmus paper test	Blue			
	Red			
Final volume (ml)				
Initial volume (ml)		0		
Volume of base used to neutralize soda pop (ml)				

a. Test the cold soda pop with litmus paper, both blue and red.

2. Record your observations in Table 1.

3. Is the soda pop acidic or basic?

Remember, litmus paper is re**d** for aci**d**
and blu**e** for bas**e**.

b. Examine the parts of the buret.

The buret is like an overgrown graduate. It is used to measure accurately small amounts of liquid. It is operated by the stopcock, as shown in the illustration.

When the stopcock handle is up and down, liquid can flow from the buret into the flask. You can use the stopcock to control the flow of the liquid from the buret.

The buret is read as a graduate is—at eye level!

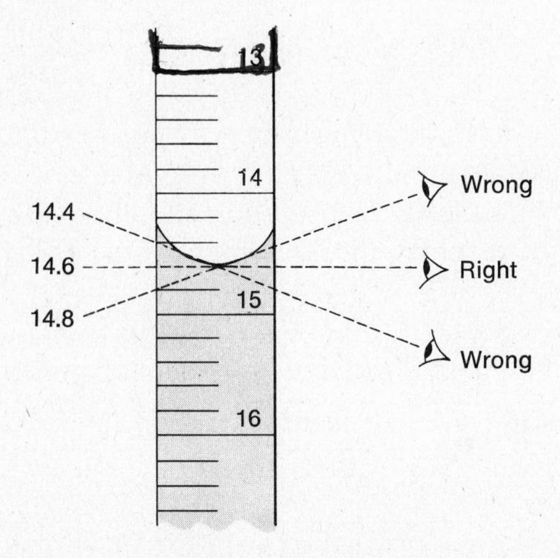

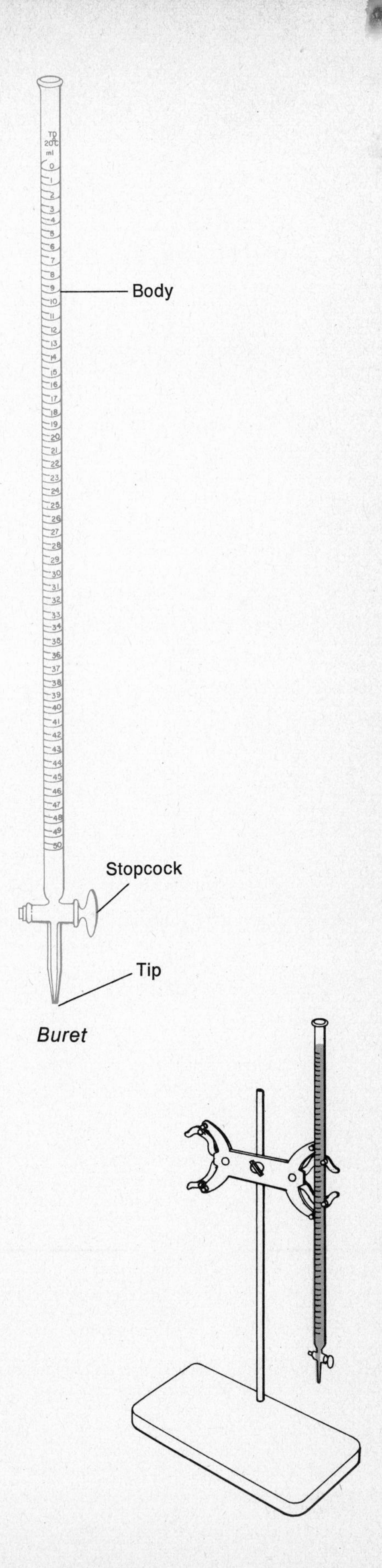

Buret

Note which way the numbers read on a buret. Zero is at the top. You fill the buret to the top. As liquid is drained out, you can tell how much you used by reading the numbers.

c. Place the buret in the buret clamp. Close the stopcock.

d. Carefully fill the buret with NaOH solution. Use a beaker and funnel for pouring into the buret. The level of the solution in the buret should be about 4 cm above the 0 ml mark.

e. Place a second beaker under the tip of the buret. It will be the waste liquid beaker.

f. Open the stopcock to release the excess solution into the waste liquid beaker. Allow liquid to flow from the buret until the bottom of the meniscus is even with the 0 ml mark. The tip of the buret should now be completely filled with liquid.

4. Why should the tip be completely filled?

You are now ready to *titrate,* or to measure, how much solution is needed to neutralize the soda pop. *Titration* is a technique frequently used in laboratories to find out concentrations of solutions.

5. Record the volume of base in the buret before titrating. Record this in Table 1 opposite "Initial volume" and under "Cold soda pop."

g. Place 25 ml of cold soda pop in the flask.

h. Add a few drops of phenolphthalein.

6. Is the solution in the flask colored? Why?

i. Place the flask on a piece of white paper under the buret.

j. Add NaOH from the buret, a drop at a time, until the soda pop is neutralized. The solution should just turn pink when you have reached the end point. Stir the contents while titrating.

7. Record the final volume of base in the buret after titrating.

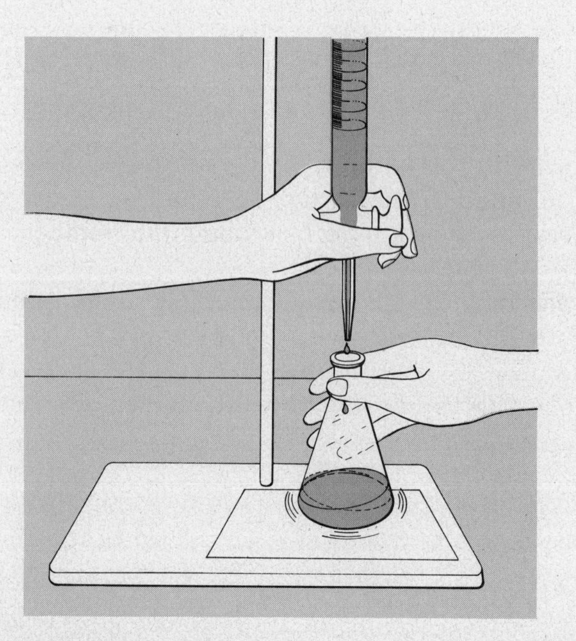

8. Calculate and record the volume of base used to neutralize the soda pop.

k. Rinse your flask, stirring rod, and graduated cylinder.
l. Repeat step **a** for room temperature soda pop and day-old soda pop left at room temperature.
m. Titrate 25 ml of room temperature soda pop with NaOH (steps **d** through **j**).
n. Titrate 25 ml of day-old soda pop left at room temperature.

9. Record all your results in Table 1.

Now examine Table 1.

10. Which soda pop needed the most base to neutralize it?

11. Which soda pop was the most acidic?

12. Which soda pop had the most CO_2 gas dissolved in it?

13. What must the CO_2 do to the soda pop when it is dissolved in it?

You now know the effect of CO_2 on soda pop. Other commercial products are worth testing by titration. Try to test the acid in vinegar.

B. Let's Stay Neutral

Vinegar is a weak solution of acetic acid. Vinegar can be made from hard apple cider, wine, or other alcoholic liquids. All vinegar bottles state how much acetic acid is in the vinegar.

You will now titrate a sample of vinegar to see how strong an acid it is. Be as careful and accurate as possible during this experiment.

Cynara

What You Need

Phenolphthalein indicator
Sodium hydroxide (NaOH), 0.1 *M* solution
Vinegar (CH_3COOH)
Water (H_2O), distilled
Beakers, 100 ml, 2
Buret
Dropper
Flask, 250 ml
Funnel
Graduated cylinder, 10 ml
Stirring rod
Buret clamp
Paper, white
Ring
Ring stand
Safety goggles

What to Do

14. Copy Table 2 into your notebook.

TABLE 2

	Trial 1	Trial 2	Trial 3
Final volume (ml)			
Initial volume (ml)			
Volume used to neutralize vinegar (ml)			
Average volume used to neutralize vinegar (ml)			

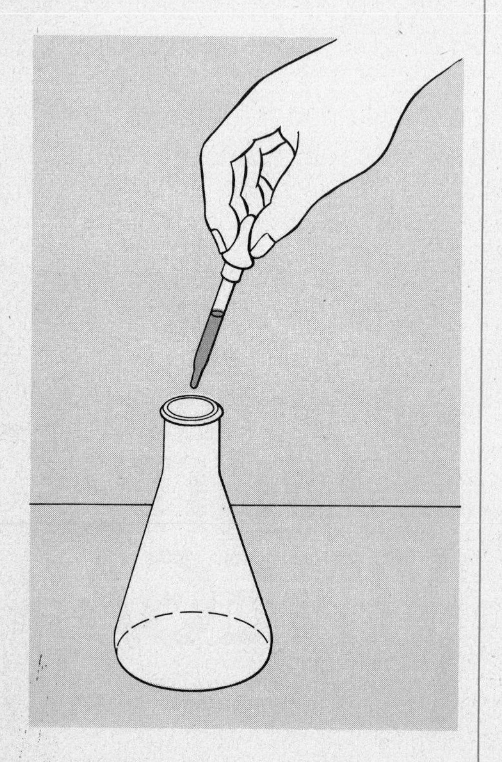

In this activity, all glassware needs to be very clean.

a. Clean all the glassware and rinse it thoroughly with distilled water.
b. Fill the buret as in Part A (steps **d** through **f**) with 0.1 *M* NaOH.
c. Place 1 ml of vinegar in your flask.
d. Add a few drops of phenolphthalein to the flask.
e. Add 5–10 ml of *distilled* water.
f. Place the flask on a piece of white paper under the buret.

15. Note the initial volume of sodium hydroxide in the buret. Record this in Table 2 under Trial 1.

g. While stirring, slowly add NaOH to the flask, a drop at a time. Add NaOH until the solution just turns pink.

16. Record the final volume of sodium hydroxide in the buret.

17. Calculate the volume of sodium hydroxide used to neutralize the vinegar. Record it in Table 2.

h. Clean and rinse the flask and stirring rod with distilled water.
i. Repeat steps **b** through **h** two more times.

18. Record the volumes of sodium hydroxide under Trial 2 and Trial 3 in Table 2.

19. Calculate the average of your three trials. Record it in the table.

j. Use Graph 1 to determine the percent of acetic acid.

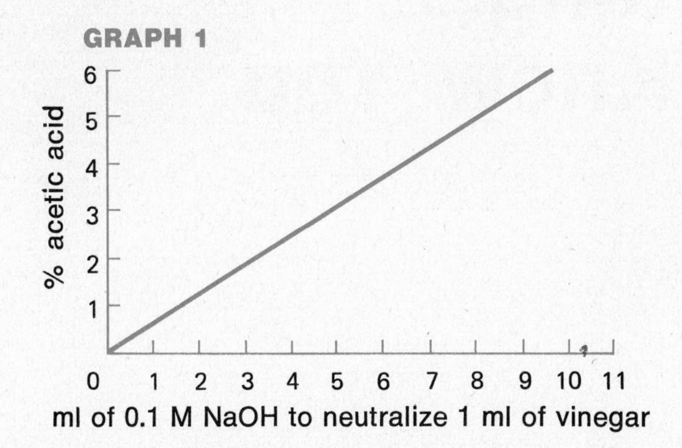

20. Record your percent acetic acid in vinegar.

21. What percent did your classmates get?

k. Check with your instructor to see how close you were to the value given on the label.

You have just completed some useful and difficult studies of acids and bases. Acids and bases are very important to us. Other substances are classified as salts and oxides. Now you will investigate some of their properties.

INVESTIGATION 10

Getting to Know You

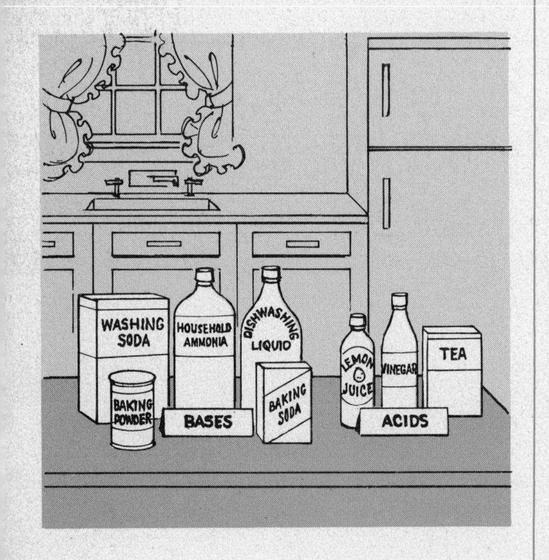

In the last two investigations you have studied the properties of acids and bases. You learned how to use these properties to classify different compounds.

Table 1 gives a list of common acids with their formulas. If you study the table; you will notice that every acid listed contains the element hydrogen. To be an acid, according to the simplest definition, a compound must have at least one hydrogen atom in it.

Bases usually contain the elements oxygen and hydrogen, joined together as the hydroxide ion, OH^{1-}. Table 2 lists a number of common bases.

Salts do not contain any one common element. This means that different salts contain different elements. Table 3 lists a number of common salts.

TABLE 1: Common Acids

Name	*Formula*
Hydrochloric acid	HCl
Sulfuric acid	H_2SO_4
Nitric acid	HNO_3
Carbonic acid	H_2CO_3
Perchloric acid	$HClO_4$

TABLE 2: Common Bases

Name	*Formula*
Sodium hydroxide	$NaOH$
Potassium hydroxide	KOH
Calcium hydroxide	$Ca(OH)_2$
Ammonium hydroxide	NH_4OH
Magnesium hydroxide	$Mg(OH)_2$

TABLE 3: Common Salts

Name	*Formula*
Sodium chloride	$NaCl$
Potassium nitrate	KNO_3
Calcium sulfate	$CaSO_4$
Ammonium chloride	NH_4Cl
Magnesium carbonate	$MgCO_3$

Salts are usually formed when acids and bases react together. For example, hydrochloric acid and sodium hydroxide react to form a salt and water.

1. Write a balanced equation for this reaction.
2. What is the salt?

Now investigate some of the properties of salt solutions.

Culver Pictures, Inc.

A. It All Comes Out in the Wash

What You Need

Copper(II) sulfate ($CuSO_4$)
Iron(II) chloride ($FeCl_2$)
Litmus paper, blue
Litmus paper, red
Magnesium sulfate ($MgSO_4$)
Potassium nitrate (KNO_3)
Sodium carbonate (Na_2CO_3)
Sodium chloride (NaCl)

Beakers, 6
Glass plate
Graduated cylinder, 100 ml
Stirring rod

Balance
Electrical conductivity apparatus
Safety goggles
Spatula
Wax pencil
Weighing paper

What to Do

3. Copy Table 4 into your laboratory notebook.

TABLE 4: Properties of Salts

Salt	*Formula*	*Ions present*	*Litmus paper test*		*Electrical conductivity*
			Blue	*Red*	
Sodium carbonate	Na_2CO_3				
Copper(II) sulfate	$CuSO_4$				
Sodium chloride	NaCl				
Potassium nitrate	KNO_3				
Iron(II) chloride	$FeCl_2$				
Magnesium sulfate	$MgSO_4$				

a. Label six clean beakers Na_2CO_3, $CuSO_4$, NaCl, KNO_3, $FeCl_2$ and $MgSO_4$.
b. Add 1 gram of the proper salt to each beaker.

c. Add 50 ml of water to each beaker. Stir to dissolve the salts. Be sure to rinse the stirring rod between uses.

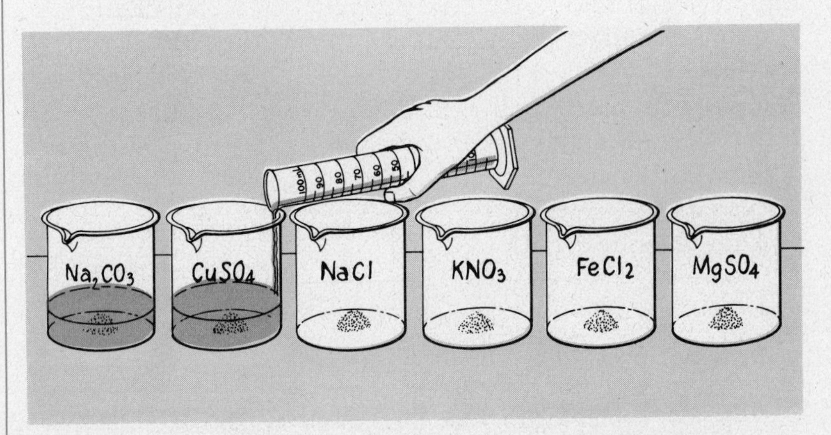

d. Separately test the reaction of each solution with blue litmus paper and red litmus paper. Remember to rinse the stirring rod between tests.

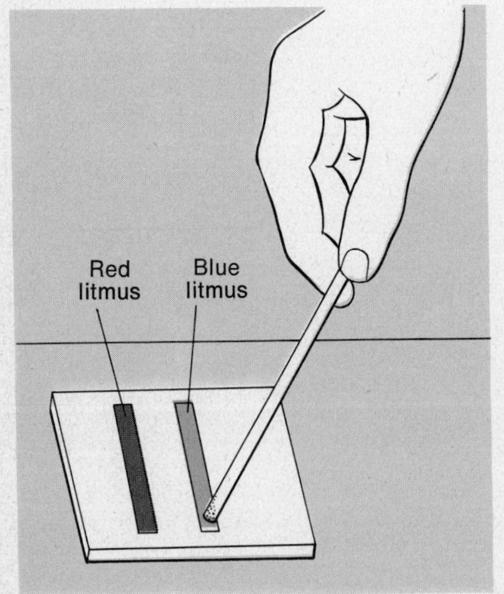

4. Record your observations in Table 4 in your notebook.

As was mentioned earlier, salts are formed from acids and bases. When an acid is dissolved in water, hydrogen ions separate from the acid molecules. The formula for these ions is H^{1+}. The hydrogen ions give acids their characteristic properties. In *strong acids,* almost all the hydrogen ions separate. In *weak acids,* not all of the hydrogen ions separate from the acid molecules.

When a base dissolves in water, hydroxide ions separate from the base molecules. The formula for these ions is OH^{1-}. In *strong bases,* almost all of the hydroxide ions separate. In *weak bases,* not all of the hydroxide ions separate from the molecules.

When acids and bases react, the H^{1+} ions and OH^{1-} ions join together to form water (HOH). The other parts of the original molecules join together to form a salt. If the salt is formed from a strong acid, its solution will give an acid reaction with litmus paper. If the salt is formed from a strong base, its solution will give a basic reaction with litmus.

5. Which of the salts you tested were formed from strong bases?

6. Which of the salts you tested were formed from strong acids?

7. What would be the reaction with litmus of a salt formed from a strong acid and a strong base?

e. Test the solution of Na_2CO_3 for electrical conductivity. **Do not plug the apparatus in until it is in place on the beaker.**

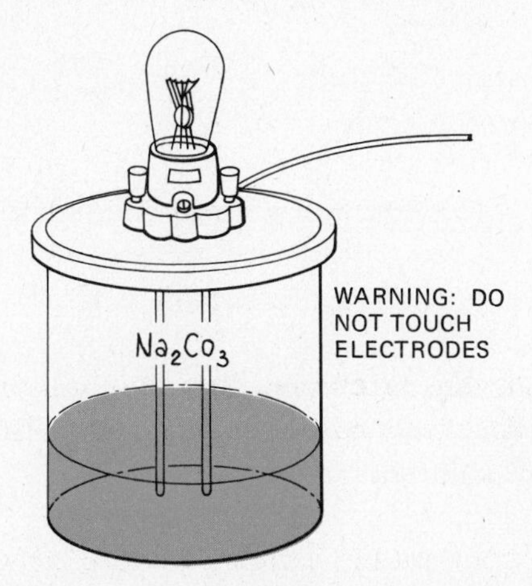

f. **Unplug** the electrical conductivity apparatus. Rinse and dry the prongs. Then test each of the other solutions in turn. **The apparatus should be plugged in only when it is sitting in a beaker. Always unplug it before removing it from a beaker.**

8. Record all of your results in Table 4.

9. Did the salt solutions conduct electricity?

A salt is a compound with ionic bonds. This means that when the compound is formed, electrons are transferred from one atom to another. In sodium chloride, an electron is transferred from the sodium atom to the chlorine atom. This results in two ions—a sodium ion with a positive (+) charge, and a chloride ion with a negative (−) charge. The formulas of these ions are Na^{1+} and Cl^{1-}.

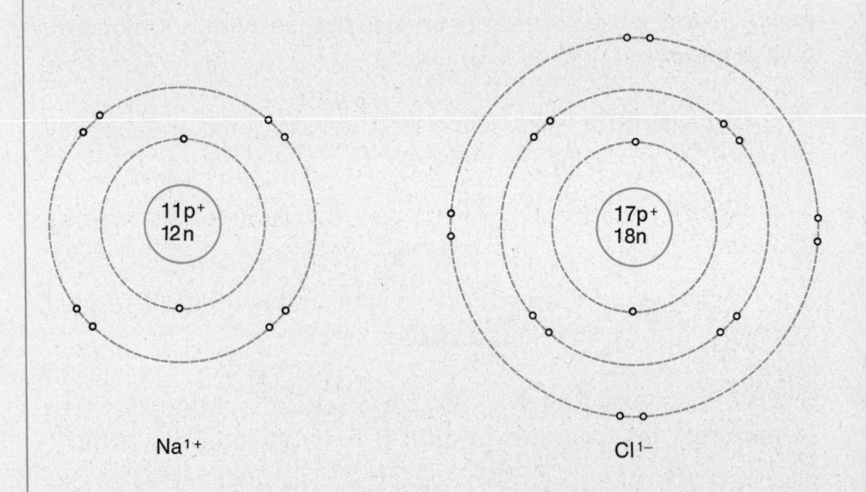

When a salt dissolves in water, the particles in solution are the ions, rather than molecules.

10. Why is a salt solution able to conduct electricity?

11. Do all salts contain a common ion?

From your work thus far, it is safe to say that salt solutions can affect litmus paper because of the acids and bases that formed them. Also, salt solutions conduct electricity.

The chemical compounds called salts are very important to everyday living. For example, the body requires salts of calcium for the proper growth of bones and teeth. The concentration of sodium chloride in the blood is critical to health. Too much or too little can be deadly. The actual sodium chloride concentration of the blood is 0.9% by mass.

Woodward/Black Star

Salts are also needed for making fertilizers, explosives, cement, and preservatives. Many useful salts are found naturally in the earth, but many more must be prepared in a laboratory.

B. Everyone Into the Pool

Now look at the properties of the oxides of some metals. The name *oxide* tells you that the compound contains oxygen. You decomposed lead(IV) oxide to lead and oxygen in Unit 3, Investigation 4.

You have seen what happens when you mix salts and water. (What happens?) Now find out what happens when you mix oxides and water.

What You Need

Aluminum oxide (Al_2O_3)
Barium peroxide (BaO_2)
Calcium oxide (CaO)
Litmus paper, blue
Litmus paper, red
Magnesium oxide (MgO)
Water (H_2O), distilled
Dropper
Stirring rod
Watch glass
Safety goggles
Spatula

What to Do

12. Copy Table 5 into your laboratory notebook.

TABLE 5: Properties of Oxides

Oxide	Formula	Reaction in water	Litmus paper test		Forms	
			Blue	Red	Base	Acid
Aluminum oxide	Al_2O_3					
Calcium oxide	CaO					
Magnesium oxide	MgO					
Barium peroxide	BaO_2					

a. Take a sample of oxide, about the size of a small pea. Put it in the center of the watch glass.

b. Add 1–2 ml of distilled water to the watch glass. Watch carefully.

13. In Table 5, record any change that takes place.

c. Stir the solution with a stirring rod. Then test it on red and blue litmus paper.

14. Record the results in your table.

d. Clean the watch glass with distilled water. Test the final rinse water with litmus to make certain that the watch glass is thoroughly clean.
e. Repeat the test with the other oxides.

15. Record your results in Table 5.

16. Complete the last column of Table 5.

17. Do the metal oxides you tested form acids or bases in water?

Generally, the oxides of metals form bases in water; but there are some exceptions. Oxides of non-metals usually form acids in water. For example, sulfur dioxide in water forms sulfurous acid. Sulfur dioxide in polluted air can combine with water in your lungs.

18. What is one disadvantage of burning fuels with a high sulfur content?

There are a great many oxide compounds. This is because oxygen is so common. Since green plants are continually renewing the supply of oxygen, there should always be oxides.

Table 6 shows some common oxides and their uses.

TABLE 6: Common Oxides

Oxide	*Formula*	*Description*	*Uses*
Sodium oxide	Na_2O	white solid	dehydrating agent
Magnesium oxide	MgO	white solid	in fire brick
Aluminum oxide	Al_2O_3	white solid	electrical insulator; filler for paints
Silicon dioxide	SiO_2	quartz	manufacture of glass
Iron(III) oxide	Fe_2O_3	red solid	red pigment; grinding and polishing lenses

19. What properties of salts have you seen in this investigation?

20. What properties of oxides have you seen in this investigation?

In general, salts and oxides are related to the acids and bases you studied in the last investigation. Salts are formed from acids and bases. Oxides can form acids or bases when dissolved in water.

Because the reactions of acids, bases, salts, and oxides can be predicted in advance, it is convenient to classify compounds in this way. For example, you know that all acids turn blue litmus paper red. You know that citric acid—even without testing it—will turn blue litmus paper red, because it is an acid.

Now that you have the means of classifying compounds, you can begin an understanding of why reactions occur.

INVESTIGATION

11

Reddy-Reddy-Redox

By now you are a pretty sophisticated chemist. You can make some educated predictions about what's going to happen when someone mixes a few chemicals together.

You are already familiar with four kinds of reactions:

a. combination reactions
b. decomposition reactions
c. single displacement reactions
d. double displacement reactions

But, even with all of these tools to help you, you still can't predict what's going to happen 100% of the time. There are still a few more tools you need in your toolbox. One such tool is a knowledge of *oxidation-reduction* (*redox*). It can give you a deeper understanding of what is happening in the kinds of reactions you have been studying. This is because it explains what is happening at the atomic level.

So that's it!

$Pb(NO_3)_2 + 2\,KI \longrightarrow 2\,KNO_3 + PbI_2$

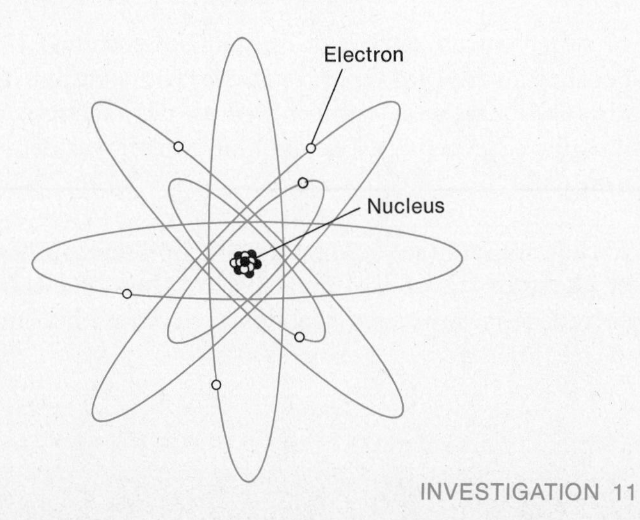

You are already familiar with the electrons, protons, and neutrons in an atom. Electrons have a negative charge, and protons have a positive charge. Oxidation-reduction is a process that describes the movement of electrons in a reaction.

A. You Lose Some; You Gain Some

In this Part, you'll re-examine an old stand-by experiment. You've done this experiment before as an example of a single displacement reaction. Now do it again, this time concentrating on what is happening to the electrons of the atoms in the reaction. This should help you make better predictions in the future.

What You Need

Copper(II) sulfate ($CuSO_4$), 0.1 *M* solution
Iron (Fe) nail
Beaker, 50 ml
Safety goggles

What to Do

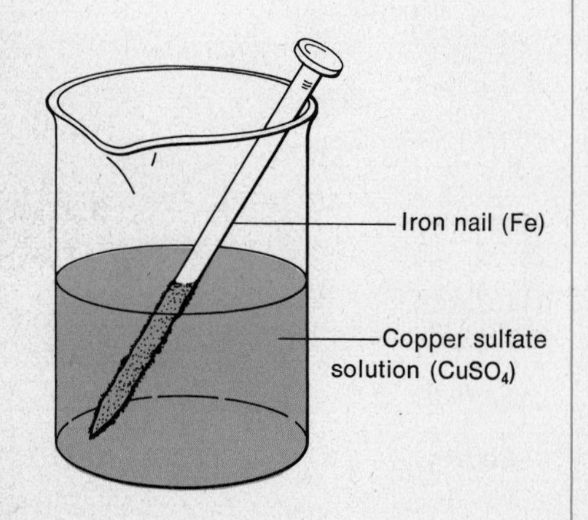

a. Place an iron nail in a beaker $\frac{1}{2}$-full of $CuSO_4$ solution.
b. After 5 minutes, look at the nail.

1. What happened?

We can write two equations to show what happened with the electrons.

$$Fe(\textit{metal}) \longrightarrow Fe^{2+}(\textit{in solution}) + 2e^-$$

$$Cu^{2+}(\textit{in solution}) + 2e^- \longrightarrow Cu(\textit{metal})$$

In other words, each iron atom (Fe) lost two electrons and became an iron ion (Fe^{2+}) dissolved in solution. Each copper ion (Cu^{2+}) in the solution picked up two electrons and became metallic copper (Cu). The metallic copper was deposited on the iron nail.

Earlier, we said that oxidation-reduction refers to the movement of electrons. *Oxidation* means that a loss of electrons has occurred. Any substance that loses electrons becomes *oxidized.*

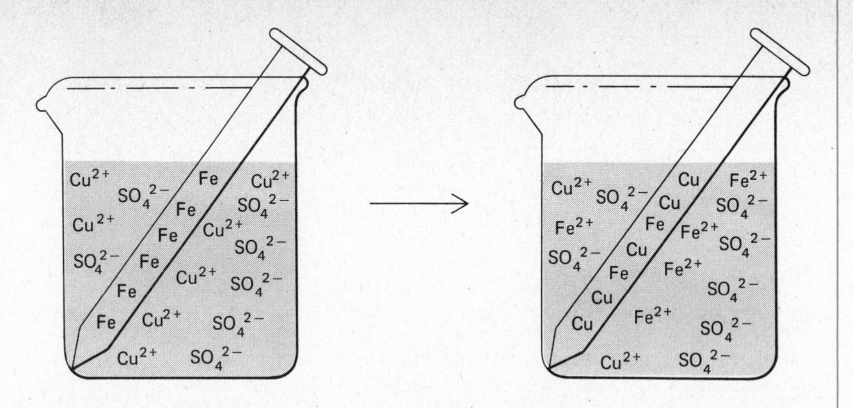

Any substance that gains electrons becomes *reduced.* This process of gaining electrons is called *reduction.*

Substances which have the greatest tendency to take electrons away from other atoms are called *oxidizing agents.* The greater this tendency is in a substance, the stronger an oxidizing agent it is. A substance which doesn't hold its own electrons tightly and has a tendency to give them away is called a *reducing agent.* The more easily it gives electrons away, the stronger a reducing agent it is.

We can rank substances on the basis of their oxidizing or reducing ability. When we have done this, we can use the information to predict whether or not a given oxidizing agent and reducing agent will react together.

When an element is oxidized, it loses electrons and gets a more positive charge. When an element is reduced, it gains electrons and gets a more negative charge. For example, in the reaction between iron and chlorine, the chlorine *gains* electrons and is reduced. The iron *loses* electrons and is oxidized.

oxidized (Fe → Fe^{3+})

$$2\,Fe + 3\,Cl_2 \longrightarrow 2\,Fe^{3+}Cl_3^{1-}$$

reduced (Cl_2 → Cl_3^{1-})

$$1e^- + Cl \longrightarrow Cl^{1-}$$

$$Fe \longrightarrow Fe^{3+} + 3e^-$$

Look at this equation:

$$Zn + 2\,H^{1+}Cl^{1-} \longrightarrow Zn^{2+}Cl_2^{1-} + H_2$$

? (H^{1+} → H_2)

2. Is the hydrogen oxidized or reduced?

Look at this equation:

$$C \longrightarrow C^{4+} + 4e^-$$

3. Is the carbon oxidized or reduced?

B. Time To Reduce

What You Need

Copper(Cu) strips, 3
Copper(II) nitrate [$Cu(NO_3)_2$], 0.1 *M* solution
Lead (Pb) strips, 3
Lead(II) nitrate [$Pb(NO_3)_2$], 0.1 *M* solution
Zinc (Zn) strips, 3
Zinc nitrate [$Zn(NO_3)_2$], 0.1 *M* solution
Test tubes, 3
Safety goggles
Sandpaper or steel wool
Test tube rack
Wax pencil

What to Do

4. Copy Table 1 into your laboratory notebook.

Rogers/Monkmeyer

TABLE 1

	SOLUTION		
Metal	*$Zn(NO_3)_2$*	*$Cu(NO_3)_2$*	*$Pb(NO_3)_2$*
Zinc			
Copper			
Lead			

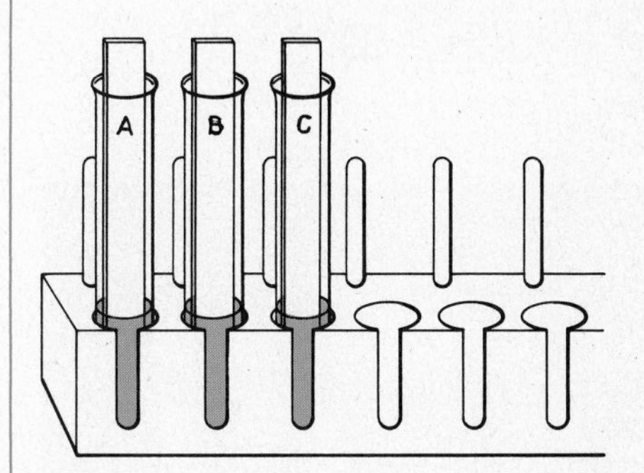

a. Label three test tubes **A, B,** and **C.** Place them in a test tube rack.
b. Fill tube **A** $\frac{1}{3}$-full with $Zn(NO_3)_2$ solution.
c. Fill tube **B** $\frac{1}{3}$-full with $Cu(NO_3)_2$ solution.
d. Fill tube **C** $\frac{1}{3}$-full with $Pb(NO_3)_2$ solution.
e. Clean the three strips of zinc with sandpaper or steel wool.
f. Place a strip of the clean zinc in each test tube.
g. Carefully observe what happens within the first 5 minutes.

5. Record your observations in Table 1 in your notebook.

h. Clean the test tubes.
i. Repeat steps **b** through **h** using strips of copper.

6. Record your observations in the table.

j. Repeat steps **b** through **h** using strips of lead.

7. Record your observations.

Series of reactions such as you have just observed can be used to rank oxidizing and reducing agents in order of their strength. Such a list can then be used to predict the tendency of given oxidizing and reducing agents to react together.

k. Examine your results in Table 1.

8. Which metal strip reacted with two solutions?

9. Which metal strip reacted with one solution?

10. Which metal strip didn't react?

11. Rank copper, lead, and zinc on the basis of their activity. Put the most active metal first (the metal that reacted with the most solutions).

12. Write a balanced chemical equation to show what happens when zinc reacts with copper nitrate.

13. Write a balanced chemical equation to show what happens when lead reacts with copper nitrate.

14. How does this activity help you to predict what will happen in certain chemical reactions?

Suppose you had to store some lead nitrate solution overnight.

15. Would you store it in a zinc or a copper container?

Sears, Roebuck & Co.

C. You'll Get a Charge Out of This One

Let's use the information you gathered in the last part to study *electrochemistry.* Electrochemistry can help you answer a number of questions. For example, "How does a battery operate?" "How do batteries produce an electric current?" "Why do batteries run down?"

What You Need

Copper (Cu) penny
Lemon (other fruits and vegetables)
Zinc (Zn) strip
Alligator clips, 2
Galvanometer
Insulated wire, ends stripped, 2
Razor blade, single edge
Safety goggles

What to Do

a. Make two parallel slits in a lemon. Each slit should be about 3 cm long.

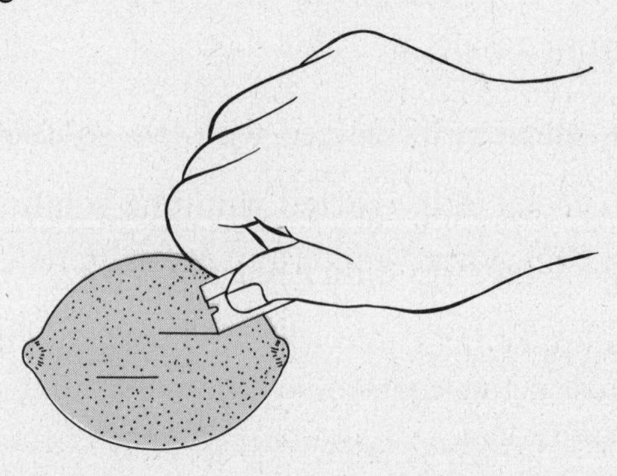

b. Carefully insert a zinc strip in one slit and a penny in the other.

c. Use a piece of wire and an alligator clip to connect the zinc strip to the negative (−) terminal of the galvanometer.

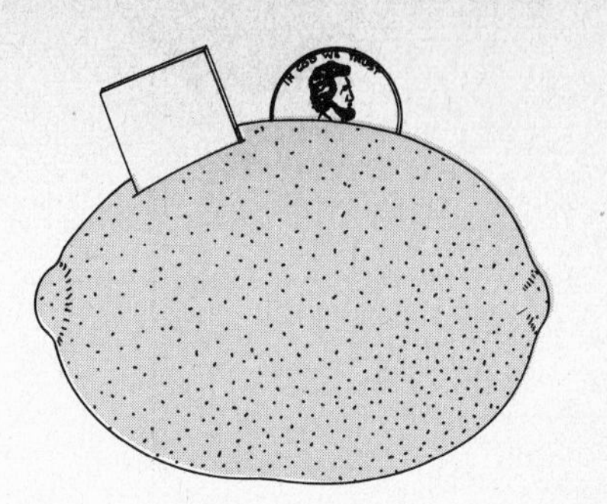

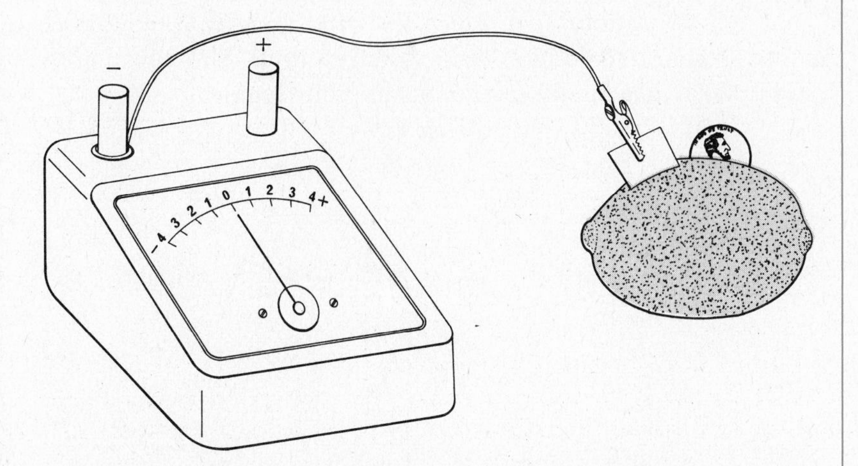

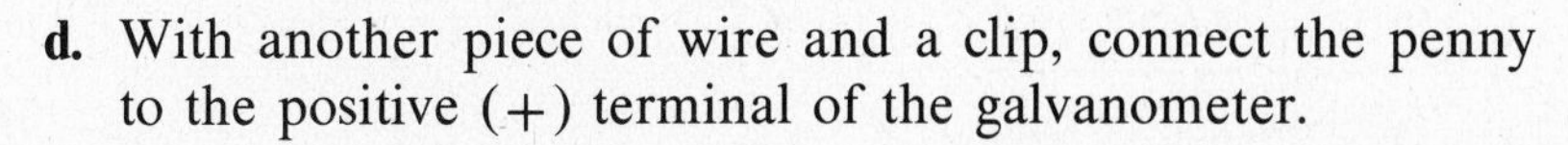

d. With another piece of wire and a clip, connect the penny to the positive (+) terminal of the galvanometer.

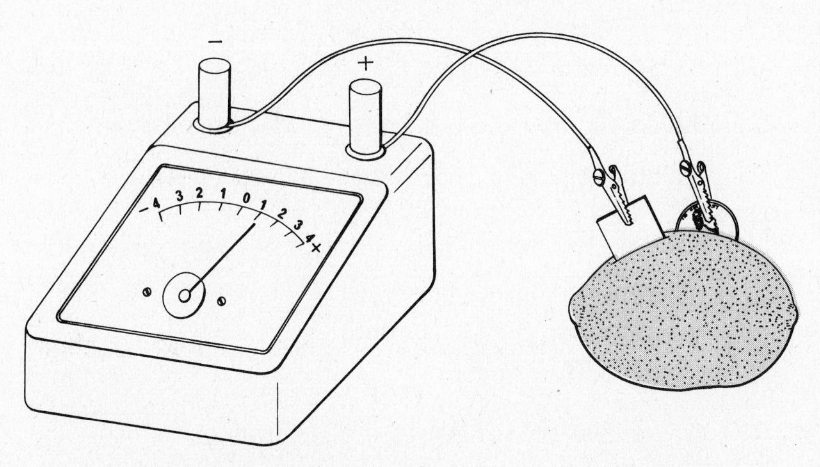

16. What happens to the needle on the galvanometer?

The needle is moving because there is a flow of electrons. The lemon, penny, zinc strip, and wires together make a battery.

A battery is an *electrochemical cell.* It produces a flow of electrons or an *electric current.* To make a battery, you need two substances with different oxidation-reduction abilities and a

solution which can conduct electricity. Such a solution is called an *electrolyte.* Electrolytes contain ions which allow a transfer of electrons.

You used the galvanometer to measure the electric current from your lemon battery. The electrons produced in your battery came from an oxidation-reduction reaction. The illustration shows what is actually happening in your lemon.

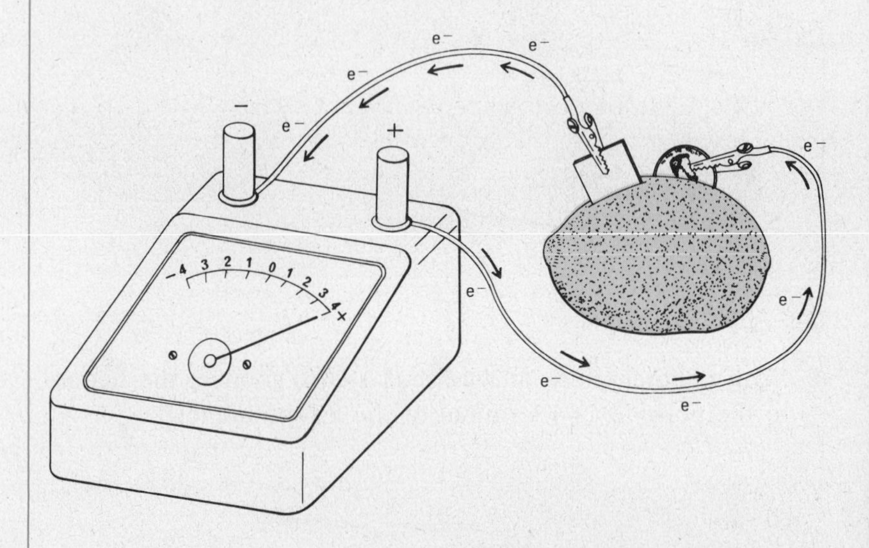

17. Which element in your battery lost electrons?

18. Which element in your battery gained electrons?

All batteries or electrochemical cells must have the same parts:

a. A reaction that loses electrons.
b. A reaction that gains electrons.
c. An electrolyte.

19. In your battery, what was the electrolyte?

Congratulations! You just made a baby brother to the car battery. But remember, it would take a pretty big lemon to run your car.

You can try using other metals and also other fruits. You can also moisten your fingers and try holding two unlike metals connected to the galvanometer.

D. Time For The Big Boys

Let's take a more detailed look at how a battery works. It's actually just an example of your new friend, the oxidation-reduction (redox) reaction.

What You Need

- Ammonium nitrate (NH_4NO_3), 0.1 *M* solution
- Copper (Cu) strip
- Copper(II) nitrate [$Cu(NO_3)_2$], 0.1 *M* solution
- Zinc (Zn) strip
- Zinc nitrate [$Zn(NO_3)_2$], 0.1 *M* solution
- Beakers, 250 ml, 2
- U-tube
- Alligator clips, 2
- Cotton
- Galvanometer
- Insulated wire, ends stripped, 2
- Safety goggles

What to Do

a. Fill a 250-ml beaker $\frac{2}{3}$-full with $Zn(NO_3)_2$ solution.
b. Place a zinc strip in the beaker. The end of the strip should extend well above the surface of the solution.
c. Fill another beaker $\frac{2}{3}$-full with $Cu(NO_3)_2$ solution. Place a copper strip in the beaker.

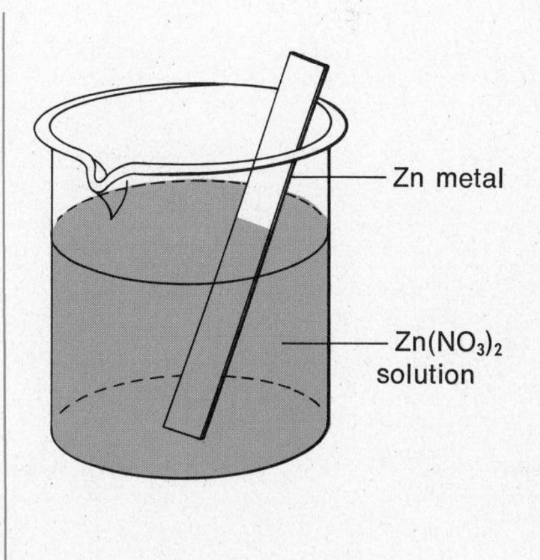

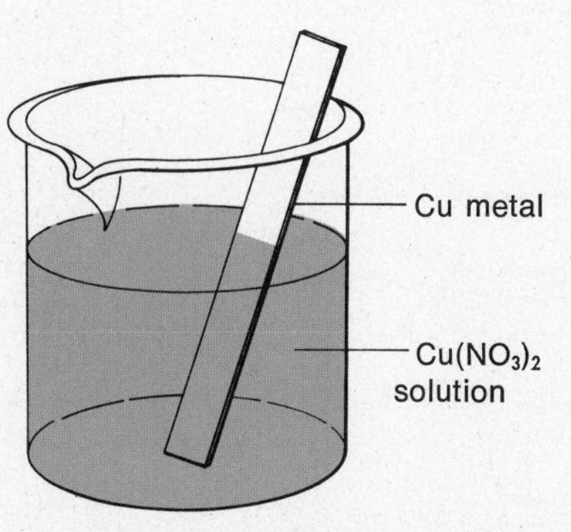

d. Use a piece of wire and an alligator clip to connect the zinc strip to the negative (−) terminal of the galvanometer.

e. With another piece of wire and a clip, connect the copper strip to the positive (+) terminal of the galvanometer.

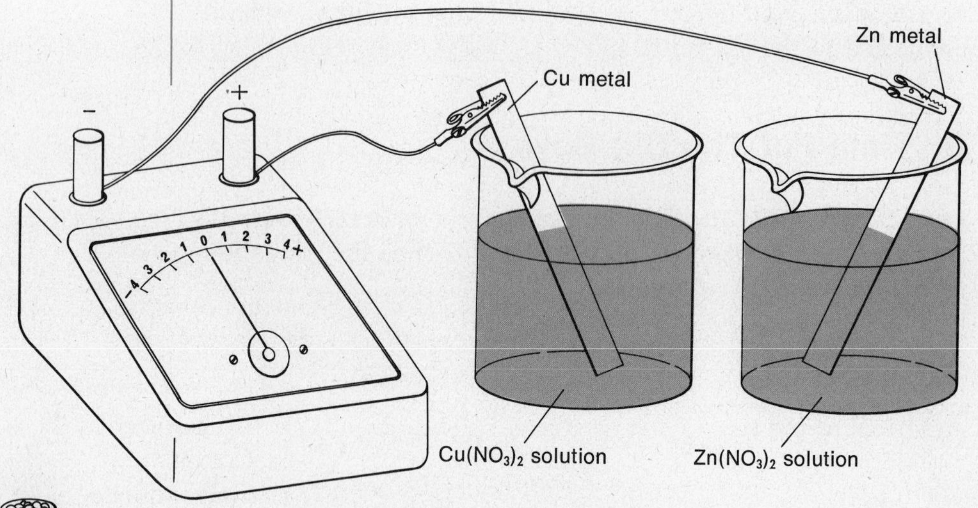

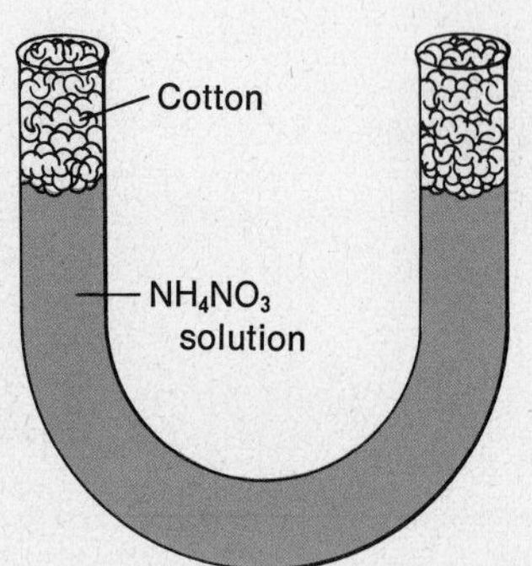

20. What happens?

f. Fill the glass U-tube with NH_4NO_3 solution. Plug the ends with cotton.

g. The device you made in step **f** is called a *salt bridge.* Use the salt bridge to connect the two beakers.

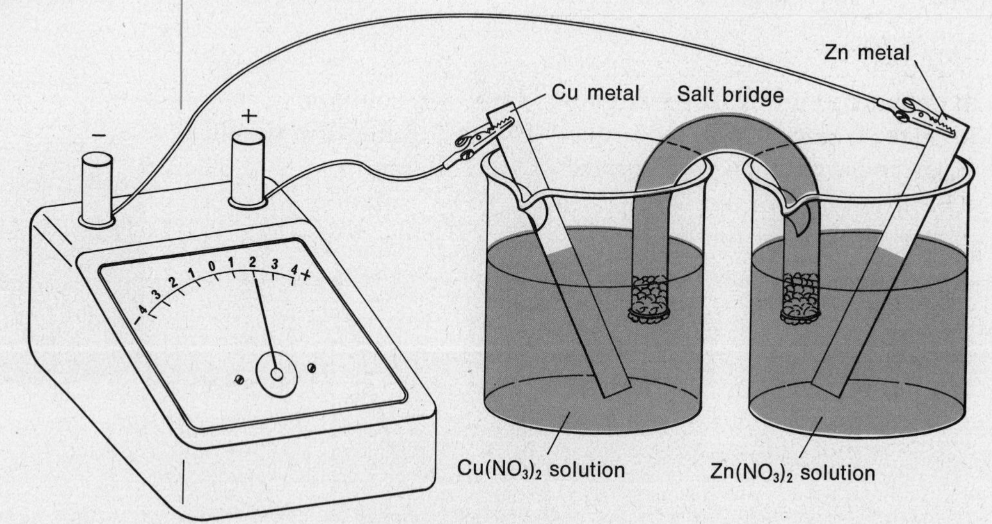

21. What happens to the needle on the galvanometer now?

h. Disconnect the wires from the galvanometer.

22. Compare your answers to questions 20 and 21. How did the salt bridge affect the readings?

In your lemon battery, the lemon juice was the electrolyte. In the battery you just made, there were several electrolytes.

23. What were they?

24. Describe the appearance of the metal strips in the beakers.

The equations in questions 25 and 26 explain what happened chemically to the metals in your electrical battery. Complete each equation in your notebook.

25. $Zn(\textit{metal}) \longrightarrow Zn^{2+}(\textit{in solution}) + \underline{\ ?\ }$

26. $Cu^{2+}(\textit{in solution}) + 2e^{-} \longrightarrow \underline{\ ?\ }$

27. Which element was oxidized?

28. Which element was reduced?

Batteries may use different elements in the oxidation-reduction reaction, but the principle of electron flow remains the same. The two most common batteries in use today are the *dry cell* and the *lead-sulfuric acid battery.*

The dry cell is the familiar battery used in flashlights and radios. It delivers a fairly small current. It is essentially a zinc case with a carbon rod in the center. The rod is surrounded by a moist paste of manganese dioxide (a filler) and ammonium chloride. The reactions in a dry cell are:

$$(1) \quad Zn \longrightarrow Zn^{2+} + 2e^{-}$$

$$(2) \quad 2\ NH_4^{+} + 2e^{-} \longrightarrow 2\ NH_3$$

29. Which substance is oxidized?

30. Which substance is reduced?

The automobile battery is called a lead-sulfuric acid storage cell. Electrical energy is stored as chemical energy in the battery through oxidation-reduction.

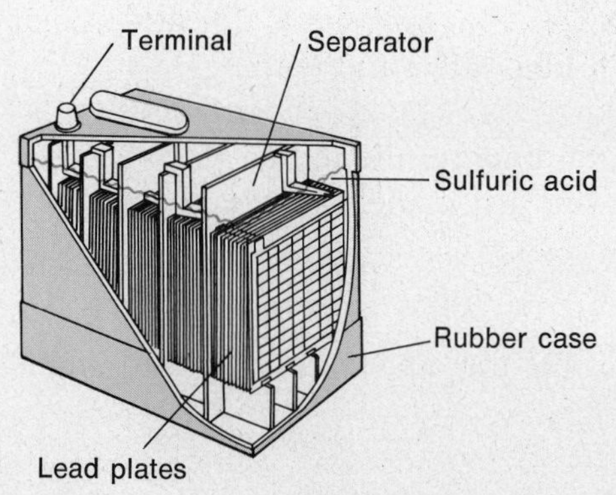

A lead-sulfuric acid storage battery.

You have grouped substances into acids, bases, salts, and oxides. There is yet another way to classify substances, which you will look at in the next investigation.

INVESTIGATION 12

Go Organic

The compounds you have studied were mostly *inorganic compounds.* These compounds are made up of combinations of many elements. There is another class of compounds. All compounds in this class contain the element carbon. The chemistry of carbon compounds is called *organic chemistry.* Organic means *from living things.*

Rogers/Monkmeyer

A. What's The Difference, Anyway?

Are organic compounds so different from inorganic compounds? Let's investigate by comparing some organic and inorganic compounds in the lab. First, compare the solubility of organic and inorganic compounds in water.

What You Need

Organic compounds
- Acetone (CH_3COCH_3)
- Citric acid ($C_6H_8O_7$)
- Isopropyl alcohol (C_3H_7OH)
- Para-dichlorobenzene ($C_6H_4Cl_2$)
- Sucrose ($C_{12}H_{22}O_{11}$)

Inorganic compounds
- Calcium hydroxide [$Ca(OH)_2$]
- Copper(II) sulfate ($CuSO_4$)
- Sodium chloride ($NaCl$)

Graduated cylinder, 10 ml
Stirring rods
Test tubes, 8

Balance
Safety goggles
Spatula
Test tube rack
Wax pencil
Weighing paper

What to Do

1. Set up a table in your notebook similar to Table 1.

TABLE 1

Compound	Organic or inorganic	Strong or weak odor	Solubility in water

a. Label your test tubes, one for each compound.
b. If the compound is a liquid, add 1 ml of it to the proper test tube. **(Caution: acetone is very flammable.)** If it is a solid, add 1 gram of it to the proper test tube.

c. Smell each compound carefully.

It isn't exactly Chanel No. 5!

2. Record your results in the table.

d. Add 5 ml of water to each test tube. Mix the contents with a stirring rod. (Be sure to rinse the stirring rod between uses.)

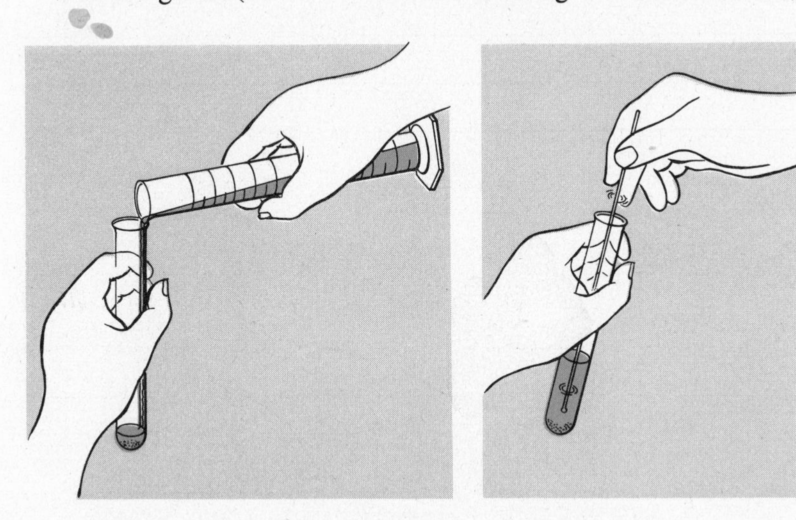

3. Record your results in Table 1.
4. Examine your data. Which class of compounds—organic or inorganic—dissolves more easily in water?
5. Which class of compounds has the stronger odor?

B. Danger—Degrading Chemical

What You Need

Organic compounds
- Citric acid ($C_6H_8O_7$)
- Sucrose ($C_{12}H_{22}O_{11}$)

Inorganic compounds
- Calcium hydroxide [$Ca(OH)_2$]
- Copper(II) sulfate ($CuSO_4$)
- Sodium chloride ($NaCl$)

Beaker, 250 ml
Graduated cylinder, 100 ml
Stirring rod
Test tubes, 5

Balance
Burner
Electrical conductivity apparatus
Matches
Safety goggles
Spatula
Test tube holder
Test tube rack
Wax pencil
Weighing paper

What to Do

6. In your laboratory notebook, set up a table similar to Table 2.

TABLE 2

Compound	*Organic or inorganic*	*Observation of decomposition*	*Electrical conductivity*

a. Label your test tubes, one for each compound.

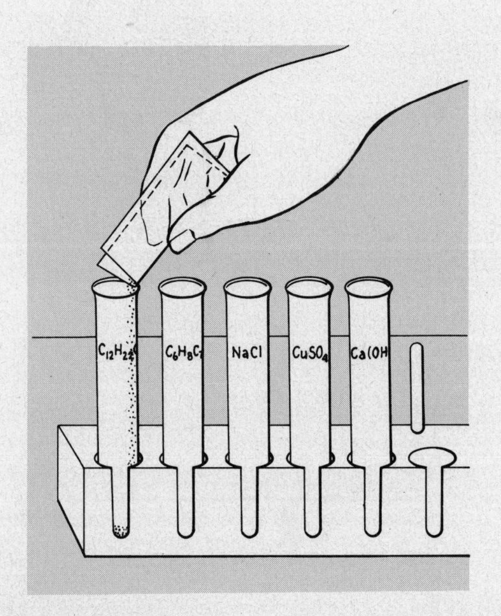

b. Add about $\frac{1}{2}$ gram of each compound to the proper test tube.
c. To test whether or not the compound is decomposed, heat it gently over the burner flame.

7. Record your results in Table 2.

8. Which class of compounds—organic or inorganic—is more easily decomposed by heat?

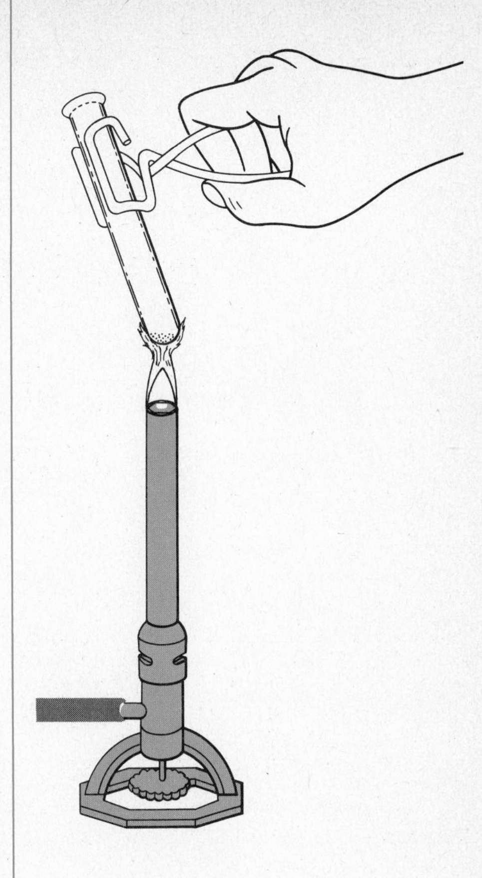

d. Place about 2 grams of the compound to be tested in a beaker. Label the beaker.
e. Add 150 ml of water to the beaker. Stir to dissolve.
f. Test the solution for electrical conductivity. **The apparatus should be plugged in only when it is in place on the beaker.** Unplug it before you remove it from the beaker.

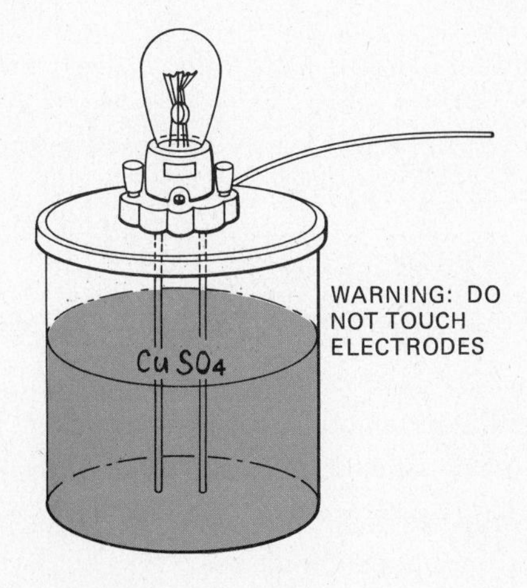

g. Repeat steps **d** through **f** for each compound. Rinse the prongs of the electrical conductivity apparatus between uses.

9. Record your results in Table 2 in your notebook.

10. Which class of compounds—organic or inorganic—seems to conduct electricity better?

11. What properties of organic compounds have you discovered in this investigation?

You have just finished a complete Unit on the properties and reactions of compounds. Now you have the tools to study the chemistry of organic compounds. On to Unit 4!

Home Activities

Investigation 1

1. What ions are in hard water?
2. What ions are in a water softener to soften the water?
3. Take a sample of water at home and test it for hardness.

Investigation 2

1. Make a list of names of chemicals and formulas in common household items (baking soda, mouthwash, deodorant, etc.).
2. Do any household items contain the same chemicals? The same elements? Which ones?

Investigation 3

1. Completely fill a glass with water. Place a piece of paper over the opening. With one hand pick up the glass. With the other hand, hold the paper. Turn the glass over. Remove the hand holding the paper. What happens? Why? (Do this over a sink.)

Investigation 6

1. Why are dust explosions a problem in coal mines and grain elevators?
2. Design different containers to slow down the melting of ice cubes. How do yours compare with those of your classmates?
3. a) Add a drop of food coloring to water at different temperatures. Which one mixes the fastest?
   b) Add drops from different heights. What do you observe?

Investigation 8

1. Take some litmus paper or *p*H paper home and test different solutions to see if they are acid or base. (Try shampoo, soap, vinegar, drain cleaners, etc.) Make a data table. Which ones are most acidic or basic?

Investigation 10

1. Make a list of different kinds of salts found in the home and of their uses.
2. Go to a grocery store and look at several salts. Complete the following table:

Brand name	Ingredients	Weight	Cost

Investigation 11

1. Look around you and identify several kinds of metals. Which ones react a lot? Which ones react very little?
2. Cut a dry cell (flashlight battery) in two. Make a sketch of its parts and describe them.
3. Take a worn out dry cell (flashlight battery) and try to get it to work again. (Warm it gently, store it in a refrigerator or freezer overnight, tap it for a while.) Can you get it to work again? How?

Investigation 12

1. List several foods you eat and the chemicals in them. What elements are in most of them?
2. Review articles of food preparation and regulations of the Federal Food and Drug Administration. Write a summary of the articles.
3. How many grams of food do you eat each day? About 1/10 of the food you eat becomes your body mass. How many grams of mass does your body keep each day?
4. What will your community's needs be for food and water in ten years?

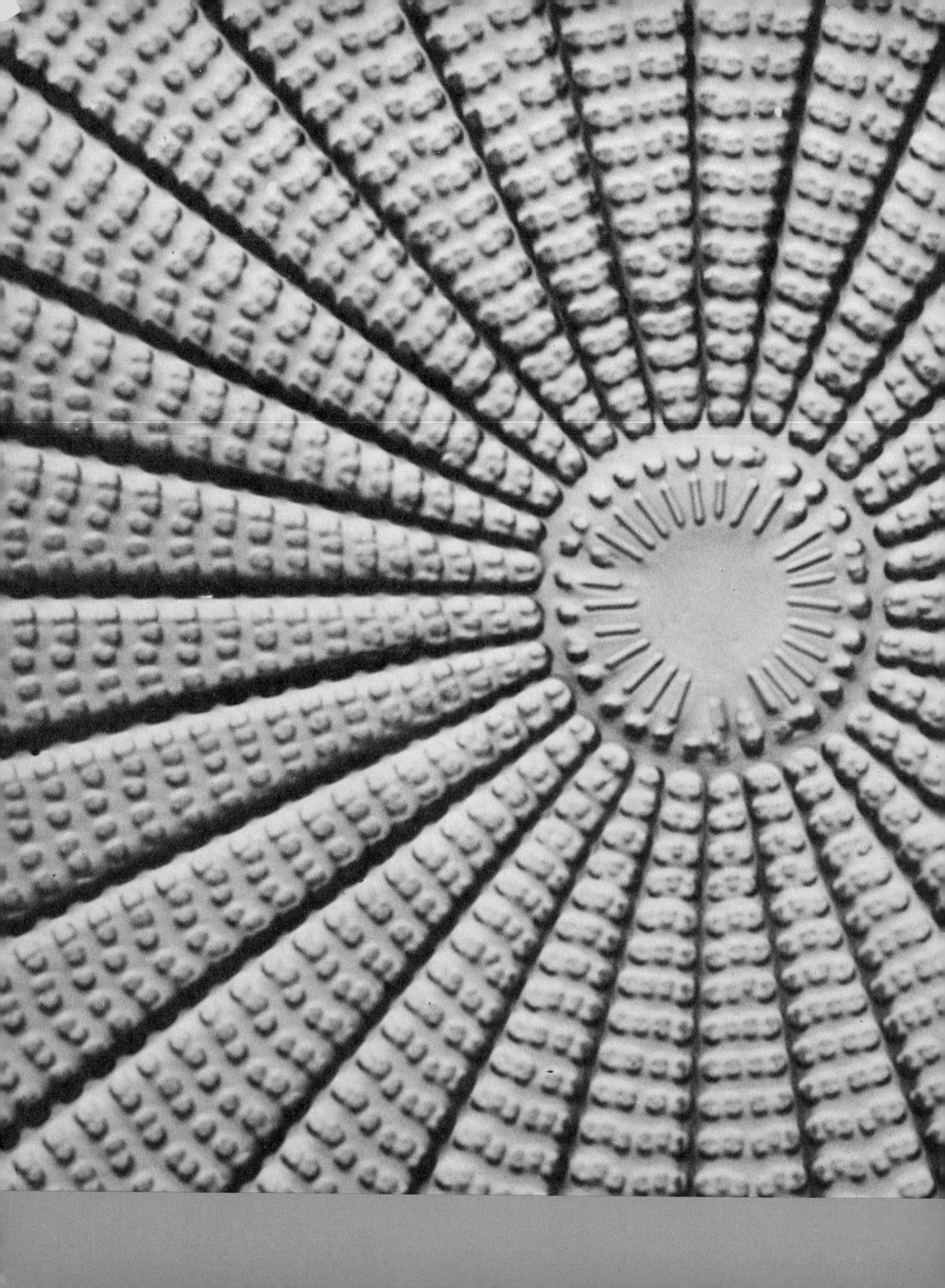

Carl Strüwe/Monkmeyer

UNIT 4

INVESTIGATION 1

Carbon, Carbon, Everywhere

What element on, in, and around the planet Earth is most important to life as we know it? It happens to be the element carbon.

The study of carbon and its compounds is so important that it is a separate branch of chemistry, called *organic chemistry.*

You will begin your study of organic chemistry by isolating the element carbon. Then you can examine some of its properties.

NASA

A. I'd Walk a Mile For a . . .

What You Need

Litmus paper, blue
Water (H_2O), distilled
Wood splints (slivers)
Glass bend, right angle
Test tube, Pyrex
Burner
Cotton ball
Matches
Ring stand
Safety goggles
Stopper, 1-hole
Test tube clamp
Test tube rack

What to Do

a. Pack the test tube $\frac{1}{2}$-full with wood slivers.
b. Push a cotton ball $\frac{1}{4}$ of the way into the test tube.

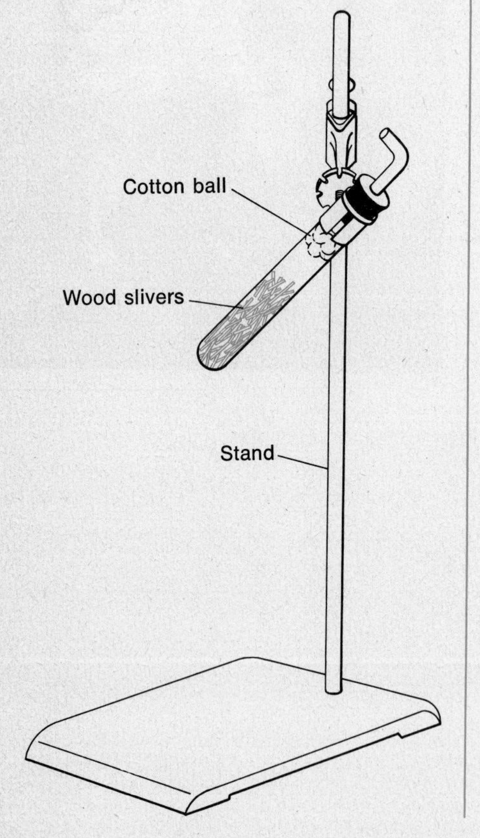

c. **Carefully** insert the glass bend into the stopper.
d. Insert the stopper with the glass tubing into the test tube, as shown below.

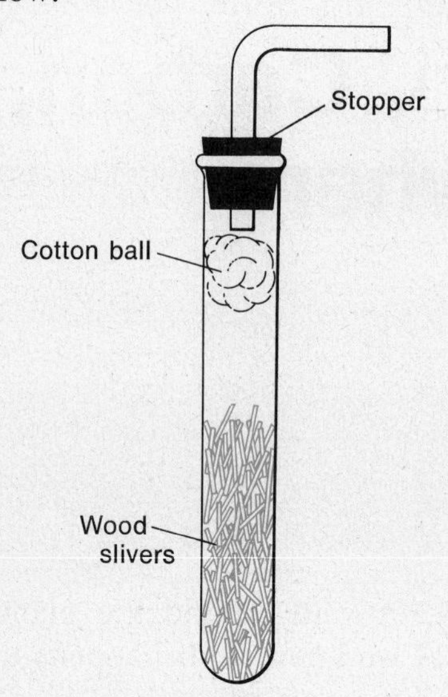

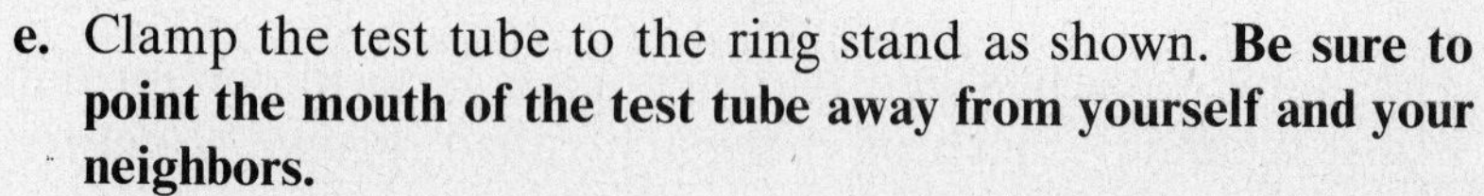
e. Clamp the test tube to the ring stand as shown. **Be sure to point the mouth of the test tube away from yourself and your neighbors.**
f. Light the burner and heat the wood slivers. The flame should be under the lower portion of the test tube. **Do not heat the cotton with the flame.**

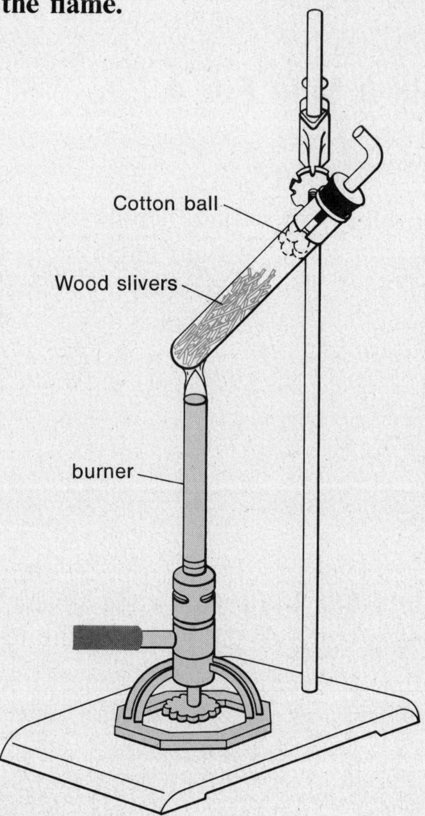

g. A dense smoke will develop inside the test tube. When you see the smoke coming out of the glass tubing, hold a lighted match to the opening of the tubing.

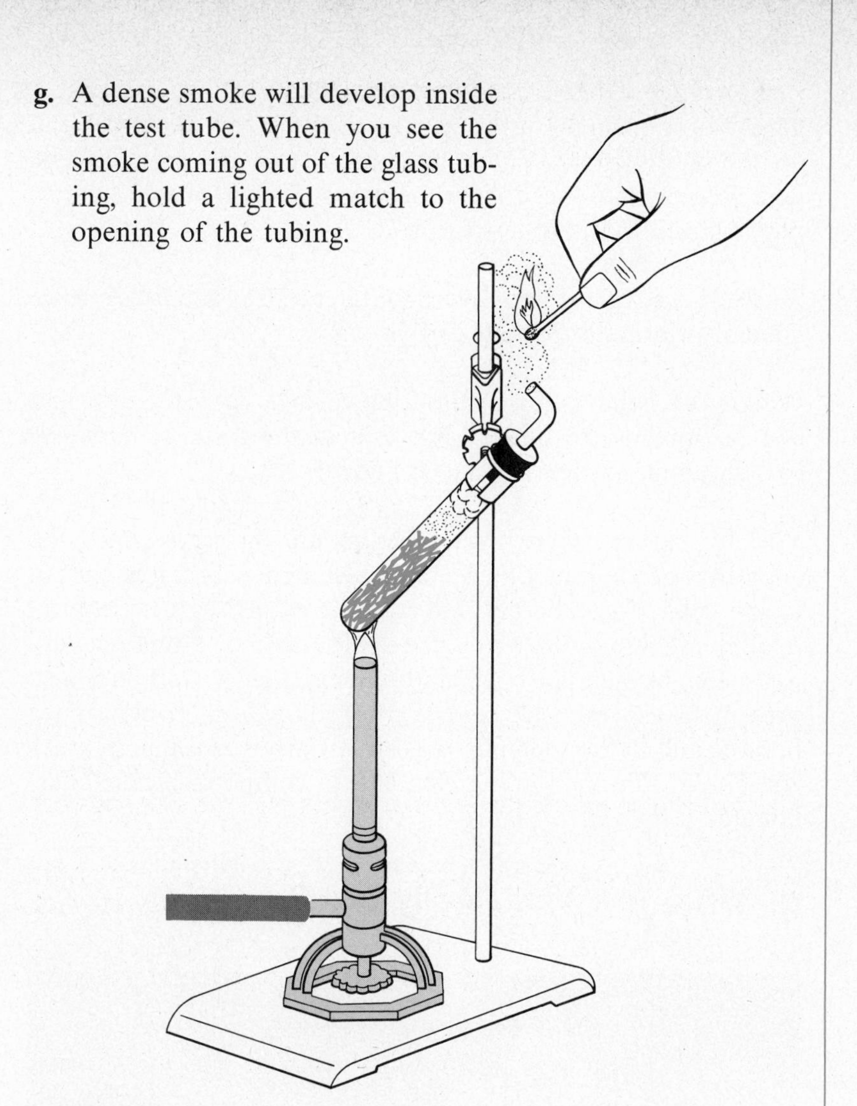

1. What happens?

h. When smoke is no longer being produced, turn off the burner.

i. Remove the stopper. (The tube will be hot.) Bring a piece of moistened blue litmus paper to the mouth of the test tube.

2. What happens to the litmus paper?

3. What has happened to the wood?

The residue in the bottom of the test tube is almost pure carbon. It was produced by slowly and carefully heating the wood in little or no air. This is how charcoal is produced.

The smoke which was produced contained various gases. These gases burned when you lit them with the match. The material on the cotton is tar, which condensed from the gases. Acetic acid was also produced from the wood. This is why the litmus paper gave a positive test for acid.

Set aside the residue and save it for later tests. First, here's some general information about carbon.

Would you believe that in the form of *soot,* carbon is soft and can be spread like butter? But that in the form of *diamond,* carbon is the hardest substance known?

Without carbon, there would be no life on earth, since the chemistry of life is based on this one element. Carbon is present in the cells of our bodies and in the food we eat. It is found in all living things. Carbon is usually found in combination with hydrogen, oxygen, nitrogen, and other elements. Carbon is also present in gas, oil, and coal. Carbon dioxide is in the air we breathe and in the soda pop we drink. Carbon is found in rocks combined with many other elements in compounds called carbonates.

B. What Are You Really Like?

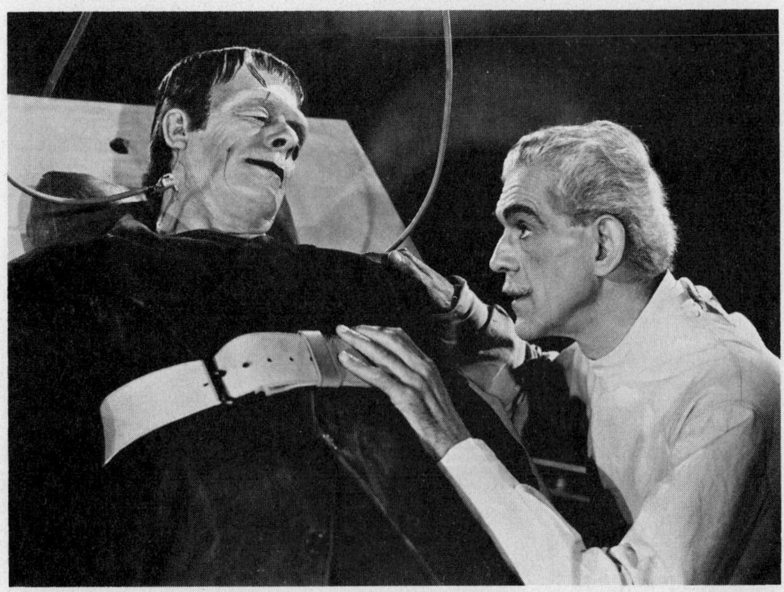

Museum of Modern Art/Film Stills Archive

Tell me about yourself.

What You Need

Carbon (C) sample from Part A
Hydrochloric acid (HCl), 0.1 *M*
Sodium hydroxide (NaOH), 0.1 *M* solution
Graduated cylinder, 10 ml
Stirring rod
Test tubes, 3
Electrical conductivity apparatus
Safety goggles
Spatula
Test tube rack
Wax pencil
Weighing paper

What to Do

a. Place the carbon sample from Part A on a piece of weighing paper.
b. **Very carefully,** as shown at the right, place the two prongs of the electrical conductivity apparatus on the sample. **Note: Do not touch the prongs with your fingers.**
c. Plug in the apparatus. If the bulb glows, the sample is said to conduct electricity. **Unplug the apparatus.**

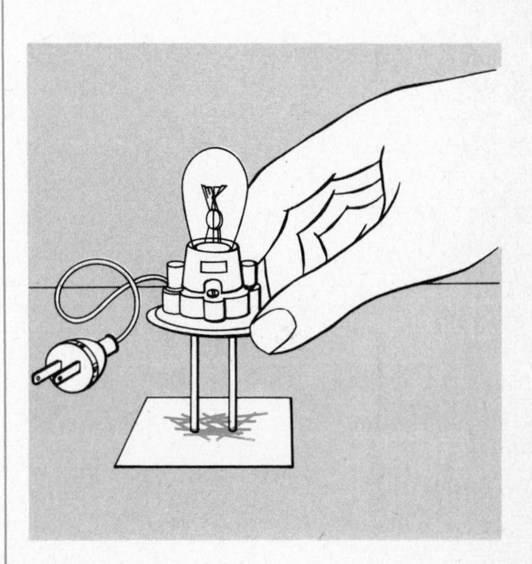

4. Is charcoal a conductor of electricity?

d. Put three test tubes in a rack. Label them as in the illustration below.

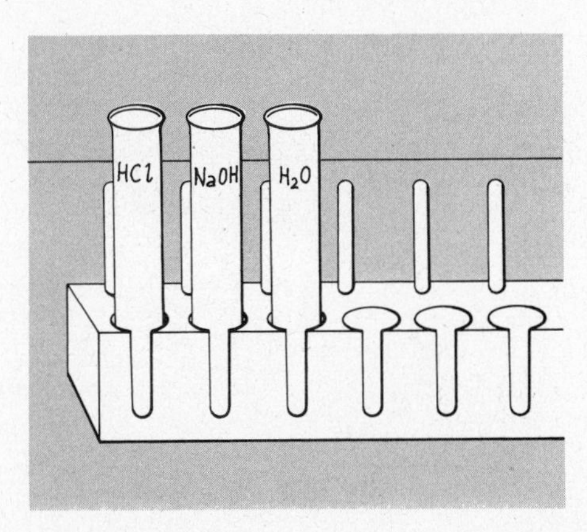

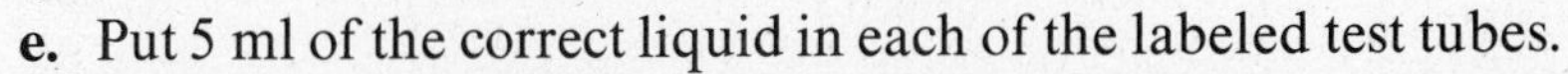

e. Put 5 ml of the correct liquid in each of the labeled test tubes.
f. Using your spatula, place a sample of carbon, about the size of a pea, into each test tube.
g. With a clean glass stirring rod, carefully stir each test tube. Rinse and dry the stirring rod after each use.

5. Describe the solubility of carbon.

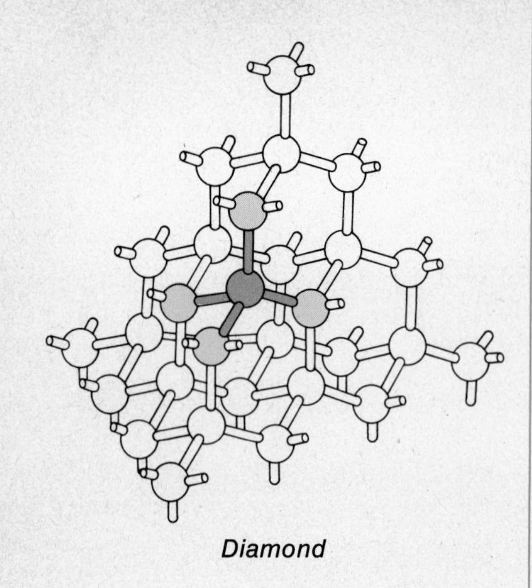

Diamond

Carbon is one of the few elements known to exist in different solid forms. Soot and diamond have already been mentioned. Diamond is clear and colorless. It is the hardest material known. In diamond, each carbon atom is bonded to four other carbon atoms.

The unit for expressing the mass of a diamond is the *carat.* A carat is equal to one-fifth of a gram.

6. What would be the mass in grams of a 15-carat diamond?

Graphite is a form of carbon which is black, shiny, soft, and quite slippery. It is often used as a lubricant. The "lead" in your pencil is actually graphite. Charcoal and soot are composed mostly of very small crystals of graphite arranged in a loose structure. Graphite is made up of layers of carbon atoms arranged in a hexagonal (6-sided) pattern.

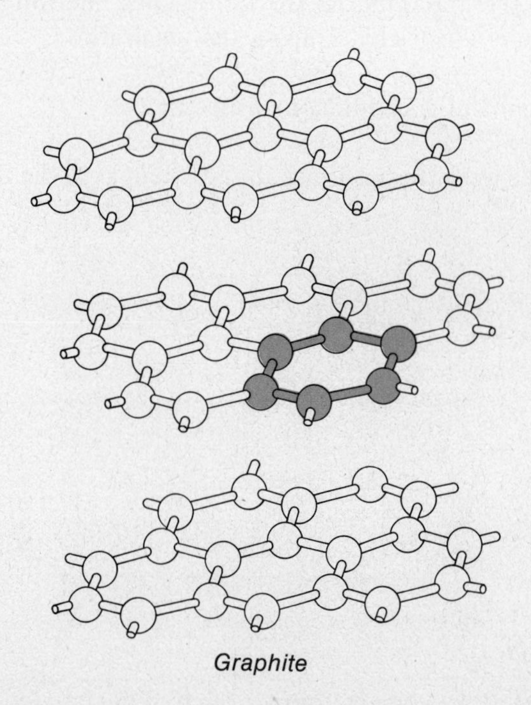

Graphite

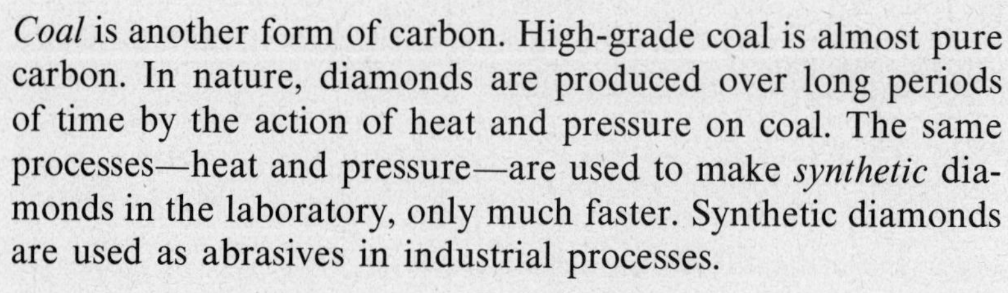

Coal is another form of carbon. High-grade coal is almost pure carbon. In nature, diamonds are produced over long periods of time by the action of heat and pressure on coal. The same processes—heat and pressure—are used to make *synthetic* diamonds in the laboratory, only much faster. Synthetic diamonds are used as abrasives in industrial processes.

C. The Character of Carbon

Along with the properties you've just investigated, carbon has other interesting properties.

What You Need

Charcoal (C)
Copper(II) oxide (CuO)
Limewater [$Ca(OH)_2$]

Glass bend, right angle
Test tubes, 2

Balance
Burner
Matches
Mortar and pestle
Ring stand
Rubber tubing
Safety goggles
Spatula
Stopper, 1-hole
Test tube clamp
Test tube rack
Weighing paper

What to Do

a. Grind several grams of charcoal to a fine powder in a mortar.
b. Get 1 gram of CuO on a piece of weighing paper.
c. Add an equal volume of powdered charcoal to the weighing paper. Carefully mix the charcoal and copper(II) oxide.
d. Place two test tubes in a test tube rack.
e. Fill one test tube $\frac{1}{3}$-full with limewater.
f. In the other test tube, place the copper(II) oxide-charcoal mixture.
g. Assemble the apparatus as shown below.
h. Light the burner and heat the test tube containing the copper(II) oxide and charcoal.

7. What happens to the limewater? What does this mean?

i. Remove the rubber tubing before turning off the burner.

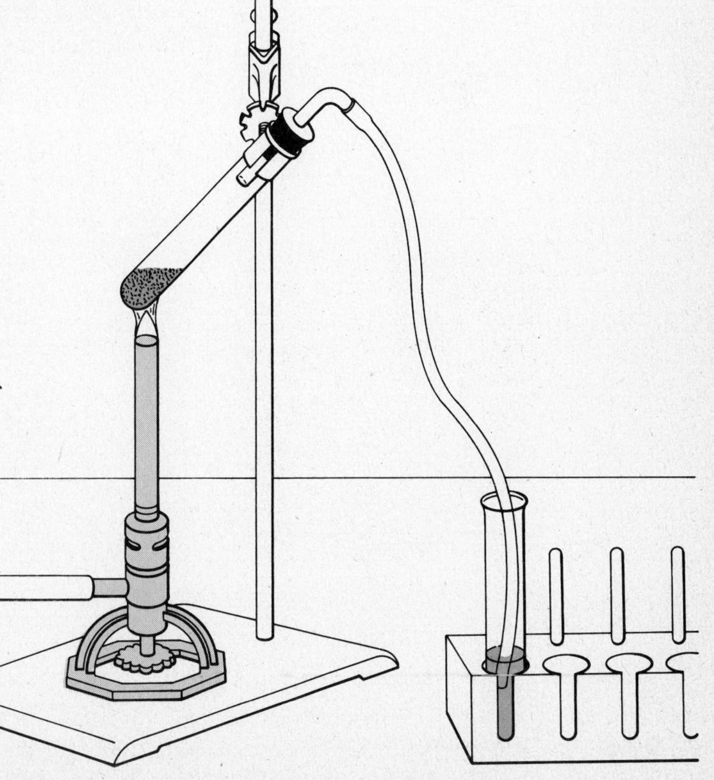

8. Write a chemical equation to show the reaction between copper(II) oxide and carbon. Be sure to check and see that it is balanced.

9. What element was produced in the test tube?

10. What type of chemical reaction is this?

As you have seen, at room temperatures charcoal conducts electricity and is not soluble in an acid, a base, or water. Many properties of carbon depend largely on temperature. Carbon when heated will combine with the oxygen in air to form carbon dioxide. Even diamonds will burn in pure oxygen.

D. Carbon, Chemistry, and You

Now think about the carbon atom.

11. How many electrons are in the outer energy level of carbon?

Let's review a little. We can say that carbon's outer level is half-filled with electrons or that it is half-empty. This means that when carbon combines with another element it could either lose four electrons or gain four electrons. Actually, it shares electrons and forms covalent bonds with other atoms.

Look what happens when the element carbon combines with the element hydrogen.

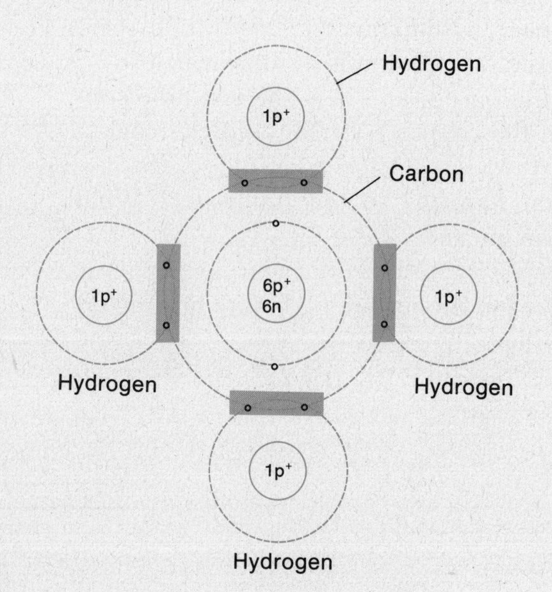

12. How many hydrogen atoms can combine with one carbon atom?

13. If two carbon atoms are joined together, how many hydrogen atoms would be needed to complete the compound?

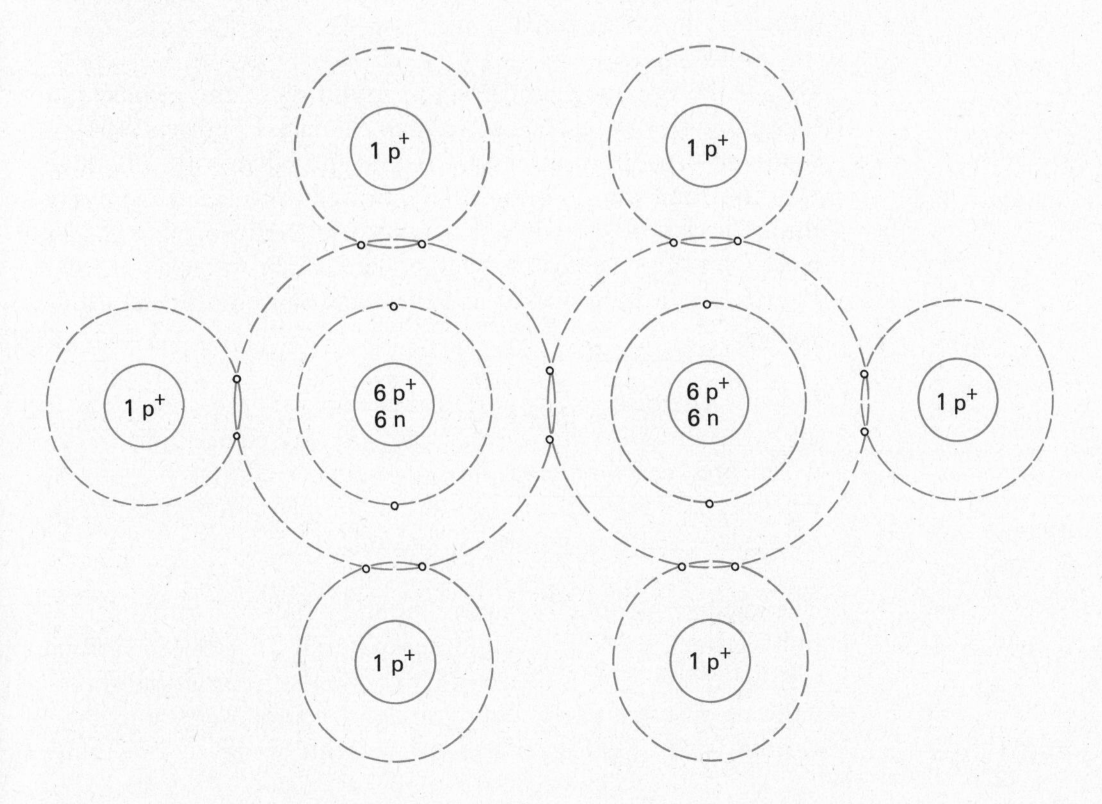

We can also use chemical shorthand to describe this compound. We can write:

```
    H  H
    |  |
 H—C—C—H
    |  |
    H  H
```

or we can write C_2H_6. This compound is called ethane.

If three carbon atoms are joined together, you get propane. Propane would be:

```
    H  H  H
    |  |  |
 H—C—C—C—H
    |  |  |
    H  H  H
```

We would write this as C_3H_8.

You could keep joining carbon atoms together; you could therefore have an almost unlimited number of different organic compounds. In fact, the number of organic compounds is about one million. In contrast, there are only about 30,000 compounds which do not contain carbon.

Originally, organic chemistry was the study of the products of living things. But in 1828, a German chemist, Friedrich Wöhler, made an organic compound, urea, in the laboratory. This was the first time that a compound normally produced by living things was synthesized in a laboratory. Because of Wöhler's work, organic chemistry became the study of *carbon compounds*—both those found in nature and those produced artificially.

Organic chemistry is important because it is the chemistry of all living things. A few of the many organic compounds in the world are: carbohydrates, fats, proteins, vitamins, hormones, enzymes, drugs, wool, silk, cotton, nylon, perfumes, dyes, soap, detergents, plastics, gasolines, and oils.

The modern contractor builds different styles of buildings by putting steel, brick, tile, and other materials together in various combinations. The modern chemist builds the invisible molecules of different useful compounds by combining elements in various ways. This is especially true with carbon compounds.

In Unit 4, you will study the chemistry of carbon compounds and their importance to life.

The list of organic materials appears endless.

INVESTIGATION

2

You Burn Me Up!

A. What's in a Name?

In Unit 3 you made a model of methane, the simplest compound of carbon and hydrogen. The molecular formula of methane is CH_4. You can also write a *structural formula* for methane:

```
    H
    |
H—C—H
    |
    H
```

This kind of formula tells how the atoms are arranged in the molecule. It shows which atoms are bonded to which other atoms.

In the last investigation, you saw how two carbon atoms can form a covalent bond by sharing a pair of electrons. When all the remaining bonds are formed with hydrogen, the resulting compound is called ethane (C_2H_6). The structural formula for ethane is:

```
    H   H
    |   |
H—C—C—H
    |   |
    H   H
```

If three carbons are bonded together, and the rest of the bonds are with hydrogen, the compound is propane (C_3H_8).

$$
\begin{array}{ccccccc}
 & & H & & H & & H & & \\
 & & | & & | & & | & & \\
H & - & C & - & C & - & C & - & H \\
 & & | & & | & & | & & \\
 & & H & & H & & H & &
\end{array}
$$

Compounds of carbon and hydrogen alone are called *hydrocarbons.* Chemists determine the names of hydrocarbons and other organic compounds by the number of carbon atoms in them. Part of the name tells how many carbon atoms are in the compound. If it contains more than four carbons, a Greek number is used. For instance, a hydrocarbon with five carbons is called pentane. The prefix, or first part of the word, *pent-*, comes from the Greek word for five. The next three hydrocarbons are called hexane, heptane, and octane. *Hex-*, *hept-*, and *oct-* are from Greek words meaning six, seven, and eight.

Table 1 will help you learn the more common prefixes used in naming organic compounds.

TABLE 1: Prefixes of Carbon Compounds

Prefix	*Number of carbons*
Meth-	1
Eth-	2
Prop-	3
But-	4
Pent-	5
Hex-	6
Hept-	7
Oct-	8
Non-	9
Dec-	10

1. Write the structural formula for butane.
2. Write the molecular formula for butane.
3. How many hydrogen atoms are there in butane?
4. Copy Table 2 into your laboratory notebook and complete it.

TABLE 2: Hydrocarbon Compounds

Compound	Carbon skeleton	Molecular formula	Number of carbons	Number of hydrogens
Methane	—C—	CH_4	1	4
Ethane	—C—C—	C_2H_6	2	6
Propane				
Butane				
Pentane				
Hexane				

5. As you add each carbon atom, how many more hydrogen atoms are needed?

B. Give It The Gas

Many hydrocarbons serve as important *fuels.* These include propane, butane, natural gas, gasoline, kerosene, diesel fuel, and fuel oil.

Where do all these products come from? They are all different *fractions,* or parts, of a mixture called *petroleum.* Petroleum, or crude oil, is a thick, brownish liquid that is pumped from rock layers in many parts of the world. Petroleum has been known for thousands of years. But only about a century ago did people discover how to separate it into various parts. They began to learn how to put each part to its best use.

Petroleum is not really one substance, but is a mixture of many different hydrocarbons. It is generally believed that petroleum was formed from the remains of organisms in ancient seas. The reduction of the carbon compounds the organisms contained has produced a mixture of hydrocarbons. Petroleum is seldom used just as it comes from the ground. Instead, it is transported

Shell Oil Co.

Exxon Corp.

A petroleum refinery.

to a chemical plant called a *refinery.* Here the different hydrocarbons are separated by a process of careful heating and cooling. This process is called *fractional distillation.*

The lightest hydrocarbons in petroleum are those with the fewest number of carbon atoms in the chain. These compounds are gases. Often they bubble out of the petroleum mixture as soon as a well is opened up. Sometimes they separate out underground so there are pockets in the rock layers which contain only these gases. The mixture of petroleum gases is called *natural gas.* In the United States, natural gas is sent around the country through pipelines. It is used for cooking, heating, and a variety of industrial uses.

Gasoline is a mixture of liquid hydrocarbons, mostly heptane and octane. It is a light, watery-looking liquid that boils, evaporates, and catches fire very easily. Gasoline vapor mixed with air explodes when it is ignited with a spark. That happens in your car's engine. The *octane rating* of a gasoline is a measure of how much engine knock the fuel will cause. The higher the rating, the less knock.

After the gasoline is separated out of petroleum, the next fuel to come from the mixture is called *kerosene.* It is denser than gasoline and less flammable. Then comes *diesel oil,* still more dense and thick. The last liquids to be separated out of petroleum are the *lubricating oils.* These are the oils used in the joints and bearings of machinery to overcome friction; they make machinery run more smoothly. The material left from the petroleum is now *tar* and *wax.* These are used in making asphalt and a great variety of useful products.

6. Which hydrocarbons are gases?
7. Name the products of the separation of petroleum.

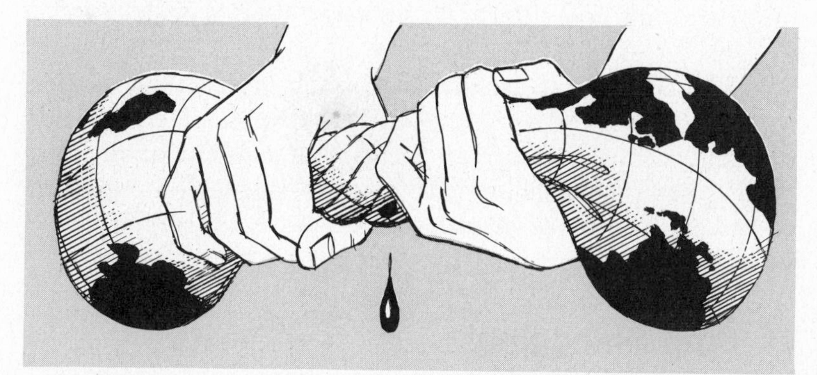

Concept courtesy of San Diego Gas and Electric Co.

C. Don't Be Fuelish

Another useful fuel found in the earth is *coal.* It was formed from plants that died and were buried millions of years ago. They were subjected to continuous heat and pressure. After many thousands of years, the plants turned into coal. Coal contains a high percentage of carbon, mixed with hydrogen, oxygen, and traces of nitrogen and sulfur.

Murray/Nancy Palmer

Petroleum, natural gas, and coal are all sources of organic compounds for industrial use. They are sources of energy when used as fuels. The energy they contain was stored up during the earth's past history through the reduction of carbon dioxide by living organisms. In a process called *photosynthesis,* plants can combine carbon dioxide and water to form *carbohydrates.* The carbohydrates contain carbon, hydrogen, and oxygen. In coal, further reduction occurred after the plants died. In hard coal, the final product is almost pure carbon. In soft coal, the reduction is incomplete. In the last investigation you decomposed wood by *destructive distillation.* Soft coal can also be destructively distilled. When that is done, many complex gases and tarry materials are given off. The residue is *coke,* a form of coal essential to the manufacture of steel.

8. Why do you think coal has been called "buried sunshine"?

Wood has always been an important fuel. Wood is either still living or was recently alive. It has not been reduced over millions of years to form carbon and hydrocarbons. But wood is an organic fuel. It is mostly a type of compound called *cellulose,* $(C_6H_{10}O_5)_n$. The n in the formula indicates that it is a long molecule with many $C_6H_{10}O_5$ units attached to each other.

When wood is burned in air, the products are carbon dioxide, water, and ash. When wood is heated without enough air, different products are formed. Some of these products are driven off as gases. The cellulose is changed to charcoal, a pure form of carbon. You made charcoal in the last investigation.

D. Stop Smogging!

The important question about fuels is "How do they react?" Their only major reaction is *combustion,* or burning. Put simply, when a fuel burns, the hydrogen and carbon in it react with

the oxygen in the air. Water and carbon dioxide are formed. The equation below sums up the burning of propane gas.

$$C_3H_8 + 5\,O_2 \longrightarrow 3\,CO_2 + 4\,H_2O$$

A problem with fuels comes when there is not enough oxygen. If the oxygen supply is limited, carbon monoxide will be formed instead of carbon dioxide. This reaction occurs in the burning of gasoline in automobiles. Carbon monoxide from automobile exhaust is dangerous. In humans and other mammals it prevents the blood from carrying oxygen.

The problem of pollution is not a new one. Long, long ago someone discovered that certain black rocks would burn.

By the Middle Ages, coal was a common fuel. But people began complaining of the smoke these rocks created in the cities. In the thirteenth century, the use of coal was forbidden in London. Later, Queen Elizabeth I stopped its use during sessions of Parliament. She stated, "The health of the knights might suffer during their abode in the metropolis."

The problem of pollution today is very much worse. Our most vital resource is the air we breathe; and we are slowly choking to death. We pollute our atmosphere daily by exhausts from cars, smoke from mills and factories, and by burning trash.

Syd Greenberg/dpi

USDA Photo

The leaf on the left is from a plant grown in polluted air.

These materials are the by-products of hydrocarbon reactions. They include carbon dioxide, carbon monoxide, nitric oxide, and sulfur dioxide. Some of these gases are poisonous. They can cause eye irritation, plant damage, respiratory illness, and even death. A polluted atmosphere destroys crops, materials, and people.

So, hydrocarbon fuels can be a mixed blessing. You have studied their background and chemistry, where they come from, and what they are made of. You may want to study their uses and effects in your area as a class project.

Air Pollution Control District, County of Los Angeles

Carbon Cliques

INVESTIGATION

3

The hydrocarbons you studied in the previous investigation are all called *alkanes.* The carbon atoms are all joined together with *single covalent bonds.* But carbon atoms can join together with two other kinds of bonds: *double covalent bonds* and *triple covalent bonds.* Get your molecular models and do the following activity to see what this is all about.

A. Model Time, Again

What You Need

Molecular model set

What to Do

a. Build a model of ethane (C_2H_6).
b. Remove one hydrogen atom and its bond from each carbon atom.

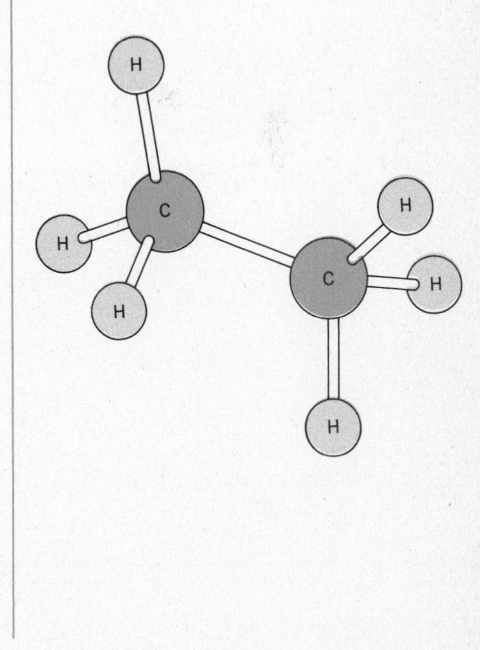

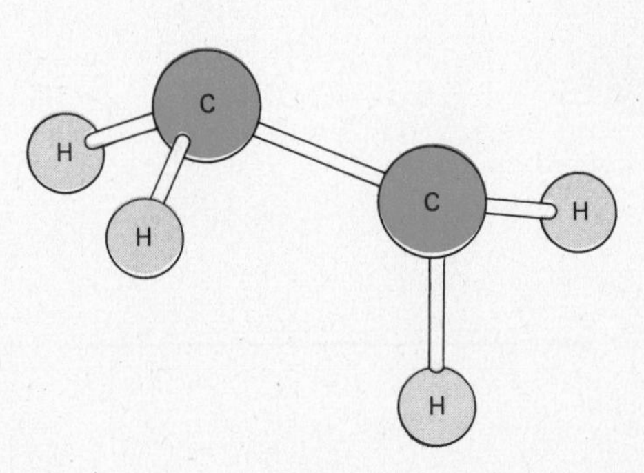

c. Use springs to joint the two carbon atoms together.

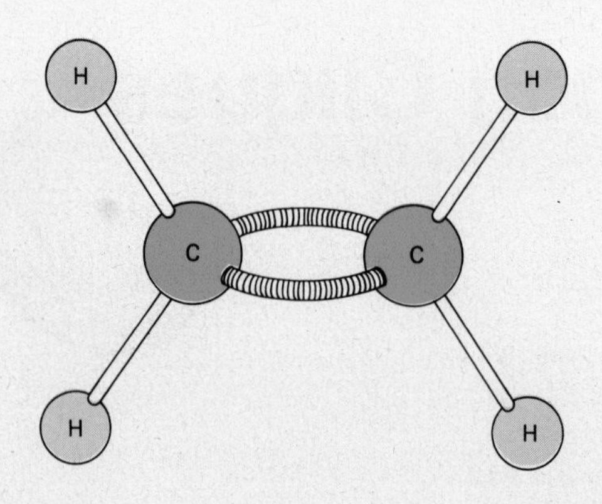

What you have now is a model of ethene (C_2H_4). The two carbon atoms are joined together by a double covalent bond, or, more simply, a *double bond.* In a double bond, two pairs of electrons are shared between two atoms. Two electrons come from one atom and two electrons come from the other atom.

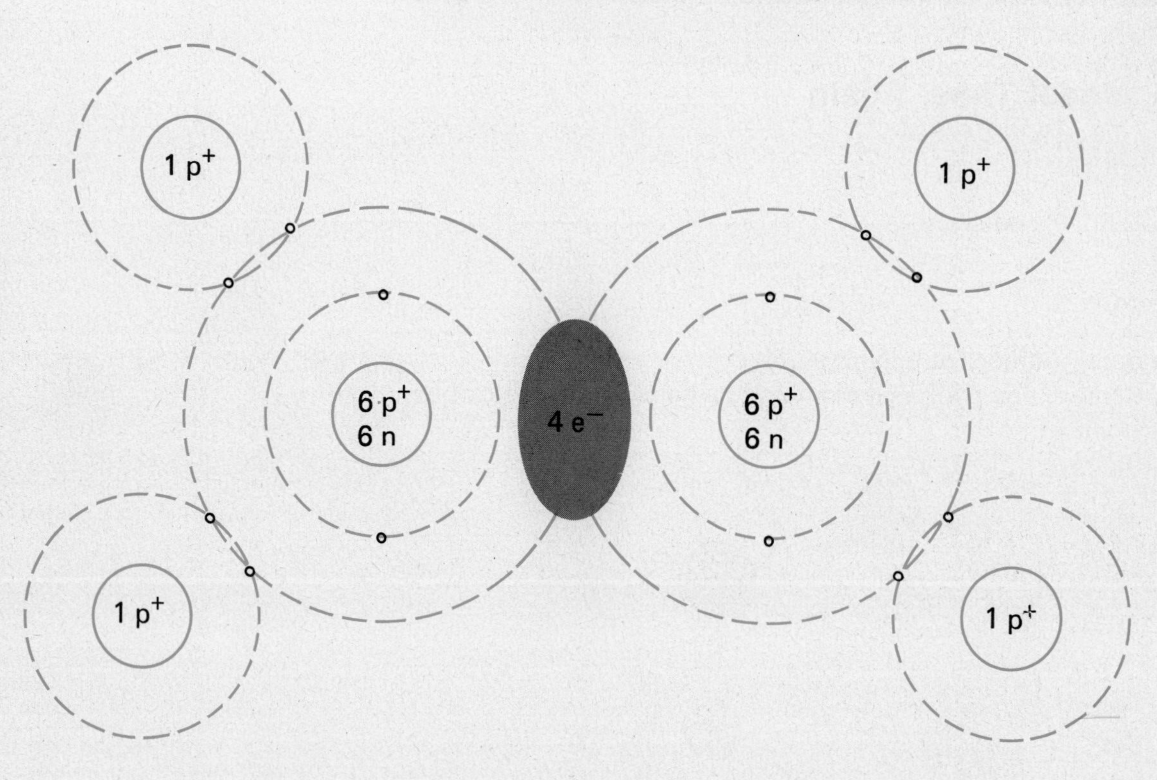

d. Remove another hydrogen atom and its bond from each carbon atom.

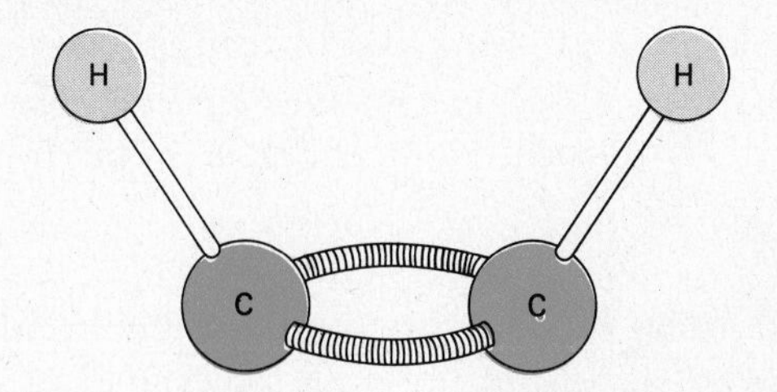

e. Join the free bonds together.

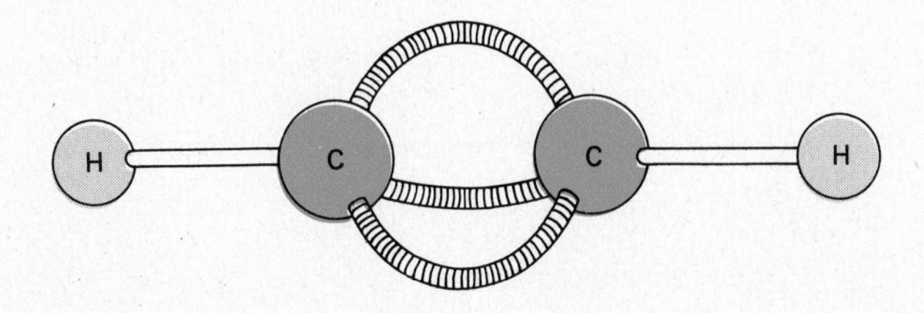

You now have a model of ethyne (C_2H_2). The common name for ethyne is acetylene. Ethyne has a triple bond; three pairs of electrons are shared between two atoms. Three electrons come from one atom and three electrons come from the other atom.

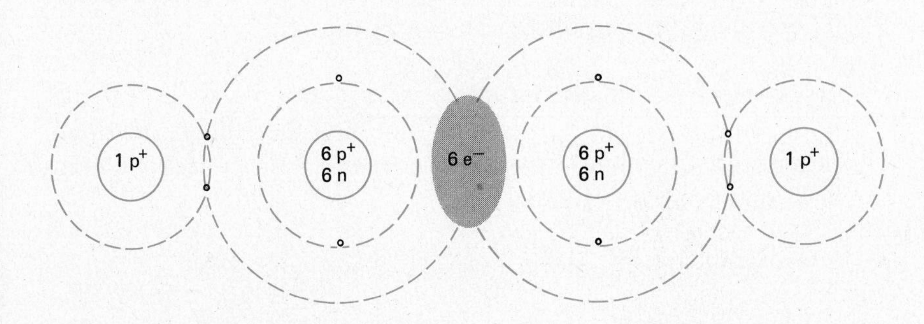

In a structural formula a double bond is written as C=C. A triple bond is written as C≡C. The structural formula for ethene is:

$$\begin{matrix} H & & & & H \\ & \diagdown & & \diagup & \\ & & C{=}C & & \\ & \diagup & & \diagdown & \\ H & & & & H \end{matrix}$$

1. Write the structural formula for ethyne. (Look at your model.)

Ethene and ethyne, and other hydrocarbons containing double or triple bonds, are called *unsaturated hydrocarbons.* They are not "saturated" with hydrogen; there is room for additional atoms.

Organic compounds which contain only single bonds are called *saturated compounds.* In the previous investigation you studied the saturated hydrocarbons. Here are some examples of saturated hydrocarbons, or alkanes:

$$\begin{array}{ccccc} & & H & & \\ & & | & & \\ H & - & C & - & H \\ & & | & & \\ & & H & & \end{array}$$

Methane, CH_4

$$\begin{array}{ccccccc} & & H & & H & & \\ & & | & & | & & \\ H & - & C & - & C & - & H \\ & & | & & | & & \\ & & H & & H & & \end{array}$$

Ethane, C_2H_6

$$\begin{array}{ccccccccc} & & H & & H & & H & & \\ & & | & & | & & | & & \\ H & - & C & - & C & - & C & - & H \\ & & | & & | & & | & & \\ & & H & & H & & H & & \end{array}$$

Propane, C_3H_8

$$\begin{array}{ccccccccccc} & & H & & H & & H & & H & & \\ & & | & & | & & | & & | & & \\ H & - & C & - & C & - & C & - & C & - & H \\ & & | & & | & & | & & | & & \\ & & H & & H & & H & & H & & \end{array}$$

Butane, C_4H_{10}

In carbon compounds, every carbon atom must form four bonds. There are four possibilities shown at the left.

$$\begin{array}{c} | \\ -C- \\ | \end{array}$$ Carbon with four single bonds.

$$=C\!\!\begin{array}{l} / \\ \backslash \end{array}$$ Carbon with one double bond and two single bonds.

$=C=$ Carbon with two double bonds.

$\equiv C-$ Carbon with one triple bond and one single bond.

Hydrocarbons with one double bond are called *alkenes.* Ethene is the simplest alkene. It is used, among other things, to ripen tomatoes artificially. Then comes propene, a three-carbon chain with one double bond.

2. Write the structural formula of propene (C_3H_6).

Note that all alkenes end in *-ene.* The prefix depends on the number of carbon atoms in the chain. Check back to Table 1 on page 278 for a list of prefixes.

Hydrocarbons with one triple bond are called *alkynes.* You just made a model of ethyne, the simplest alkyne. Note that all alkynes end in *-yne.* As with the alkanes and alkenes, the prefix depends on the number of carbon atoms.

3. What would you call a three-carbon chain with one triple bond?

4. What would you call a four-carbon chain with one triple bond?

5. Write the structural formula of propyne.

6. What is the name of a four-carbon chain with one double bond?

7. Why are there no compounds called methene or methyne?

B. A Hydrocycle Built for Two

In addition to the *open*-chain hydrocarbons you have been studying, there is another class of carbon-hydrogen compounds. These are the *closed*-chain or *ring* compounds. In these compounds, the carbon atoms form a closed ring. One example is cyclopropane:

```
      H   H
       \ /
        C
       / \
  H—C——C—H
     |    |
     H    H
```

Other examples are cyclobutane and cyclopentane:

```
    H   H                   H     H
    |   |                    \   /
 H—C—C—H              H     C     H
    |   |                   \  / \  /
 H—C—C—H              H—C       C—H
    |   |                    \     /
    H   H                H—C—C—H
                             |   |
                             H   H
```

Notice how these compounds are named. The prefix *cyclo*- indicates a ring compound. From there on, it's familiar ground. Ring compounds can be either saturated or unsaturated. If you had a five-carbon ring compound with one double bond, it would be called cyclopentene. Cyclopentene would look like this:

```
          H
          |
          C     H
        //  \  /
   H—C       C—H
       \     /
    H—C—C—H
        |   |
        H   H
```

The illustration below shows a saturated and an unsaturated 6-carbon cyclic compound.

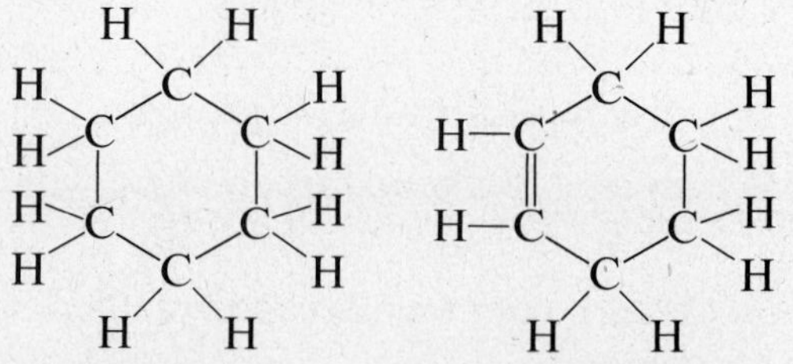

Saturated   Unsaturated

8. What is the name of the compound on the left?

9. What is the name of the compound on the right?

Do the following activities to see if these compounds react in the same way.

What You Need

Bromine (Br_2) solution in TTE
Cyclohexane (C_6H_{12})
Cyclohexene (C_6H_{10})
Potassium permanganate ($KMnO_4$), 0.1 *M* alkaline solution
Dropper
Graduated cylinder, 10 ml
Test tubes, 2
Safety goggles
Stoppers, solid, 2
Test tube rack
Wax pencil

What to Do

a. Place two clean, dry test tubes in the test tube rack.
b. Label one for cyclohex**ane** and the other for cyclohex**ene**.
c. Add 10 drops of cyclohex**ane** to the proper tube.
d. Add 10 drops of cyclohex**ene** to the proper tube.
e. Add 2 drops of alkaline $KMnO_4$ solution to each tube.
f. Stopper both tubes. Then shake them gently.

10. Describe the contents of each test tube.

g. Over the next several minutes, shake the tubes occasionally.

11. Describe the contents of each tube at the end of 5 minutes.

h. Carefully rinse the test tubes and stoppers. Then dry them thoroughly.

i. Repeat steps **a** through **d.**

j. **Do the following under the hood.** Add 2–3 drops of bromine solution, drop by drop, to each test tube. **Do not allow the bromine solution to touch your skin. It can cause bad burns.** After adding the bromine, stopper the tubes and shake them **gently.**

12. Describe what happens in each tube.

13. Which compound—saturated or unsaturated—is more reactive?

Look at the reaction of cyclohexene with bromine.

```
      H   H                         H   H
       \ /                           \ /
 H      C                      H      C       H
  \   /   \                     \   /   \   /
H—C        C—H     Br          H—C        C—Br
   |       ||   +  |    ⟶         |        |
H—C        C—H     Br          H—C        C—Br
  /   \   /                     /   \   /   \
 H      C                      H      C       H
       / \                           / \
      H   H                         H   H
```

Cyclohexene is an unsaturated compound. When it reacts with bromine, the bromine atoms are *added* to the molecule. Unsaturated compounds can react by *addition.* The double or triple bonds can split. Other atoms can then be added to the compound.

14. Why is cyclohexene unsaturated?

15. Is the product of the reaction above saturated or unsaturated? Why?

16. Why doesn't bromine add to cyclohexane?

One of the most useful cyclic compounds is benzene (C_6H_6). Benzene is the starting point in the manufacture of many organic dyes, drugs, and explosives. Its structure is often drawn with a double bond between every other pair of carbon atoms:

```
        H
        |
        C
      /   \\
H—C         C—H
   ||        |
H—C         C—H
      \   //
        C
        |
        H
```

Benzene is often abbreviated as:

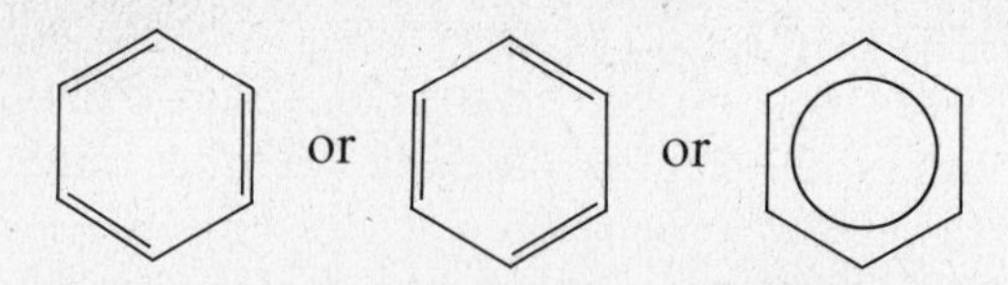

C. Food for Thought

Both saturated and unsaturated compounds are present in many common foods. Medical research suggests that a high intake of saturated fats could lead to a high level of cholesterol in the blood. This in turn can lead to heart disease, strokes, and gallstone formation. *Polyunsaturated* fats and oils are those which have double bonds in their carbon chains. It's time to play detective and identify saturated and unsaturated oils.

What You Need

Bromine (Br_2) solution in TTE
Unknown oils
Dropper
Test tubes
Safety goggles
Stoppers, solid
Test tube rack
Wax pencil

Some oil and fat products.

What to Do

a. Label a test tube with the name of each oil you will test.
b. Add 10 drops of the proper oil to each test tube.
c. **Work in the hood.** Add 2–3 drops of bromine solution, drop by drop, to each tube. **Remember that bromine will burn your skin.** After adding the bromine, stopper the tubes and shake them **gently.**

Table

17. Which of the oils you tested were unsaturated?

18. Write a report on this experiment.

You're well on your way to becoming an organic chemist! You already know a great deal about the structure and reactions of organic compounds. How about some more?

INVEST

Meet Your Neighborhood Organic Compounds

You've undoubtedly noticed that various organic compounds have characteristic odors. Organic compounds which contain similar groups tend to have similar odors. These odors can sometimes be used to help identify organic compounds.

Your teacher has placed several labeled bottles of organic compounds on a table. Smell each sample carefully. Try to remember the odors. During this investigation you may want to refer back to these compounds to help identify unknown compounds.

1. Describe the odors. Are any similar?

A. Form and Function

In the last two investigations you've been studying hydrocarbons—the compounds of hydrogen and carbon. You're now going to look at some compounds which contain a third element—oxygen. You've already used methanol, also known as methyl alcohol. The structural formula of methyl alcohol (CH_3OH) is:

```
      H  ¦
      |  ¦
   H—C—+—OH
      |  ¦
      H  ¦
```

It's like methane, but one hydrogen has been replaced by a *hydroxide group,* —OH.

Look at the structural formula for ethyl alcohol (CH_3CH_2OH):

$$\begin{array}{ccccc} & H & & H & \\ & | & & | & \\ H- & C & - & C & -OH \\ & | & & | & \\ & H & & H & \end{array}$$

It's like ethane, but with a hydroxide group instead of a hydrogen. The right-hand parts of these alcohol molecules are the same. They are both —OH, which is the *functional group* of the alcohols. All alcohols contain the hydroxide group. In this investigation, you'll learn about two other functional groups.

A non-functional group.

The left-hand part of the molecules shown above is called the *alkyl group.* The alkyl group is named, as you've probably guessed, from the number of carbon atoms. Table 1 below is derived from Table 1 on page 278.

TABLE 1: Alkyl Groups

Alkyl group	*Formula*	*R*
Methyl	CH_3—	CH_3—
Ethyl	CH_3CH_2—	C_2H_5—
Propyl	$CH_3CH_2CH_2$—	C_3H_7—
Butyl	$CH_3CH_2CH_2CH_2$—	
Pentyl (Amyl)	$CH_3CH_2CH_2CH_2CH_2$—	
Hexyl	$CH_3CH_2CH_2CH_2CH_2CH_2$—	

"R" in Table 1 is shorthand. In formulas it can mean any alkyl group.

2. Write the shorthand notation for the formulas for *butyl, pentyl,* and *hexyl.*

B. Don't Get Pickled

What You Need

Copper (Cu) wire
Ethyl alcohol (ethanol—CH_3CH_2OH)
Methyl alcohol (methanol—CH_3OH)
Graduated cylinder, 10 ml
Test tubes, 2
Burner
Forceps
Matches
Safety goggles
Test tube rack

What to Do

a. Add 5 ml of methyl alcohol to a clean, dry test tube.
b. **Carefully** note the odor of methyl alcohol. Then place the tube in a test tube rack.
c. Make a spiral of fine copper wire. Coil the wire around a pencil several times. The spiral should be about 1 cm in length. Leave about 10 cm of wire for a handle to hook over the lip of the test tube.

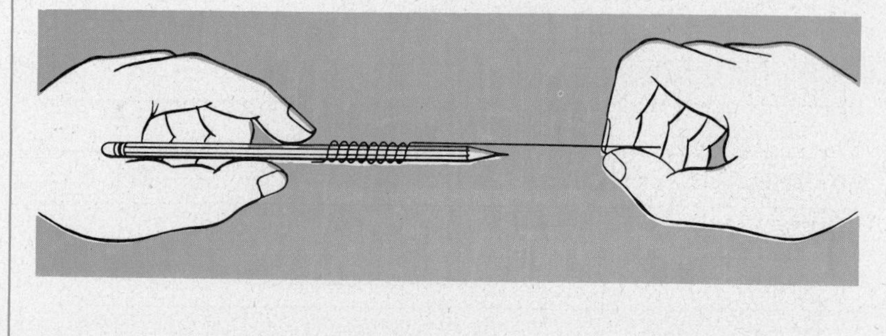

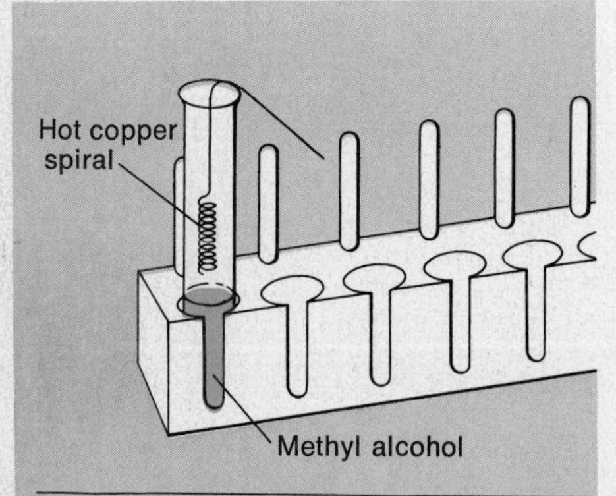

d. Hold the spiral with forceps. Heat the spiral in a burner flame. **Be sure the burner is away from the methyl alcohol. Alcohol is flammable.**
e. While the spiral is hot, hook it on the lip of the test tube.
f. Observe the copper wire during this process. Note any odor produced.

3. What changes, if any, did you observe in the wire?
4. Describe the odor produced in this reaction.

You have just made an *aldehyde* from an alcohol. Try to make another.

g. Add 5 ml of ethyl alcohol to a clean, dry test tube.
h. Carefully note the odor of ethyl alcohol. Then place the tube in the test tube rack.
i. Heat the copper spiral as before.
j. While the spiral is hot, hook it on the lip of the test tube.
k. **Carefully** smell the contents.

5. Describe the odor of the contents of the tube.

6. What kind of substance was probably formed in this reaction of ethyl alcohol?

In both of these reactions, alcohols were oxidized. In Unit 3 you learned that when atoms are oxidized, they lose electrons. This can happen with compounds as well. When organic compounds are oxidized, they generally lose hydrogen atoms. That means they lose at least one electron and one proton.

From the odors produced in these two reactions, you know that aldehydes were formed. But what are aldehydes?

C. The Smelly Old Aldehydes

All alcohols contain the group —OH. The number of carbon atoms in the compound may vary, but this particular group is what makes an alcohol a special kind of compound. Such a group is called a *functional group.* Aldehydes also have their own special functional group. It is:

$$-\mathrm{C}{\overset{\displaystyle \mathrm{O}}{\underset{\displaystyle \mathrm{H}}{\lessgtr}}}$$

If you start with an alcohol, and oxidize it so it loses two hydrogen atoms, you get an aldehyde.

$$-\overset{\displaystyle \mathrm{H}}{\underset{\displaystyle \mathrm{H}}{\overset{|}{\underset{|}{\mathrm{C}}}}}-\mathrm{OH} \longrightarrow -\mathrm{C}{\overset{\displaystyle \mathrm{O}}{\underset{\displaystyle \mathrm{H}}{\lessgtr}}}$$

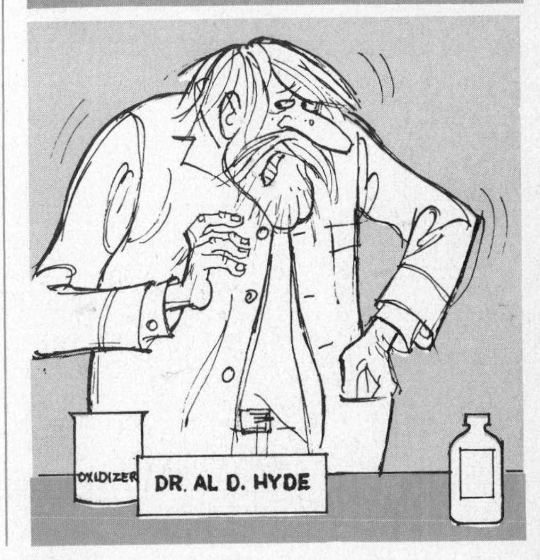

Feels as if I lost an electron!

In the reaction of methyl alcohol with the hot copper wire, two hydrogen atoms were removed from the methyl alcohol. This left a compound called formaldehyde. The reaction is a bit complicated. First, the hot copper wire reacted with oxygen in the air to form copper oxide (CuO). Then the hot copper oxide reacted with the alcohol fumes to produce the aldehyde. The reaction looks like this:

$$\text{(1)}\ \underset{\text{Copper}}{2\,Cu} + \underset{\text{Oxygen}}{O_2} \longrightarrow \underset{\text{Copper oxide}}{2\,CuO}$$

$$\text{(2)}\ \underset{\text{Methyl alcohol}}{H-\overset{H}{\underset{H}{C}}-OH} + \underset{\text{Copper oxide}}{CuO} \longrightarrow \underset{\text{Formaldehyde (or Methaldehyde)}}{H-C{\overset{O}{\underset{H}{\lessgtr}}}} + \underset{\text{Water}}{H_2O} + \underset{\text{Copper}}{Cu}$$

The reaction of ethyl alcohol with potassium dichromate and sulfuric acid produces acetaldehyde. The overall equation is:

$$3\,\underset{\text{Ethyl alcohol}}{CH_3CH_2OH} + \underset{\text{Potassium dichromate}}{K_2Cr_2O_7} + 4\,\underset{\text{Sulfuric acid}}{H_2SO_4} \longrightarrow$$

$$3\,\underset{\text{Acetaldehyde (or Ethaldehyde)}}{CH_3C{\overset{O}{\underset{H}{\lessgtr}}}} + 7\,\underset{\text{Water}}{H_2O} + \underset{\text{Potassium sulfate}}{K_2SO_4} + \underset{\text{Chromium sulfate}}{Cr_2(SO_4)_3}$$

Any aldehyde can be produced by taking two hydrogen atoms away from an alcohol. This is called *dehydrogenating* the alcohol. In fact, that is how the word aldehyde arose. It is a combination of *al*cohol and *dehyd*rogenate.

Aldehydes are important commercially. Acetaldehyde (CH_3CHO) is a flavoring agent. Formaldehyde (HCHO) is widely used as a preservative, especially for biology specimens.

When we oxidize alcohols in a different way, or when we oxidize aldehydes, we get compounds with another functional group. Back to the lab!

D. Have a Permanganate Special

What You Need

Ethyl alcohol (CH_3CH_2OH)
Potassium permanganate ($KMnO_4$), 0.1 *M* solution
Sulfuric acid (H_2SO_4), 6 *M*
Dropper
Graduated cylinder, 10 ml
Stirring rod
Test tube
Safety goggles
Test tube rack

What to Do

a. Place 2 ml of $KMnO_4$ solution in the test tube.
b. Carefully add 2 ml of H_2SO_4 to the tube.

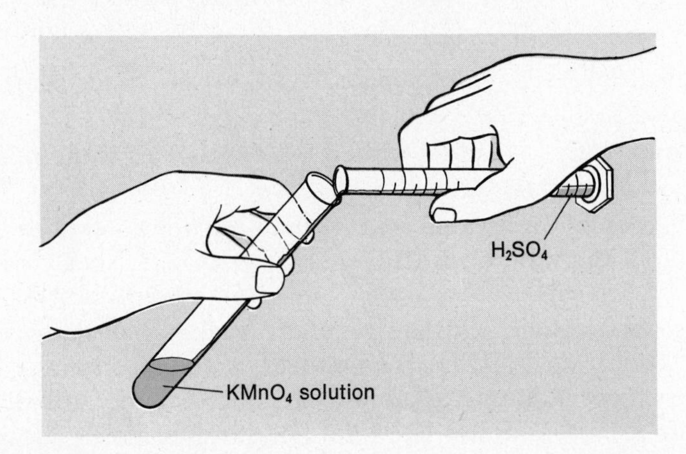

c. Add 2 drops of ethyl alcohol. Carefully stir the mixture.

7. Do you notice any change in the permanganate solution?

d. Add another drop or two of ethyl alcohol to the tube. Let the tube stand for 5 minutes. (This is important!)

8. Are there any further changes in the permanganate solution?

The color of a solution containing the permanganate ion (MnO_4^{2-}) is green. When permanganate is reduced, manganese dioxide (MnO_2), a brown precipitate, forms. Further reduction creates a light pink solution of Mn^{2+} ions.

9. What is the color of the solution after 5 minutes?

10. Was the permanganate reduced once or twice?

11. If one substance in a reaction is reduced, what must happen to another substance?

When an alcohol is oxidized, the first step produces an aldehyde, as you saw earlier. If you keep on oxidizing the alcohol, you get an *acid:*

$$\underset{\text{Alcohol}}{R-\overset{\displaystyle H}{\underset{\displaystyle H}{\overset{|}{\underset{|}{C}}}}-OH} \xrightarrow[\text{[O]}]{\text{oxidation}} \underset{\text{Aldehyde}}{R-C\begin{matrix}\nearrow\!\!\!\!\!= O\\ \searrow H\end{matrix}} \xrightarrow[\text{[O]}]{\text{oxidation}} \underset{\text{Acid}}{R-C\begin{matrix}= O\\ \searrow OH\end{matrix}}$$

The final product is an organic acid. All organic acids contain the group $-C\begin{matrix}= O\\ \searrow OH\end{matrix}$, often written —COOH. This is the functional group of an organic acid. Permanganate can oxidize an alcohol all the way to an acid.

As a group, organic acids are generally weaker than inorganic acids. Formic acid (HCOOH), however, is quite a strong acid. It is ten times as strong as almost any other organic acid. The venom, or poison, of ant bites and bee stings is mostly formic acid. You know acetic acid (CH_3COOH) as vinegar. Like ethyl alcohol, acetic acid has been known since prehistoric times.

The —COOH group can be attached to a carbon chain of any length. In naturally occurring organic acids, however, the number of carbon atoms is almost always even. This is because living organisms use acetic acid (two carbons) to make the organic acids they need. They make acetic acid from sugar or starch. They keep joining molecules of acetic acid together. Since they start out with a 2-carbon compound and add more 2-carbon compounds, they wind up with an even number of carbon atoms.

Foods which are sour or tart are sure to contain some organic acid. Fruits are good examples. Malic acid is found in unripe pears, apples, and other fruits. The amount of acid grows less, and the sugar content increases, as a fruit ripens. This is why ripe fruits taste so much better than unripe fruits.

Table 2 summarizes the functional groups you have studied in this investigation.

I guess I had too much malic acid!

TABLE 2: Functional Groups

Functional group	*Name*	*Typical compound*
—OH	alcohol	CH_3CH_2OH ethyl alcohol
$-C(=O)H$ or —CHO	aldehyde	$CH_3C(=O)H$ acetaldehyde
$-C(=O)OH$ or —COOH	acid	$CH_3C(=O)OH$ acetic acid

You have just investigated the reactions of alcohols in a test tube. Ethyl alcohol undergoes the same reactions in your body.

All alcohols in a laboratory are poisonous, including ethyl alcohol. Laboratory alcohol is *denatured,* or treated, with other compounds so that it cannot be consumed. The ethyl alcohol used in beverages is produced by the fermentation of grains. It then undergoes an involved distillation process to remove all of the byproducts.

When consumed in beverages, a small quantity of alcohol ($\frac{1}{3}$ of a shot) dulls the higher nerve centers of the brain. Judgment, attention, memory, and self-control are affected. There is usually a slight decrease in blood pressure and an increase in pulse rate. With larger quantities, the heart action is slowed and the blood pressure is increased. Alcohol also slows the reaction time, which is important in avoiding or causing accidents.

The next investigation will give you another functional group to work with. Onward!

INVESTIGATION

5

More Friendly Neighbors

In the last investigation you met three types of organic compounds: alcohols, aldehydes, and acids. Each is made of an *alkyl* group and a *functional* group. As you know, you can abbreviate the alkyl group with the capital letter R. Thus:

$$\text{alcohol} = \text{R—OH}$$

$$\text{aldehyde} = \text{R—C} \begin{matrix} \diagup\!\!\!\!\diagup \text{O} \\ \diagdown \text{H} \end{matrix}$$

$$\text{acid} = \text{R—C} \begin{matrix} \diagup\!\!\!\!\diagup \text{O} \\ \diagdown \text{OH} \end{matrix}$$

Now meet another class of compounds—the *esters.* A typical ester is ethyl acetate. The molecular formula for it is $CH_3COOCH_2CH_3$. The structural formula is written as:

$$
\begin{array}{ccccccccc}
 & & H & & & O & & & \\
 & & | & & /\!/ & & H & H & \\
H & - & C & - & C & & | & | & \\
 & & | & & & \backslash & & & \\
 & & H & & & O & -C & -C & -H \\
 & & & & & & | & | & \\
 & & & & & & H & H &
\end{array}
$$

The part inside the box is the functional group for esters:

$$
\begin{array}{ccc}
 & & O \\
 & /\!/ & \\
-C & & \\
 & \backslash & \\
 & & O-
\end{array}
$$

An ester has *two* alkyl groups. It can be written in the general form as:

$$
\begin{array}{ccc}
 & & O \\
 & /\!/ & \\
R-C & & \\
 & \backslash & \\
 & & O-R'
\end{array}
$$

where R and R′ may be the same or different alkyl groups. An ester is prepared from an alcohol and an acid. Give it a go in the lab.

A. Have a Whiff

What You Need

Acetic acid (CH_3COOH), glacial
Amyl alcohol [$CH_3(CH_2)_3CH_2OH$]
Ethyl alcohol (CH_3CH_2OH)
Methyl alcohol (CH_3OH)
Salicylic acid (HOC_6H_4COOH)
Sulfuric acid (H_2SO_4), concentrated

Beaker, 400 ml
Dropper
Graduated cylinder, 10 ml
Stirring rod
Test tubes, 3

Asbestos pad
Balance
Burner
Matches
Ring
Ring stand
Safety goggles
Spatula
Test tube rack
Wax pencil
Weighing paper

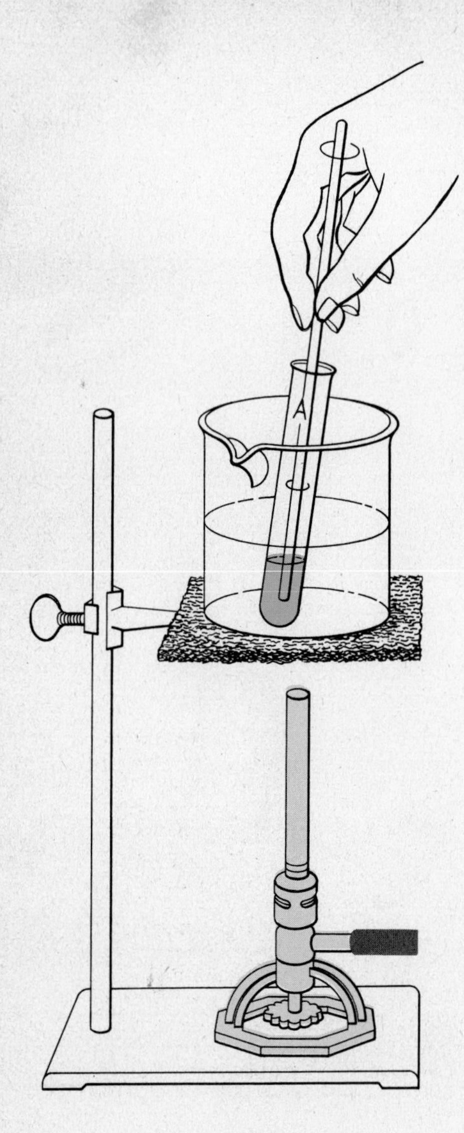

What to Do

a. Place three test tubes in a test tube rack.
b. Label the test tubes **A, B,** and **C.**
c. Fill a 400 ml beaker $\frac{1}{2}$-full with water; place it on a ring stand.
d. Heat the water in the beaker, but do not boil it.
e. Turn off the burner.
f. Into test tube **A,** place 2 ml of ethyl alcohol, 2 ml of glacial acetic acid, and 6–8 drops of concentrated H_2SO_4. **Be very careful handling the acids!**
g. Place the test tube in the beaker of warm water. Stir gently for 4 minutes with a clean stirring rod.

When you heat something in water, it is called heating in a *water bath.*

h. Carefully sniff the contents of the test tube. Remove the tube from the water bath and place it in the test tube rack.

   1. What odor do you smell?

i. Into test tube **B,** place 2 ml of amyl alcohol, 2 ml of acetic acid, and 6–8 drops of concentrated H_2SO_4.
j. Repeat steps **g** and **h.**

   2. What is the odor?

k. Into test tube **C,** place 2 ml of methyl alcohol, 0.5 gram of salicylic acid, and 6–8 drops of concentrated H_2SO_4.
l. Repeat steps **g** and **h.**

   3. What is the odor produced?

B. Ester is Terrific!

You have just prepared three different esters. Esters are organic compounds formed from the reaction of an alcohol with an organic acid. The process is called *esterification.*

The general equation is:

$$R'\text{—}OH + R\text{—}C(=O)\text{—}OH \longrightarrow R\text{—}C(=O)\text{—}O\text{—}R' + HOH$$

$$\text{Alcohol} + \text{Acid} \longrightarrow \text{Ester} + \text{Water}$$

But the reaction can also go in the opposite direction:

$$\text{Ester} + \text{Water} \longrightarrow \text{Alcohol} + \text{Acid}$$

This reaction is called *hydrolysis,* because water (*hydro-*) is used to split apart the ester.

We can combine these two equations by the use of a double arrow:

$$\text{Alcohol} + \text{Acid} \underset{\text{hydrolysis}}{\overset{\text{esterification}}{\rightleftharpoons}} \text{Ester} + \text{Water}$$

The reaction, then, is *reversible.* It can go either way. Why did it go to the right (esterification) for you? Esterification is favored by the presence of concentrated sulfuric acid, which is a powerful *dehydrating,* or water-removing agent. If you add water to an ester, the reaction will reverse; the ester will *hydrolyze* and form an alcohol and an acid.

As you have seen, esters are generally pleasing to the smell. They are used in making beverages, candies, and cosmetics. Table 1 shows some familiar esters:

TABLE 1: Familiar Esters

Odor	*Ester*
Banana	amyl acetate
Orange	octyl acetate
Pear	isoamyl acetate
Rum	ethyl formate
Wintergreen	methyl salicylate
Jasmine	benzyl acetate

Here is the equation for the reaction between ethyl alcohol and acetic acid, which you just performed:

$$\text{H—}\underset{\text{H}}{\overset{\text{H}}{\text{C}}}\text{—}\underset{\text{H}}{\overset{\text{H}}{\text{C}}}\text{—OH} + \text{H—}\underset{\text{H}}{\overset{\text{H}}{\text{C}}}\text{—C}\begin{smallmatrix}\text{//O}\\ \text{\\OH}\end{smallmatrix} \longrightarrow$$

$$\text{H—}\underset{\text{H}}{\overset{\text{H}}{\text{C}}}\text{—C}\begin{smallmatrix}\text{//O}\\ \text{\\}\end{smallmatrix}\text{O—}\underset{\text{H}}{\overset{\text{H}}{\text{C}}}\text{—}\underset{\text{H}}{\overset{\text{H}}{\text{C}}}\text{—H} + \text{H—O—H}$$

4. Write the chemical equation for the formation of amyl acetate.

5. What products would you get by adding water to amyl acetate?

You now have a familiarity with basic organic chemistry. The next three investigations will help you apply what you've learned to daily living.

TIGATION

6

B Is for Biochemistry

Biochemistry is the study of chemical processes in living things. *Bio-* means *life*. Biochemists study the chemistry of people, animals, and plants.

The study of biochemistry can tell us many things. One of these things is which foods we should eat to be healthy. Biochemistry also tells us what happens to foods once we eat them.

Most of the chemicals and chemical processes within a living thing are organic. Biochemistry is therefore studied under the heading of organic chemistry.

The organic compounds which you studied in the last several investigations have only one functional group per molecule. They are therefore rather simple molecules.

There are other kinds of organic compounds with more complicated molecules. In this investigation, you will look into *fats*, *carbohydrates*, and *proteins*. These compounds are among the principal constituents of food, and therefore of living things. Their chemical reactions are the basis of biochemistry.

First you will learn certain chemical tests that identify fats, carbohydrates, and proteins. Then you will examine the structures of their molecules to find out what characteristic groups are present. Finally, you will test a number of foods for fats, carbohydrates, and proteins.

A. Fats and Food

What You Need

Margarine	Beaker, 250 ml	Asbestos pad
Oil	Dropper	Burner
Peanuts	Evaporating dish	Matches
Trichlorotrifluoroethane (TTE)	Graduated cylinder, 10 ml	Paper, white
	Mortar and pestle	Ring
	Stirring rod	Ring stand
	Test tube	Safety goggles
		Spatula

What to Do

a. Place a piece of margarine, the size of a pea, in an evaporating dish. Heat strongly over the burner.

1. Describe the odor.

b. When the evaporating dish has cooled completely, rinse and dry it.
c. Then repeat the test with a few drops of oil.

2. Describe the odor.

The smell is due to a compound called acrolein. Its formation during burning helps identify fats.

$$H_2C=CH-CHO$$

Acrolein

d. Rub a bit of fat, such as margarine or oil, on a piece of white paper.
e. Hold the paper in front of a light.

3. How does the paper appear?

f. Crush a peanut in a mortar with a pestle.
g. With a spatula, place the crushed peanut in a test tube. Then add 5 ml of TTE to the test tube. **Avoid breathing the vapors.**

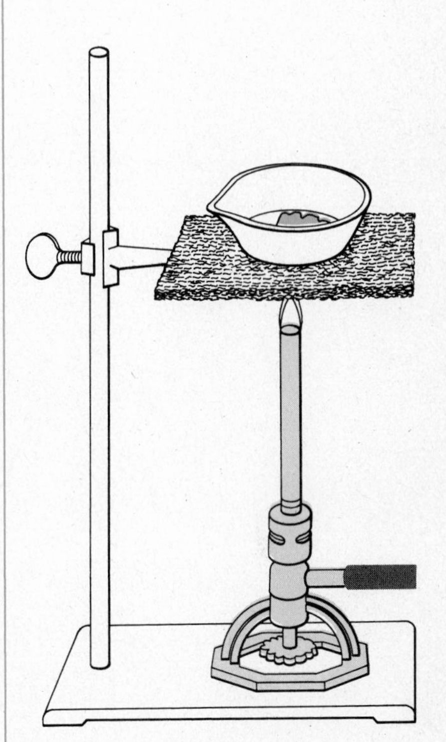

h. Stir the mixture of peanut and TTE.
i. Place the test tube in a beaker of hot water for a few minutes.
j. Using the dropper, transfer a few drops of the liquid from the test tube to a piece of white paper. Take the liquid from just under the surface.
k. Allow the spot on the paper to dry. Then hold the paper in front of a light.

4. How does the paper appear?

5. What component of peanut dissolved in the TTE?

From these tests you can conclude that fats form acrolein when heated, leave an oily spot on paper, and dissolve in TTE.

Fats are compounds of high molecular weight, composed of carbon, hydrogen, and oxygen. They provide energy for body processes. When digested, fats provide certain unsaturated organic acids which appear to be necessary for good health. Fats are found in such things as butter, lard, tallow, and animal and vegetable oils. They are an important class of food, but one can have too much of a good thing.

In order to understand the chemical nature of a fat, you need to recall what you learned about esters. Remember that an ester is formed when an alcohol reacts with an acid.

$$\underset{\text{Alcohol}}{R{-}OH} + \underset{\text{Acid}}{R'{-}C{\overset{\displaystyle O}{\diagup\!\!\diagup}}\!\!\diagdown_{OH}} \longrightarrow \underset{\text{Ester}}{R'{-}C{\overset{\displaystyle O}{\diagup\!\!\diagup}}\!\!\diagdown_{O-R}} + \underset{\text{Water}}{H_2O}$$

Esterification

Fats are large molecules which contain the functional group of an ester. Therefore, fats are esters. The illustration shows the formula of a fat molecule.

$$\begin{array}{l} R{-}\overset{O}{\overset{\|}{C}}{-}O{-}CH_2 \\ \qquad\qquad\quad | \\ R'{-}\overset{O}{\overset{\|}{C}}{-}O{-}CH \\ \qquad\qquad\quad | \\ R''{-}\overset{O}{\overset{\|}{C}}{-}O{-}CH_2 \end{array}$$

The structure of a fat molecule

6. How many ester groups are in a fat molecule?

7. Write the reaction between butyl alcohol and acetic acid.

B. Carbohydrates Are Corny

Carbohydrates are probably the most abundant organic compounds in the plant kingdom. They include a great variety of materials. Sugars, starches, and various forms of cellulose such as wood, paper, and cotton are all carbohydrates.

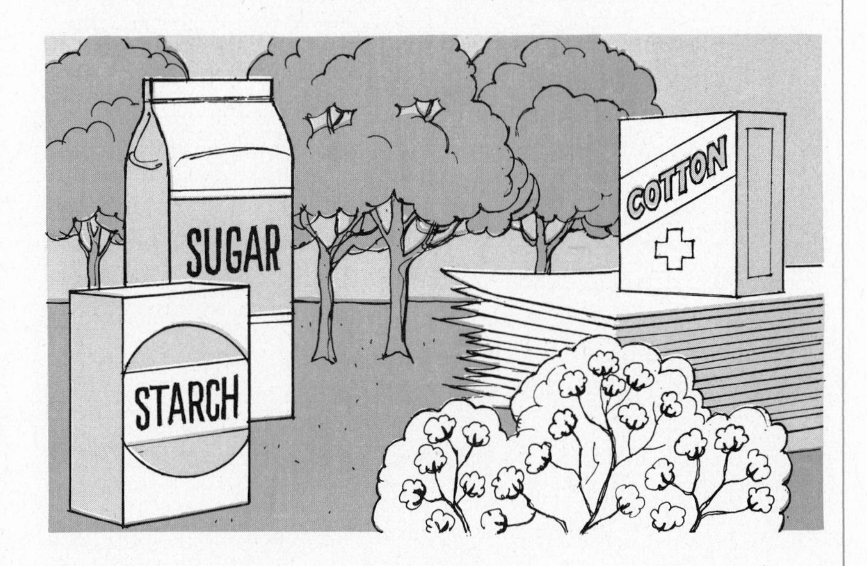

What You Need

Apple juice
Benedict's solution
Corn syrup
Glucose ($C_6H_{12}O_6$), 2 *M* solution
Iodine (I_2), 0.1 *M* solution
Orange juice
Starch [$(C_6H_{10}O_5)_n$]
Beaker, 50 ml
Dropper
Graduated cylinder, 10 ml
Stirring rod
Test tube
Burner
Matches
Safety goggles
Spatula
Test tube holder
Test tube rack

What to Do

a. Place enough starch in the bottom of a test tube to fill the curved portion.

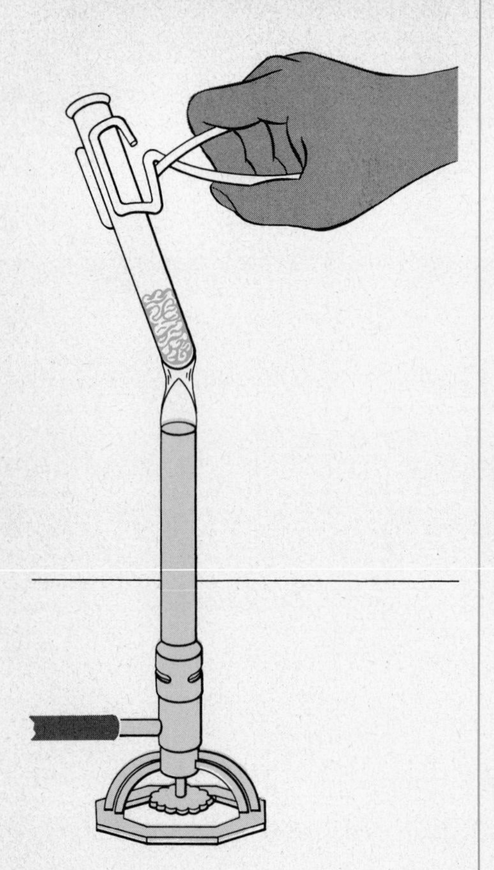

b. Use the test tube holder to hold the test tube over the burner flame. Heat the starch strongly.

8. What substance collects on the cooler portions of the tube?

9. What two elements does this test show to be present in the starch?

10. What color is the residue in the test tube?

11. To what element in the starch is this color due?

12. What three elements are present in starch?

c. Place a small sample of starch, about the size of a pea, in a beaker. Add about 5 ml of water.
d. Boil this mixture until a jellylike mass forms.
e. Dilute the jellylike material with about 5 ml of water. Stir the mixture.
f. Add a few drops of iodine solution to the mixture.

13. What happens when you add iodine to starch?

The color you observed is a positive test for starch. You tested for starch in Unit 1, Investigation 1.

g. Add 5 ml of Benedict's solution to a clean beaker. Carefully heat the solution to boiling.
h. Add 3 ml of glucose solution to the hot Benedict's solution.

14. What color is produced when you add glucose to hot Benedict's solution?

This test is used to identify glucose sugar. The structure of the glucose molecule is shown below.

```
     H    H    H    OH   H
     |    |    |    |    |      O
  H—C————C————C————C————C—C  //
     |    |    |    |    |      \
     OH   OH   OH   H    OH      H
```

The structure of glucose

i. Use Benedict's solution to find out if glucose is present in orange juice, apple juice, corn syrup, and starch.

15. Which of these substances contain glucose sugar?

Most carbohydrates in nature are formed in green plants from carbon dioxide and water by photosynthesis. Carbohydrates contain carbon, hydrogen, and oxygen. In the most common members of the group, there are two atoms of hydrogen for each oxygen atom in the molecule. This is the same proportion in which hydrogen and oxygen combine to form water. The name carbohydrate comes from *carbon* plus water (*hydrate*).

Carbohydrates are probably the original sources of all our body energy and fuel energy. They are also the starting materials from which plants and animals make practically all of the organic compounds in nature.

C. Proteins are Progressive

Proteins are the third major class of foods. They make up the muscle tissue, skin, hair, nails or claws, and the bulk of the body cells of animals. Here are two tests for identifying proteins.

What You Need

Ammonium hydroxide (NH_4OH), 2 *M* solution
Copper(II) sulfate ($CuSO_4$), 0.5 *M* solution
Egg white, fresh
Egg white, hard boiled
Nitric acid (HNO_3), 1 *M*
Sodium hydroxide (NaOH), 3 *M* solution
Dropper
Graduated cylinder, 10 ml
Stirring rod
Test tubes, 2
Burner
Matches
Safety goggles
Test tube holder
Test tube rack

What to Do

a. Place a piece of boiled egg white in a test tube. Add 5 ml of HNO_3.
b. Warm the contents ot the tube gently. **Do not inhale the fumes.**

16. What do you observe?

c. Carefully pour out the HNO_3, leaving the boiled egg white in the test tube. Then rinse off the egg white with water.
d. Add 5 ml of NH_4OH to the egg white in the test tube. Warm the contents gently. **Do not inhale the fumes.**

17. Is there any further change in color?

The orange-yellow color is one positive test for protein. Here's the other.

e. Add 5 ml of liquid egg white to a clean test tube.
f. Add 5 ml of NaOH solution to the egg white. Mix thoroughly.
g. Add 2 or 3 drops of $CuSO_4$ solution.

18. Describe any color produced.

Of the three major classes of foods, proteins are the most complex. Proteins always contain carbon, hydrogen, oxygen, and nitrogen. They may also contain sulfur. Proteins are made up of units called *amino acids.* Amino acids contain two functional groups—the amino group ($—NH_2$) and the acid group (—COOH).

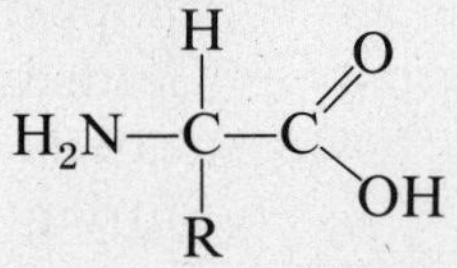

```
        H    O
        |   //
  H2N—C—C
        |   \
        R    OH
```

The structure of an amino acid

Protein molecules are long-chain molecules made up of amino acids joined together. An example is shown below:

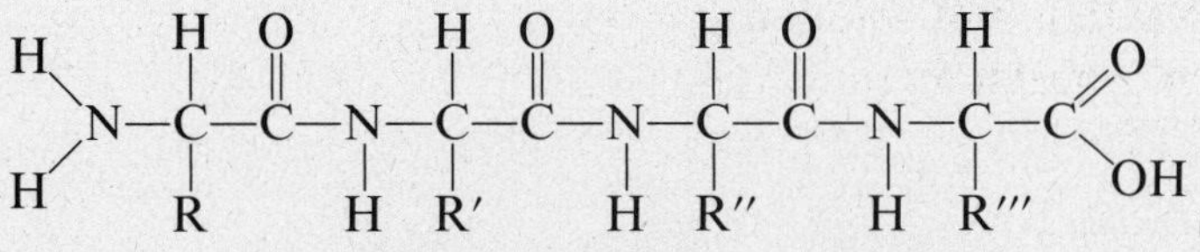

```
        H  O     H  O     H  O     H
 H      |  ||    |  ||    |  ||    |   O
  \     |  ||    |  ||    |  ||    |  //
   N—C—C—N—C—C—N—C—C—N—C—C
  /     |     |  |     |  |     |  |   \
 H      R     H  R'    H  R''   H  R'''  OH
```

The structure of a protein molecule

19. What are the general properties of proteins?

20. What use is made of fat in the body?

21. What are carbohydrates?

D. Now for the Grand Finale

What You Need

Ammonium hydroxide (NH_4OH), 2 *M* solution
Benedict's solution
Copper(II) sulfate ($CuSO_4$), 0.5 *M* solution
Food samples
Iodine (I_2), 0.1 *M* solution
Nitric acid (HNO_3), 1 *M*
Sodium hydroxide ($NaOH$), 3 *M* solution
Trichlorotrifluoroethane (TTE)

Beaker, 50 ml
Beaker, 250 ml
Dropper
Evaporating dish
Graduated cylinder, 10 ml
Mortar and pestle
Stirring rod
Test tubes

Asbestos pad
Burner
Matches
Paper, white
Ring
Ring stand
Safety goggles
Spatula
Test tube holder
Test tube rack

What to Do

22. Set up a table like Table 1 in your laboratory notebook.

a. Select various samples of food. Examples might be bread, meat, potatoes, beans, cereals, and chocolate.
b. Test each sample for fat, carbohydrate, and protein.

TABLE 1

Food	TEST USED AND RESULTS			
	Fat	Carbohydrates		Protein
		Starch	Glucose	

23. Record your data in the table.

In this investigation you did *qualitative* tests. This means you identified *what* was in a substance. In the next investigation you will do a *quantitative* test. You will find out *how much* of a substance is present.

INVESTIGATION
7

Give Me a V. . .
Give Me a C. . .

De Wys Inc.

A. The Nitty Gritty

You have just studied fats, carbohydrates, and proteins. In order to live, human beings and other animals must eat foods containing these substances. In addition, they must eat foods containing chemical substances called *vitamins* and *minerals.*

To a geologist, minerals mean rocks. But minerals in foods are the 12 or more elements required by the body in addition to carbon, hydrogen, oxygen, and nitrogen. The following elements are regarded as essential for health: calcium, phosphorus, potassium, sulfur, sodium, zinc, chlorine, magnesium, iron, manganese, copper, and iodine. We get our minerals directly from plants, or indirectly from animals that eat plants. The plants get their minerals from the soil.

Vitamins are organic compounds necessary for the normal functioning of cells. Without vitamins, humans and other animals become sick and may even die. The human body can manufacture some vitamins by itself, but often in amounts too small for its needs. Other vitamins are not formed within the body and must be supplied by outside sources.

Although the importance of all of the vitamins is not known, we do know that each vitamin has specific functions. One vitamin cannot replace another. The continued lack of one vitamin in an otherwise balanced diet will result in a *deficiency disease.*

Some of the more common vitamins and their functions are listed in Table 1.

TABLE 1: Vitamins

Vitamin	*Soluble in*	*Function*	*Source*	*Deficiency Diseases*
A	fat	aids in the building and growth of body cells; builds resistance to infection; aids eyes in adjusting to varying light intensity	carrots, milk, egg yolk, green and yellow vegetables	night blindness
B complex (includes 15 known groups)	water	converts carbohydrates to energy; promotes body's use of oxygen; prevents skin problems; aids in prevention of certain types of anemia	meat, yeast, liver,. grain	beriberi; skin sores; convulsions (in infants)
C	water	healthy blood vessels; sound bones and teeth (The body cannot store this vitamin, so it must be supplied daily.)	citrus fruits	scurvy
D (includes 10 known groups)	fat	bone development	sunlight, fish liver oil	rickets
K	fat	aids in blood clotting	cabbage, kale, cauliflower	bleeding

In the laboratory, you will examine a method to measure the amount of one of these vitamins.

B. Buy the C

The average high school student probably needs a daily supply of vitamin C equal in mass to the common pin—100 mg or 0.1 gram. This vitamin helps form teeth, bones, red blood cells, and connective tissue. It is essential for the normal healing of wounds. Vitamin C is used to prevent or cure scurvy, a disease which causes bleeding under the skin and weakness.

Only small amounts of vitamin C can be stored in the body, so you need to get a fresh supply each day. Where do you get it? There are many sources. You can buy vitamin C tablets in the drug store; or you can drink a glass of fruit juice each day.

Natural fruit juices differ in vitamin C content. The juice of the acerola, a cherry native to Puerto Rico, is 80 times richer in vitamin C than is orange juice. Acerola juice is sometimes added to other fruit juices to increase the vitamin C content.

By now, you are pretty good chemists; so you can test different kinds of fruit juice for their vitamin C content. Vitamin C reacts with an iodine solution. When the reaction is complete, excess iodine will react with starch to give a blue color. This shows the end point, or the end of the reaction. Starch is used in this reaction as an indicator.

To test for vitamin C, first determine the amount of iodine solution needed to react with a known amount of vitamin. Then measure the amount of iodine solution required to react with an unknown amount of vitamin C. From this you can calculate the amount of the vitamin present in the unknown.

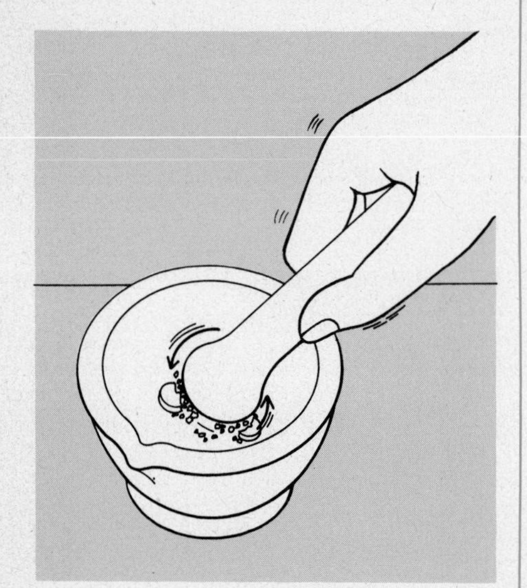

What You Need

Lugol's iodine solution
Starch solution
Vitamin C tablet of known strength
Beakers, 250 ml, 2
Buret
Dropper
Flask, 250 ml
Graduated cylinder, 10 ml
Mortar and pestle
Buret clamp
Paper, white
Ring stand
Safety goggles
Weighing paper

What to Do

a. With a pestle, grind a vitamin C tablet of known strength in a mortar.
b. Transfer the powdered tablet to a piece of weighing paper. Then transfer it into a flask.

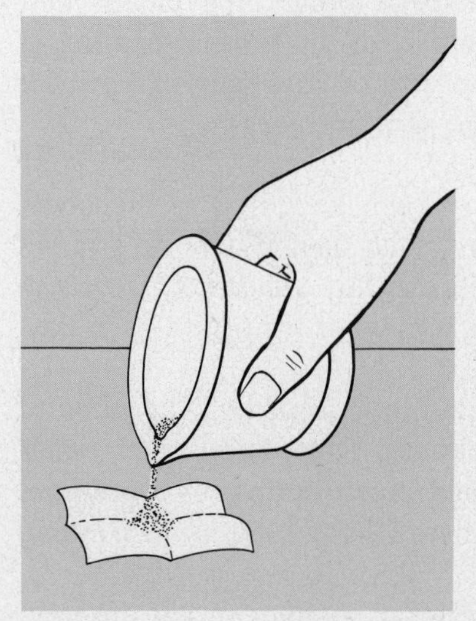

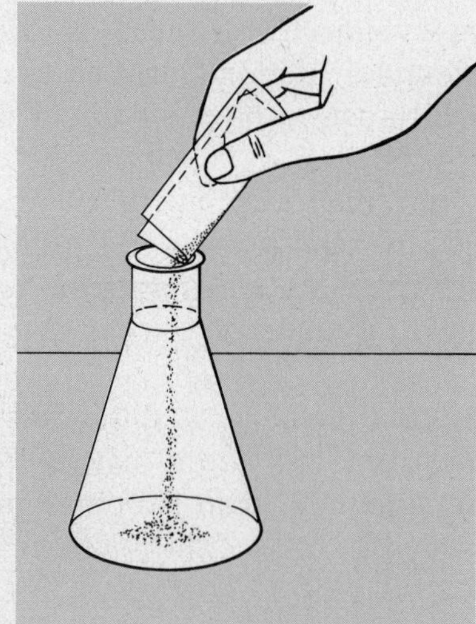

c. Add 5 ml of water to the flask. Then add 1 ml of starch solution. Swirl the flask to mix the contents.
d. Place the buret clamp on the ring stand. Place the buret in the buret clamp.

e. Carefully fill the buret with Lugol's iodine solution. Use a beaker for pouring into the buret. The level of the solution in the buret should be about 4 cm above the 0 ml mark.
f. Place a second beaker under the tip of the buret. It will be the waste liquid beaker.
g. Open the stopcock to release the excess solution into the waste liquid beaker. Allow liquid to flow from the buret until the bottom of the meniscus is at the 0 ml mark. The tip of the buret should be completely filled with liquid.

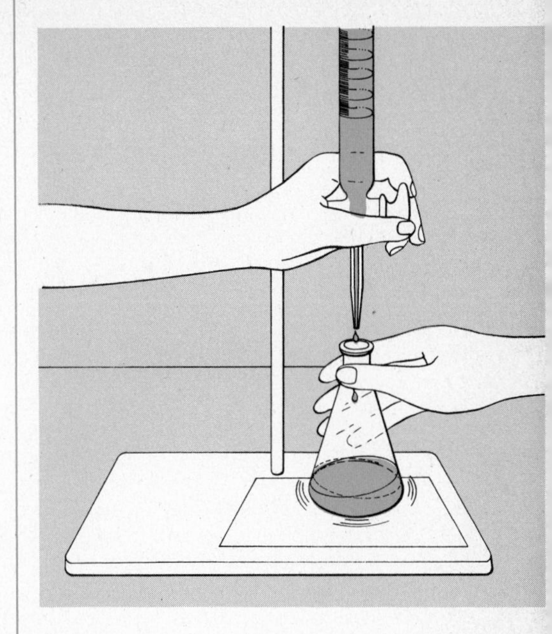

You are now ready to titrate.

1. In your notebook, record the initial reading of the buret.

h. Swirl the flask gently while holding it under the tip of the buret.
i. Slowly add the iodine solution from the buret, drop by drop, until a faint blue color persists in the flask.
j. After the faint blue color appears, wait a few minutes before you read the buret. This lets the liquid drain down the side of the buret so that the reading is more accurate.

2. Record the final reading.

3. Subtract the initial reading from the final reading and record your answer.

Now you need to find out how many milligrams of vitamin C are equivalent to 1 ml of iodine solution. Suppose, for example, that 25 ml of iodine solution were used to titrate 100 mg of vitamin C. You would then be able to say that 1 ml of iodine solution was equivalent to 4 mg of vitamin C.

How can you do this? Simply by dividing the number of milligrams of vitamin C by the number of milliliters of iodine solution.

$$\frac{\text{Milligrams of vitamin C (amount in tablet)}}{\text{Milliliters of iodine solution (amount used in titration)}}$$

In this example you get

$$\frac{\text{100 mg of vitamin C}}{\text{25 ml of iodine solution}} = \text{4 mg of vitamin C/ml of iodine}$$

4. How much vitamin C was in the tablet you used?

5. How many ml of iodine solution did you use in your titration?

6. In your titration, how many mg of vitamin C were equivalent to 1 ml of iodine solution? Set up your problem like this:

$$\frac{\_\_\_\text{ mg of vitamin C}}{\_\_\_\text{ ml of iodine solution}} = ?$$

This value is known as the *factor* of your solution. You will use this number in all your calculations.

7. Is it possible for different groups in the class to have different factors? Why?

C. Let's Test Brand "X"

Now different groups will test different kinds or brands of juice for their vitamin C content.

What You Need

Fruit, fresh
Juices, canned, bottled, or frozen, *e.g.*,
- Grapefruit juice
- Hi-C
- Lemonade
- Orange juice

Lugol's iodine solution
Starch solution
Beakers, 250 ml, 2
Buret
Dropper
Flask, 250 ml
Funnel
Graduated cylinder, 10 ml
Buret clamp
Clay triangle
Filter paper
Paper, white
Ring stand
Safety goggles

What to Do

a. Shake the container of juice you are going to use.
b. Put 10 ml of the juice in a clean flask.
c. Add 1 ml of starch solution. Swirl the flask.
d. Fill the buret with iodine solution.

8. Record the initial reading in your notebook.

e. Titrate with iodine solution until a faint blue color persists in the flask. Remember to swirl the flask constantly.

f. When you reach the end point, wait a few minutes and then make the final reading.

9. Record the final reading in your notebook.

10. Subtract the initial reading from the final reading and record your answer.

11. Calculate the amount of vitamin C in your sample of juice. To do this, multiply the amount of iodine solution used by the factor you determined in Part B. The answer will be the number of mg of vitamin C in your sample.

12. Set up a table like Table 2 in your notebook.

TABLE 2: Vitamin C Content of Juices

Sample	Amount of iodine solution used	Factor (Question #6)	Mg of vitamin C in a 10 ml sample

g. Titrate other samples of juice. You may want to compare fresh, frozen, day-old, and warm samples.

13. Record your results in Table 2.

14. What kind of juice has the most vitamin C?

15. What kind of juice has the least vitamin C?

16. What kind of fresh fruit has the most vitamin C?

17. What kind of fresh fruit has the least vitamin C?

18. Of all samples tested, which had the most vitamin C?

If you're going to buy fruit or fruit juice for the vitamin C, you might as well get the most for your money. That doesn't apply to vitamin C only. A smart shopper is a healthy shopper. In the next investigation, we will look more at foods and what may be found in them.

INVESTIGATION

8

Whom Can You Trust?

You have studied about the food you eat and what foods you should eat. But that's not the whole story.

You know that food will spoil after it has been exposed to air for some time. You have all seen moldy fruit or bread and spoiled meat. This spoilage is due to the growth of *microorganisms* in the food. Microorganisms are tiny plants and animals that can be seen only with a microscope. Sometimes the spoiled food contains dangerous *toxins,* or poisons, which are harmful to the body.

The processes used to preserve food slow down spoilage. They also make it possible for us to enjoy foods from all parts of the world. Fruits and vegetables grown in season can be preserved and then eaten at any time of the year.

There are many ways of preserving food. The problem starts when manufacturers are over enthusiastic in loading their products with chemicals. They can also use preservatives to disguise a poor product. With so many preservatives in use, food doesn't rot any more; it slowly mummifies.

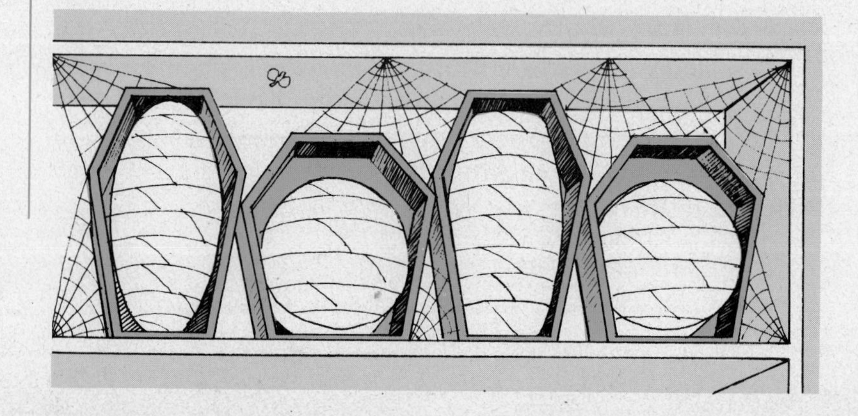

A. The Nasty Nitrites

Nitrite compounds contain the NO_2^{1-} group. They have been widely used in food preservation. Nitrites are used to color meats such as corned beef and hot dogs.

You will test for nitrite, using sodium nitrite as the sample. This will show you a positive test.

What You Need

Potassium iodide (KI), 1 *M* solution
Sodium nitrite ($NaNO_2$)
Starch [$(C_6H_{10}O_5)_n$] solution
Sulfuric acid (H_2SO_4), 0.5 *M*
Dropper
Graduated cylinder, 10 ml
Stirring rod
Test tubes, 2
Safety goggles
Spatula
Test tube rack

What to Do

a. Put a pinch of $NaNO_2$ in a test tube.
b. Add 5 ml of water.
c. Add 3 drops of KI solution. Then add 3 drops of H_2SO_4. Stir.
d. Half-fill another test tube with starch solution.
e. Add a few drops of the mixture from step **c** to the starch solution.

1. What do you observe?

A blue-black color is a positive test for nitrite. In this reaction, the potassium iodide and sulfuric acid react with the nitrite ions to give free iodine (I_2). The iodine reacts with the starch to give the blue-black color.

B. The Benzoate Bomb

Sodium benzoate has been a favorite preservative. It is added to coffee cake, tomato products, and soft drinks. You will test for benzoates, using sodium benzoate. This will show you a positive test.

THE SMITH FAMILY By Mr. and Mrs. George Smi

What You Need

- Cereal box
- Iron(III) chloride ($FeCl_3$), 2 *M* solution
- Sodium benzoate (C_6H_5COONa)
- Sulfuric acid (H_2SO_4), 0.5 *M*
- Trichlorotrifluoroethane (TTE)
- Dropper
- Graduated cylinder, 10 ml
- Stirring rod
- Test tube
- Balance
- Safety goggles
- Spatula
- Test tube rack
- Weighing paper

What to Do

a. Put about 0.5 gram of sodium benzoate in a test tube.
b. Add 5 ml of water.
c. Add 3 drops of H_2SO_4. Then add 5 ml of TTE. **Do not inhale** the fumes.
d. Carefully stir the contents of the tube.
e. Add 3 drops of $FeCl_3$ solution.

2. What do you observe?

The salmon-colored precipitate indicates a positive test for benzoate.

You can give a lot of thought to what you should eat; and even then you're not safe. Almost every week you can read about some food product being pulled off the market because it is contaminated. You pay good money for food, and then you have to worry about it hurting you.

f. Get a box of cereal and look at the list of ingredients.

3. Which ingredients are natural foods?
4. Which ingredients are chemical preservatives?
5. Can you do anything about chemicals in your food?

All is not darkness and gloom in the food situation. The Food and Drug Administration checks for the presence of harmful chemicals in our foods. The improper use of chemicals in food often results from out-of-date laws and uneducated consumers. To change the laws, you have to educate the consumers. This means you will need to continue to educate yourselves.

C. You Are What You Eat

What is food? Why do we have to eat?

Our bodies are very complicated. We can live and perform our normal activities only within a limited body temperature range. The normal body temperature of an adult human being is 37°C. This temperature must be maintained in spite of external temperature changes.

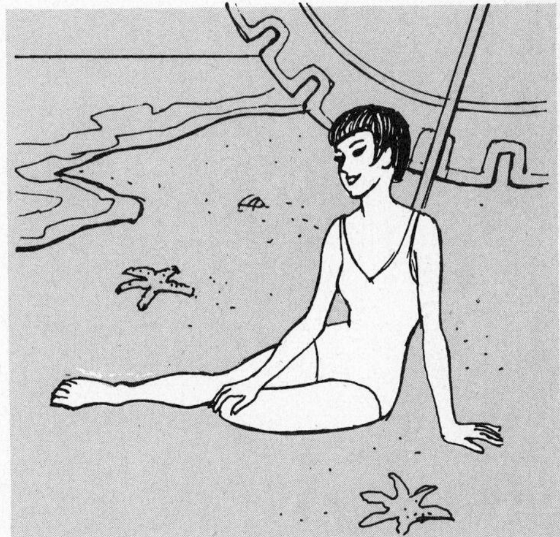

The body must generate heat to maintain its temperature. It must also produce enough additional energy for such important processes as digestion, breathing, blood circulation, and muscular motion. To obtain the necessary energy for all of these activities, everyone eats.

Practically all of the foods that we eat have been analyzed. They contain five classes of substances: fats, carbohydrates, proteins, vitamins, and minerals. Most foods also contain water. Table 1 shows the function of these substances in the body.

TABLE 1: Functions of Foods

Functions in the body	*Fats*	*Carbo-hydrates*	*Proteins*	*Vitamins*	*Minerals*	*Water*
To provide energy	X	X	X			
To regulate body processes				X	X	X
To build, maintain, and repair tissue	X	X	X		X	

How do you make sure that your diet is well balanced? The trick is to eat foods from four basic food groups each day. If you do this, your daily nutritional needs will be met.

The "Basic Four" are the milk group, the meat group, the bread-cereal group, and the fruit-vegetable group.

By this time, you realize that food is an important factor in the growth of a strong, healthy body. Food is used to build muscle, bone, teeth, and hair—in fact, every tissue in the body. Food is used to produce energy and to repair body tissues. But it has to be the right kind and the right amount of food.

Since most of us know too little about the foods we ought to eat, everyone tries to give us advice. We are surrounded by announcements of health-giving foods and of diets from faddists.

Which advice should you accept? Which advice should you reject? You need to learn the answers. After all, it's your body; and in a lifetime, you will probably consume some 50 tons of food!

Fifty tons?!?

Glance through the newspapers and magazines; listen to the radio; watch the TV. You are being "sold" on the idea that it is fashionable to have a "slim and graceful figure." A lot of people are willing to go to any extreme of dieting to lose weight. Don't fall for any miracle or wonder diet. The needs of the body are based upon size, sex, age, occupation, and activities. No one diet is adequate and safe for everyone. Using drugs is the most dangerous of the so-called "scientific" methods of reducing. Some reducing drugs contain thyroxine, a hormone secreted by the thyroid gland. Thyroxine regulates the rate at which the body burns food. It should be given only by a doctor.

Other drugs are sold as pills to be taken before or after meals. Those taken before meals slow down the movements of the digestive tract, so you don't feel hungry. Those taken after meals speed up the movement of the food through the digestive tract. The food moves so fast that the nutrients in the food cannot be used. Since these drugs act on the nerves in the digestive tract, they can permanently damage them.

But why should you go to all this trouble to figure out what to eat? Because if you don't, nobody else will. It's your body and it's the only one you've got. You've got it for the rest of your life. So you'd better take care of it.

Do some reading in newspapers and magazines. Look at the articles and advertisements.

6. Are the people who advertise fad diets trying to con you? If so, how?

D. Let's See Where We Are

Okay, where do we stand now? In this Unit and in the first three Units, you have been taking in a lot of information. You've come a long way from identifying the parts of a balance. You now have a set of basic chemical skills, similar to those of any chemist in any laboratory.

Here's a partial list of what you know and what lab skills you have.

What You Know

- Metric system
- Graphing
- Chemical symbols
- Structure of the atom
- Properties of the elements
- Uses of the periodic table
- Combining numbers
- The mole
- Ionic and covalent bonding
- Reactions of acids, bases, salts, and oxides
- How to balance chemical equations
- The different types of reactions
- Oxidation-reduction reactions
- Classification of organic compounds
- Reactions of functional groups
- Factors influencing the rates of reactions

What You Can Do

- Bend glass
- Recognize chemical reactions
- Make reports
- Determine solubility
- Construct a battery
- Make toothpaste
- Consumer testing
- Qualitative tests
- Quantitative tests
- Classify substances
- Chemical tests to identify:
  - fats
  - carbohydrates
  - proteins
  - vitamin C
  - nitrites
  - benzoates
- Titrate

This is a long and impressive list; and it doesn't even cover everything you know and can do. But, so what? Unless you can put this knowledge to work for you, you've gone to a lot of trouble for nothing.

In Unit 5, you will see how chemists take the same information you have and put it to work for them. Chemists work in the fields of nursing, nutrition, metallurgy, oils, rubber, paper, plastics, and many others.

Now is your opportunity to let chemistry go to work for you.

Home Activities

Investigation 1

1. Be a science fiction author and write a short story about a planet where life is based not on carbon but on silicon.
2. Describe the natural process by which the element carbon becomes a diamond.
3. If water is called "the universal solvent," would carbon be considered the "universal element"? Defend your answer.

Investigation 2

1. Research the subject "Oil" and write a report on any one or a combination of the following:
   a) Where and how is oil found?
   b) How is oil brought up from the earth?
   c) How is oil refined?
   d) Petroleum derivatives are used in the manufacture of what products?
   e) Oil and pollution.
   f) Oil *vs.* coal as a source of heat.

Investigation 3

1. Using pipe cleaners, polystyrene balls, and paint, make models of:
   a) an alkane
   b) an alkene
   c) an alkyne
   d) an alcohol
   e) benzene
   f) a saturated oil
   g) an unsaturated oil
2. Write a report summarizing the latest findings on cholesterol.

Investigation 4

1. Why does alcohol make a poor antifreeze?
2. You have heard the saying "Drinking and driving do not mix." As a chemist, explain this statement.

Investigation 5

1. Visit various stores in your neighborhood and list the odor-bearing chemicals found in cosmetics.

Investigation 6

1. Using your pipe cleaners, polystyrene balls, and paint, make models of:
   a) a protein
   b) a fat
   c) a carbohydrate
2. How are fats, carbohydrates, and proteins digested in the body?

Investigation 7

1. What are given as the minimum daily requirements of vitamins and minerals as determined by the United States Government?
2. Examine several brands of vitamins and minerals. Taking into account the cost, determine which brand gives the most for the money.

Investigation 8

1. Prepare a list of food preservatives in various packaged foods.
2. Design a well-balanced diet, including vitamins, minerals, fats, carbohydrates, and proteins. Make use of the four basic food groups.

Monkmeyer

UNIT 5

Let's Enroll At Good Old Health U

INVESTIGATION 1

You have learned a lot of chemistry this year. But how does it apply to you and to your life style? What can you do with all of the knowledge you have collected?

This Unit will show you some of the many different ways you can use the information you now have at your fingertips. You can start by looking into how chemistry is related to the health sciences. The health sciences include dietetics, dental hygiene, hospital chemistry, medical technology, pharmacy, and nursing.

A chemist working in the health sciences usually performs tests on urine, blood, and other body fluids. These fluids are not

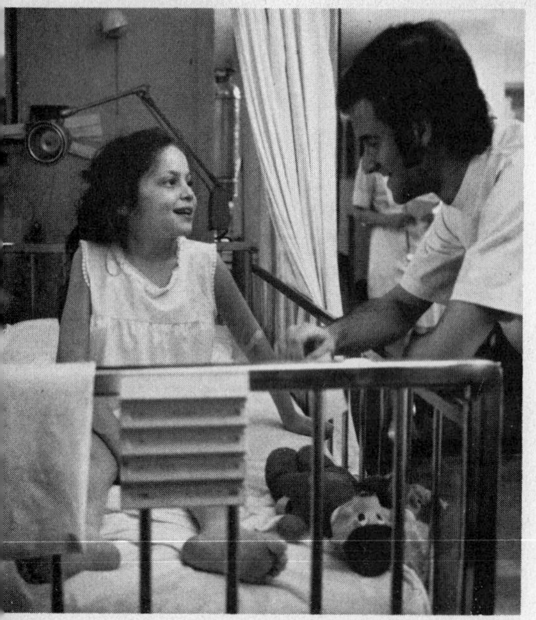
Lucy B. Lazzopina/Columbia-Presbyterian Medical Center

magical or mysterious. They are composed of the same elements found in the compounds you have studied. They undergo the same kinds of reactions that you have already studied.

Nursing includes both the licensed practical nurse (L.P.N.) and the registered nurse (R.N.). An R.N. usually attends a college or nursing school for two to four years after high school. An L.P.N. trains at a special school for one or more years. More and more men are entering the field of nursing and finding it a rewarding career.

Both R.N.'s and L.P.N.'s do a lot of different things in a hospital or nursing home. They have the responsibility for the well-being of their patients. They must watch and keep track of their patient's temperature, blood pressure, and medication. You can investigate the chemistry of blood and also learn how solutions are made.

A. You Have a Lovely Neck, My Dear

Photo courtesy of Freelance, Inc., Lansdale, Pa.

Blood is an essential body fluid. It carries dissolved food, waste products, oxygen, carbon dioxide, hormones, and antibodies to and from cells. (Hormones are regulatory substances produced by glands; antibodies help fight against disease.)

Blood is about 7%, by mass, of the normal adult human body. It is made up of plasma—a complex, watery fluid—and different types of cells. These include red blood cells, white blood cells, and platelets.

One of the more important things you can know about blood is its type. There are four basic blood types: A, B, AB, and O. A blood transfusion from one person to another with a different blood type can result in fatal reactions. Thus, it is necessary to test for blood type before a transfusion. This process is called *blood typing.* Blood typing is done by mixing red blood cells with two solutions. The solutions contain known chemical compounds. If a blood sample clumps, or clots, with Anti-A serum (solution), the blood is type A.

If the blood sample clots with Anti-B serum, the blood type is B.

If the blood clots with both solutions, the blood type is AB. If it does not clot with either solution, the person has blood type O.

With the aid of your teacher, you can find out the blood type of your lab partner.

What you need

Alcohol
Anti-A serum
Anti-B serum
Anti-Rh serum
Glass slides, 2
Cotton balls (swabs)
Hemolets, sterile, 4
Toothpicks
Wax pencil

What to do

a. Read the experiment through step **j.**
b. Label the glass slide as shown in the illustration.

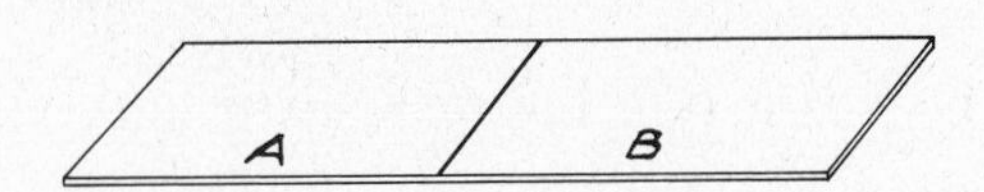

c. Place a large drop of Anti-A serum on the left side of the slide. Place a large drop of Anti-B serum on the right side.
d. With an alcohol-soaked swab, scrub the end of the third finger of your lab partner's hand.

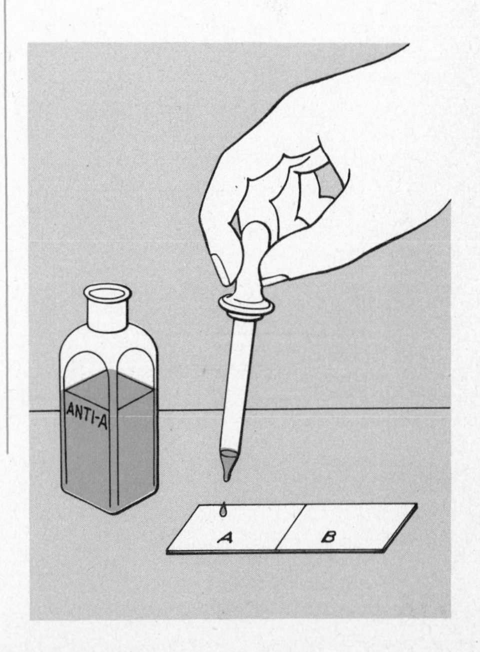

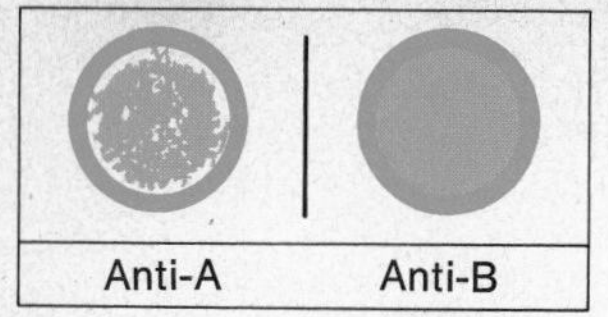

Blood type A

Clotting with Anti-A

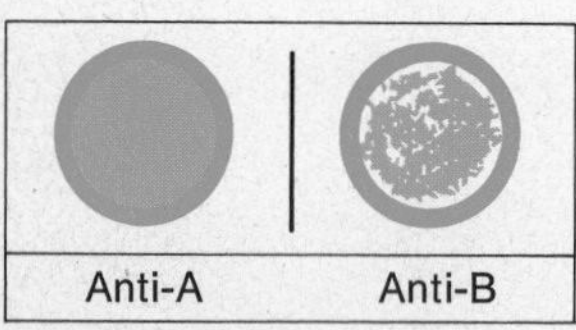

Blood type B

Clotting with Anti-B

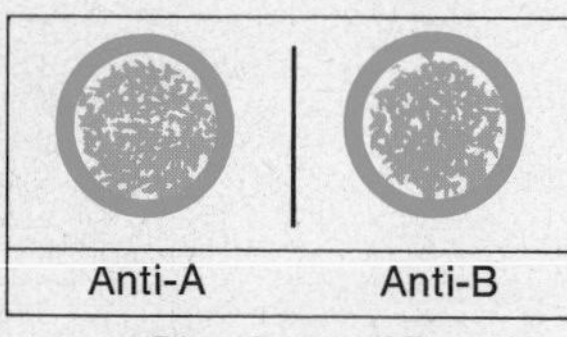

Blood type AB

Clotting with both

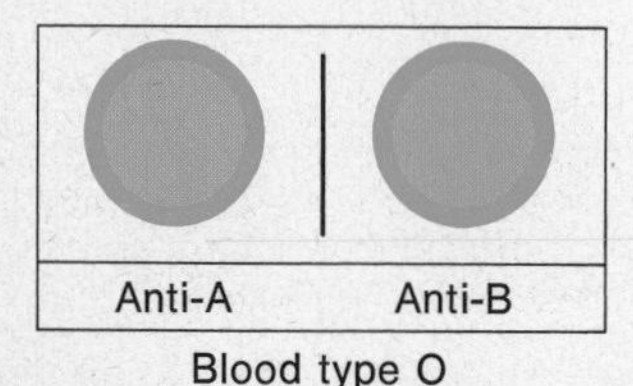

Blood type O

No clotting

e. Prick this finger with a sterile hemolet (a disposable blood lancet).
f. Turn the finger over and hold it just above the slide.
g. Squeeze out a drop of blood. Let it fall on the Anti-A serum on the slide.
h. Let another drop fall on the Anti-B serum. Discard the lancet. **Never reuse a lancet.**
i. Quickly stir each mixture with a toothpick. Be sure to use a different toothpick for each mixture. **Break and discard the toothpicks.**
j. Look carefully at the slide. Compare it to the drawing.

1. What is your lab partner's blood type?

k. Have your lab partner test a blood sample from your finger.

2. What is your blood type?
3. Make a survey of the students in the class. How many of your classmates have type O, A, B, or AB blood?
4. Which blood type is most common in your class?
5. Which blood type is least common in your class?

On the average, 46% of the American population has type O blood, 42% has type A, 9% has type B, and 3% has type AB.

6. Do these percentages agree with the results in your class? If not, why not?

Another substance found in blood is the Rh factor. An Rh factor is present in about 85% of the population. If Rh positive (Rh+) blood is injected into a human whose blood is Rh negative (Rh−), then clotting can take place. This could be fatal. If a woman who is Rh− bears a child who inherited an Rh+ factor from the father, the child's red blood cells could be fatally damaged.

You can easily see if your lab partner's blood contains an Rh factor.

l. Get a clean, dry slide.
m. With a wax pencil, draw a single circle in the center of the slide.
n. Place a drop of Anti-Rh serum in the circle.

o. Scrub the tip of the little finger with an alcohol-soaked swab.
p. Use a sterile hemolet to prick the cleaned finger. Discard the lancet. **Never reuse a lancet.**
q. Place a drop of blood on the Anti-Rh serum.
r. Mix thoroughly with a clean toothpick. **Break and discard the toothpick.**
s. Tilt the slide back and forth gently for about 2 minutes.

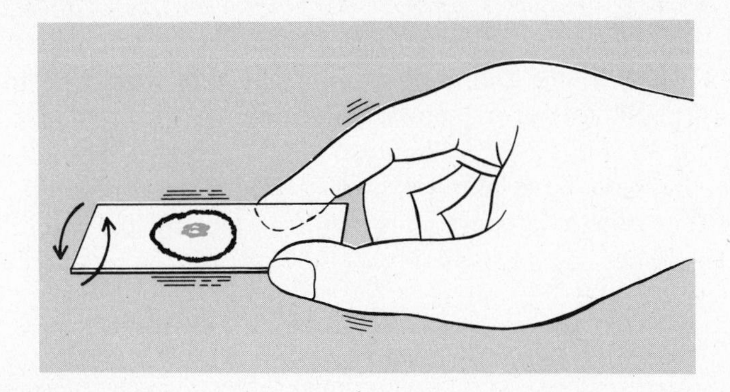

t. Examine the mixture for clotting, as before.

7. Is your partner's blood Rh positive or negative?

u. Have your lab partner test your blood.

8. Is your blood Rh positive or negative?

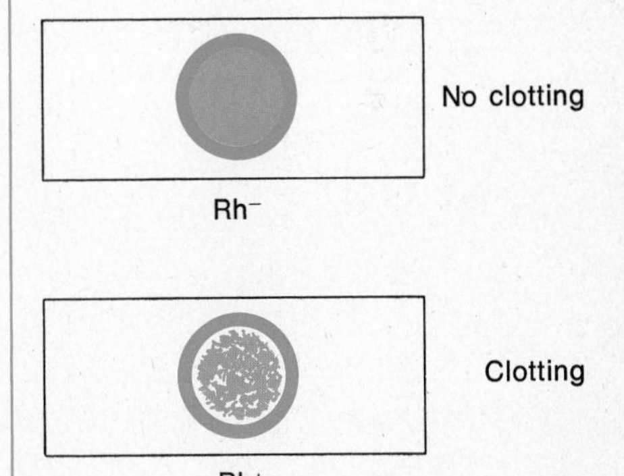

B. Time to Concentrate

Now you can investigate how solutions are made. Most solutions used in the lab are *molar solutions.* You are already familiar with moles. When you dissolve 1 mole of a substance in water to get a final volume of 1 liter of solution, you have a *1 molar solution.*

Remember that 1 mole of any substance is its molecular weight expressed in grams. For example, the molecular weight of sodium chloride is 58.

Atomic weight

Na 23
Cl 35
58 = molecular weight

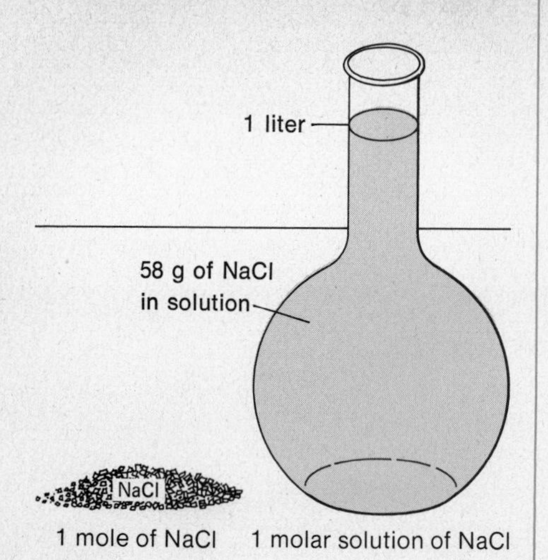

Therefore, 1 mole of sodium chloride is 58 grams. A 1 molar solution of sodium chloride is 58 grams dissolved in water to make 1 liter of solution.

9. How many grams of sodium chloride would be in 1 liter of a 0.5 molar solution?

10. How many grams of sodium chloride would be in 1 liter of a 2 molar solution?

11. How many grams of sodium chloride would be in 0.5 liters (500 ml) of a 1 molar solution?

Now that you have learned how solutions are prepared, you should be able to prepare your own solutions for the rest of this Unit.

INVESTIGATION

2

Time for Your 15,000 km Check-Up

Have you thought about how chemistry is related to your car? Chemists have a great many jobs in the automobile industry. They have to answer many questions like, "Which makes a better car body—steel or plastic?" "Does an aluminum engine have any advantages over a steel engine?" "How do you make a rubber tire last longer?" "What makes a more efficient automobile battery?"

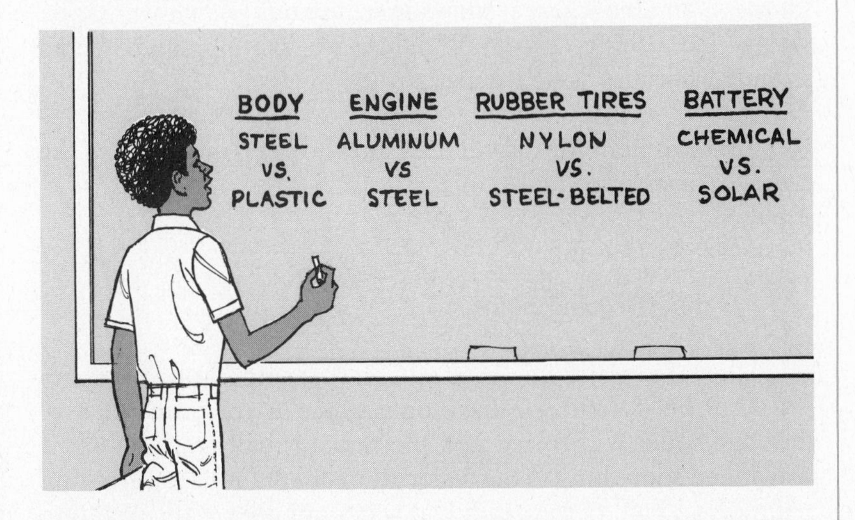

When checking the antifreeze in your car, what is the mechanic really doing? What principles of chemistry is he using?

You have already learned enough chemistry to answer some of these questions. You have also learned how to think problems through. Put your knowledge to work in this investigation.

Compix/UPI

A. Baby, It's Cold Outside

Living in an area where it snows in winter has great advantages: skiing, ice fishing, snowball fights, ice skating, and—being stranded because your car freezes up.

Here is a problem for the chemist: how to keep the water in the radiator from freezing.

1. Define freezing.
2. Define freezing point.

You know that water freezes at 0°C. If the temperature outdoors is 0°C or below and you have only water in your car radiator, then the water will freeze and the radiator will crack. What if you added something to the water? Pure methanol has a freezing point of −97°C. If methanol and water are mixed, the resulting solution will have a freezing point less than 0°C. That would keep your radiator water from freezing in most places. But the solution of methanol in water also has a lower boiling point than pure water. When the car engine is running, the temperature of the liquid in the radiator can go as high as 95°C. At this temperature the water-methanol solution would boil off.

So alcohol is not the ideal *antifreeze.* Chemists have discovered that another compound, ethylene glycol, does the trick. Here's how:

What You Need

Ethylene glycol (CH_2OHCH_2OH)
Water (H_2O)
Cooling bath
Graduated cylinder, 10 ml
Test tubes, 4
Thermometer
Flashlight
Graph paper
Safety goggles
Test tube rack
Wax pencil

What to Do

a. Label four test tubes: 10%, 20%, 30%, and 40%.
b. In the tube marked 10%, mix 9 ml of water and 1 ml of ethylene glycol.
c. In the tube marked 20%, mix 8 ml of water and 2 ml of ethylene glycol.

3. What mixtures would you put into the 30% and 40% tubes?

d. Prepare the 30% and 40% tubes.

Your teacher has a special cooling bath set up for you. It uses dry ice (solid CO_2) and alcohol, and it will cool down to $-72°C$. You will place each of your tubes in this bath, in turn, and determine at what temperature its contents begin to freeze. The first sign of a solid beginning to form is called the "slush temperature." You will then determine the temperature at which the contents become completely solid.

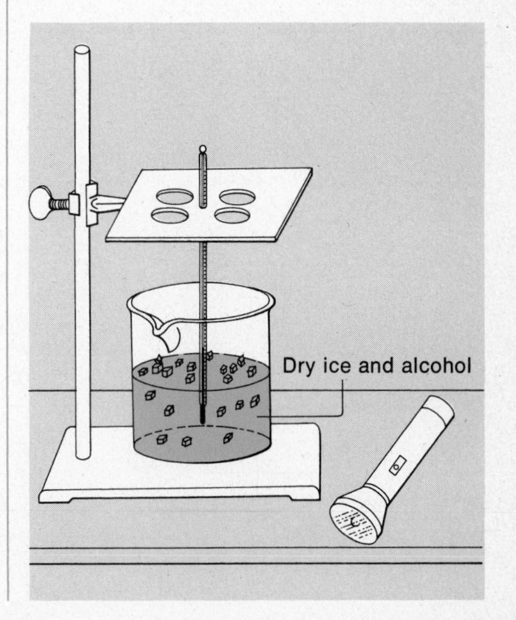

e. Place the 10% tube in the cooling bath, following your teacher's instructions. With the aid of a flashlight, determine the slush temperature.

4. Record the slush temperature.

f. Allow the mixture to become completely solid. Note the temperature.

5. Record this as the freezing point.

g. Repeat steps **e** and **f** with the other three tubes.

6. Record your results in a table.

7. Prepare a graph of your results.

8. As the concentration of ethylene glycol increases, what happens to the freezing point?

9. What would be the slush temperature of 25% ethylene glycol in water?

10. What would be the freezing point of 50% ethylene glycol in water?

11. Write an experimental procedure to show how you would determine the boiling point of your mixtures.

B. I'd Rather Remain Neutral

You will now look at another aspect of automobile chemistry—the testing of motor oil. It is estimated that over 2,000,000,000 liters of motor oils are used each year in the United States. Many chemists are hired to run tests on and study the properties of oil.

There are two principal reasons for analyzing used oils. The first is to determine the oil's suitability as a lubricant. The second is to show up problems that might be occurring in the engine. Oil analysis can often spot trouble before it starts. Analysis can point to conditions in the engine which cannot be seen. For example, the presence of water in used oil can indicate a leaking seal or a cracked engine block.

When oil is heated due to the friction produced in an engine, it breaks down. There are several signs that a lubricant is no longer good, such as a sharp smell or an increase or decrease in thickness. One product of the chemical breakdown is an organic acid called carbolic acid. Other acids formed are inorganic acids.

The extent of oil breakdown can be measured by the neutralization number test. The *neutralization number* is defined as the number of milligrams of potassium hydroxide required to neutralize the acids present in one gram of oil. The more acid found in the oil, the higher is its neutralization number.

You're going to make your own solution in this investigation; so you might want to review your work on solutions in Investigation 1. You will prepare 100 ml of 0.1 *M* KOH in a volumetric flask.

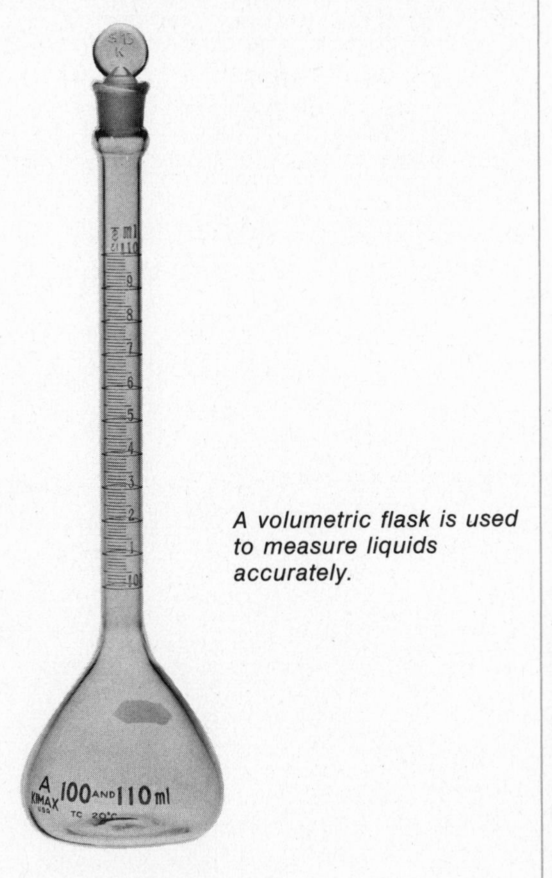

A volumetric flask is used to measure liquids accurately.

Kimble Division of Owens-Illinois, Inc.

You also need some oil. So, drain some out of your crankcase and bring it into the laboratory.

What You Need

Benzene (C_6H_6)
Isopropyl alcohol (C_3H_7OH)
Methyl orange indicator
Oil sample, unused
Oil sample, used
Potassium hydroxide (KOH)
Water (H_2O), distilled

Buret
Flask, 250 ml
Graduated cylinder, 25 ml
Stirring rod
Volumetric flask, 100 ml

Balance
Buret clamp
Ring stand
Safety goggles
Spatula
Weighing paper

What to Do

12. How many grams of KOH do you need to make up 1 liter of a 1.0 *M* solution?

13. How many grams of KOH do you need to make up 1 liter of a 0.1 *M* solution?

14. How many grams of KOH do you need to make up 100 ml (0.1 liter) of a 0.1 *M* solution?

a. Get the amount of KOH that you calculated in question 14. Place it in a 100 ml volumetric flask.

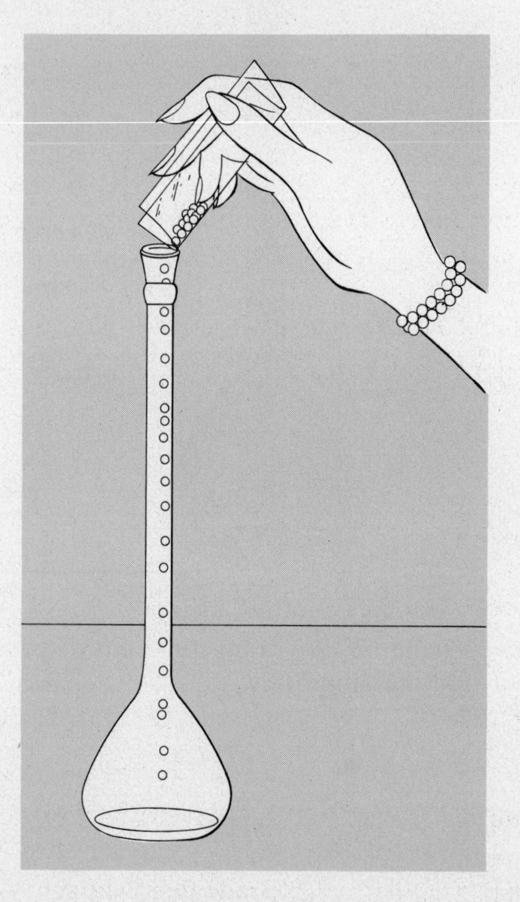

b. Add about 25 ml of distilled water to the flask. Stopper and shake well.

c. Carefully add distilled water to the level of the line in the neck of the flask. There are now exactly 100 ml of solution in the flask.

d. Stopper the flask. Shake it well.

You have now prepared 100 ml of a 0.1 M KOH solution.

e. Get 1 gram of used or unused oil in a 250 ml flask.
f. Add 25 ml of benzene and 25 ml of isopropyl alcohol. Stir well. **Caution: benzene vapors are harmful.**
g. Add a few drops of methyl orange indicator. Mix well.
h. Fill a buret with 0.1 M KOH solution.

15. Record the initial reading of the buret.

i. Titrate the sample. The end point has been reached when you get a light yellow color.

16. Record the final reading.

17. How much 0.1 M KOH did it take to neutralize the oil?

The neutralization number is defined as the number of milligrams of potassium hydroxide required to neutralize a one gram sample of oil.

As you calculated in question 13, there are 5.6 grams of KOH in 1000 ml (1 liter) of a 0.1 M solution.

5.6 g in 1000 ml
or 5600 mg in 1000 ml

Therefore there are 5.6 mg in 1 ml.

This gives you your factor (5.6) for determining the neutralization number.

j. Multiply the number of milliliters of KOH used to neutralize the oil by your factor.

$$(\text{ml of oil used}) \times 5.6 = \text{neutralization number}$$

18. Record the neutralization number of your sample.

k. Repeat the procedure with other oil samples.

19. Prepare a report on the testing of oils.

Now slide along to the next investigation.

INVESTIGATION 3

So You Want To Be in Pictures

Courtesy of Eastman Kodak Co.

Firms such as Kodak, Polaroid, Ansco, Agfa-Gevaert, and Fuji employ hundreds of thousands of people to make and process films. Photography is a major business in this country.

Photography also plays an important part in many other industries. In scientific work, photography can be used to help keep accurate records of investigations. It can show all the details of a laboratory set-up.

Even if you don't want photography as a profession, it makes a great hobby. To make first-rate pictures, however, requires very careful technique. Here goes!

A. Say Cheese, Please

What You Need

Developing solution
Fixing solution
Stop bath solution
Clothespin or clip
Developing bag, black
Developing tank (tank, cover, film rack, thermometer, and stirrer)
Film, roll, exposed black and white
Safety goggles

What to Do

a. Examine the materials shown in the photograph. They are your darkroom, and you will use them to develop your film.

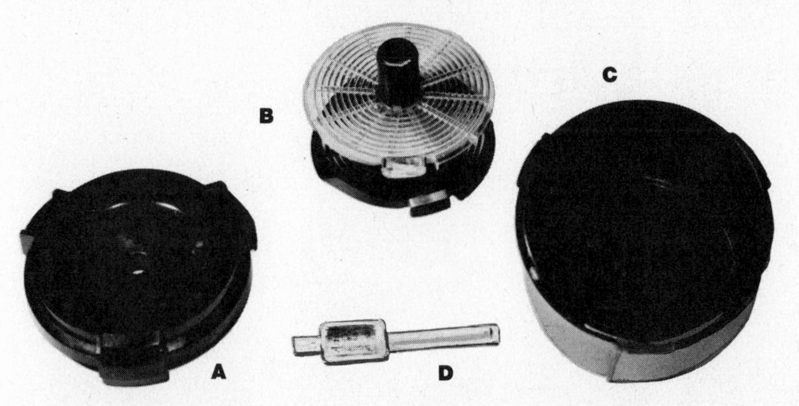

A is the cover to the developing tank. B is the film rack which fits in the developing tank. C is the developing tank. D is a thermometer and stirrer which fits into the top of the tank.

Exposed film is placed in the developing bag along with the developing tank. The trick now is to close the bag and transfer the film to the tank without looking. This is how it's done:

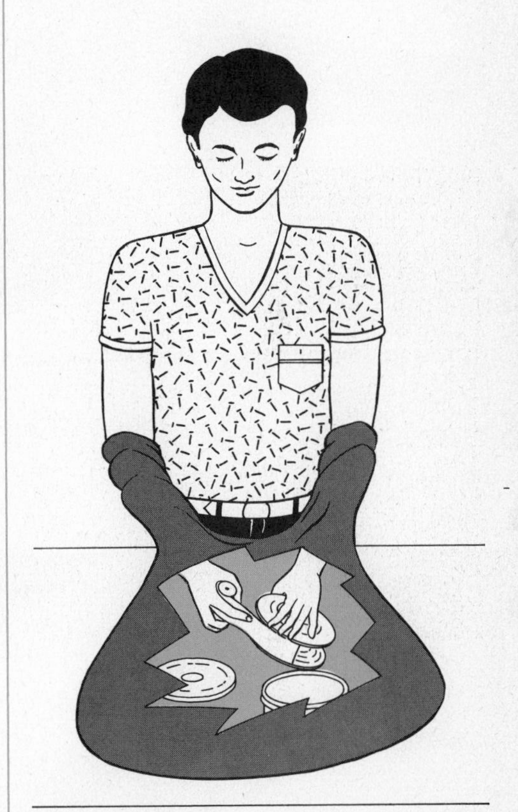

1. Why must you handle the exposed film inside a black bag?

b. Read through the rest of this investigation before you begin to develop your film. Gather all of the materials and solutions necessary before you begin.

c. Place the parts of the developing tank and the **wrapped** roll of exposed film in the black developing bag. Close the bag to all light.

d. In the light-tight developing bag, unwrap the exposed roll of film. The film is the smooth inner strip. You can find it by touch. **Handle the film by the edges only.**

This photograph shows the right *way to handle film.*

This photograph shows the wrong *way to handle film. It results in fingerprints on your negatives and prints.*

e. Thread the film into the film rack.

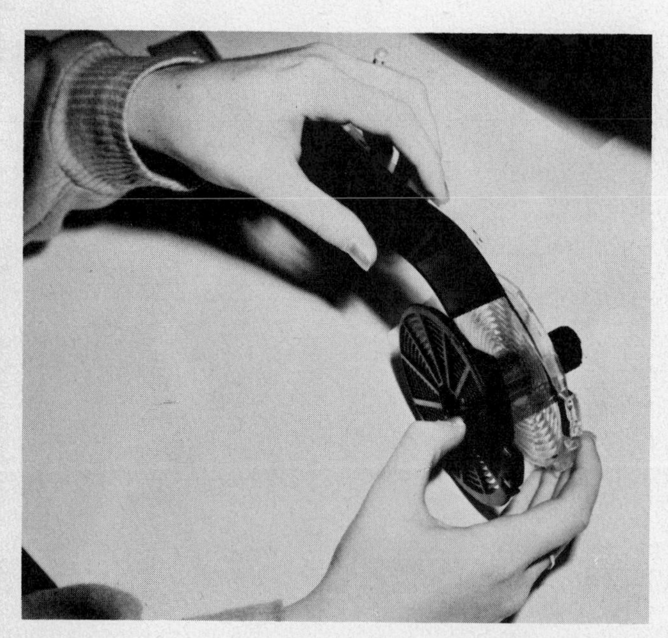

Threading the film into the film rack. Remember that you will be doing this inside *the developing bag.*

f. Place the threaded film rack into the developing tank.

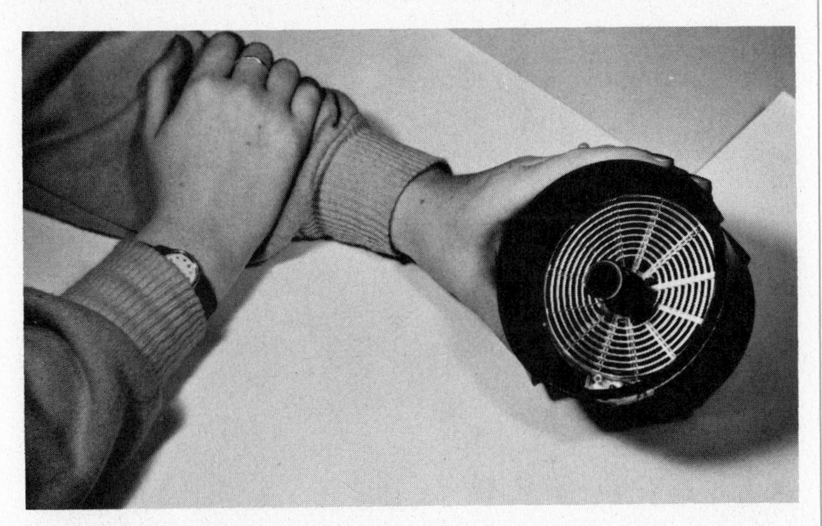

The film rack in place in the developing tank.

g. Place the cover tightly on the developing tank.

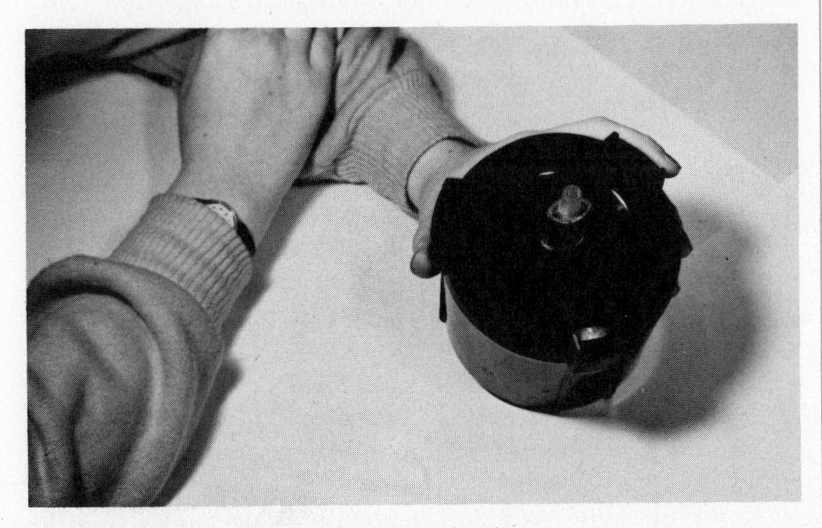

The developing tank with the cover in place.

h. Double check to be sure the cover is on tightly. Then remove the closed developing tank from the black developing bag.

i. Place the loaded tank in the sink. Carefully pour developing solution into the tank through the center opening in the cover. Pour in enough so that it just overflows.

j. Gently turn the knob on the cover of the tank. Keep turning for 6–8 minutes.

2. What does the developing solution do to the film?

k. Carefully pour the developing solution back into its original bottle. **Do not open the developing tank.** The developing solution may be used again.

l. Immediately pour the stop bath solution into the developing tank using the same procedure as in step **i.** Gently turn the knob on the cover of the tank for 1 minute.

This stops the developing process.

3. What happens if you delay adding stop bath after step **k?**

m. Pour the stop bath solution into its original bottle. This solution may also be used again.

n. Carefully pour the fixing solution into the tank using the same procedure as in step **i.** Gently turn the knob on the cover of the tank for 6–7 minutes.

4. What do you think is the purpose of the fixing solution?

o. Pour the fixing solution into its original bottle. This solution may be used again.

p. Remove the cover from the developing tank. Place the open tank containing the film under a tap of **gently running** cold water. Leave it there for about 10 minutes.

5. Why should the film be washed?

q. Carefully handling the wet film by the edges, hang it up to dry. Use a clothespin or clip to hold it.

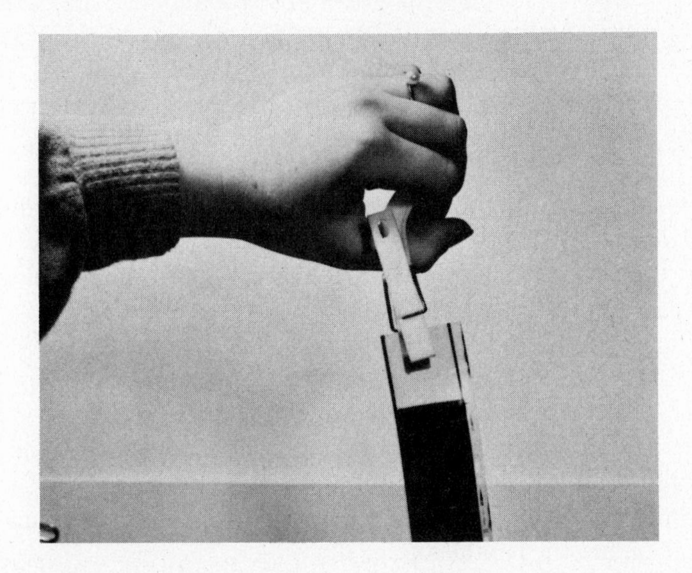

You have developed a roll of film, and you now have a set of negatives. For a look into the chemistry behind the films, read on.

B. Hollywood, Chemistry, and You

The whole business of photography is possible because of the halogen compounds of silver. Silver chloride ($AgCl$), silver bromide ($AgBr$), and silver iodide (AgI) are sensitive to light. These compounds are called *halides.*

Modern photography is based on the fact that silver halide crystals on film begin to change to metallic silver after a brief exposure to light. The process is completed when the film is placed in a weak reducing solution called a *developer.*

6. What does the term reducing, or reduction, mean in chemistry?

Kodak

In the developing process, the silver gains an electron.

$$AgCl + e^- \longrightarrow Ag + Cl^-$$

7. Write an equation for the reduction of silver bromide.

When you want to take a picture, the first thing you do is load the camera with film. Photographic film is a strip of plastic covered on one side by an *emulsion.* The emulsion is a thin layer of a jellylike substance with grains of silver halide (usually silver bromide) embedded in it. When the film is exposed to light, the silver halide crystals start to change to grains of metallic silver. A pattern of light and dark areas is formed. This process is speeded up by putting the film into the developer. A picture, or *image,* is produced on the film by grains of metallic silver. The silver grains are so small that they appear black. The image is called a *negative* at this stage. The dark areas in the original subject are light areas on the negative. The light areas in the original subject are dark on the negative.

Obviously there is something more to be done before you can put the pictures in your scrapbook. The next step is to make some *contact prints.* Contact prints, which are finished photographs, are made by placing a negative in contact with photographic paper. Light shining through the negative forms an image on another emulsion on the paper. The negative controls the amount of light that can pass through it on to the emulsion. Light areas on the negative allow more light to pass through than dark areas. Dark areas in the negative stop light from reaching parts of the emulsion. The pattern of light and dark areas is reversed on the emulsion. This emulsion is called a *positive,* and it is your photograph.

When you place this paper in developer, stop bath, and fixer, you wind up with a finished print. This procedure is a little too complicated for the classroom. You need a completely dark room. If you wish to make some prints as an extra project, your teacher has some instructions for this. You might be able to use a darkroom equipped with an *enlarger.* Then you can make enlargements from your negatives.

Once you have mastered the skills of processing film and making prints, you can do a number of useful things. For example, you can make prints to illustrate reports; or you can make your own greeting cards; or you can decorate your classroom!

INVESTIGATION 4

Mine Ore Yours?

The ancients knew about metals. Gold was used in ornaments as early as 3500 B.C. Iron and copper tools go back to before 3000 B.C. Silver was used as early as 2400 B.C.

You said this necklace was the latest fashion!

Metallurgy is the study of metals. Metals are part of our daily lives. Look around and note all of the things that are made of metals. Chemistry plays an important part in making metals useful to us.

Many different metals are found in many forms within the earth's crust. While in the crust they are known as *ores.* An ore is a combination of a metal with other elements, either as a compound or as a mixture. One field of metallurgy studies the separation of metals from their ores. This is known as *extractive metallurgy.* The aim of extractive metallurgy is to refine a metal to a pure state.

Physical metallurgy studies how to convert refined metals to human use. It involves combining different pure metals into mixtures called *alloys.* Many things you take for granted are really alloys, like brass, bronze, and steel. Even a penny is a mixture, or alloy, of copper and zinc. Pure metals, such as silver, copper, iron, and gold, are rather weak. They have little or no strength by themselves. Therefore alloys are made. Table 1 lists some common alloys.

TABLE 1: Common Alloys

Alloy	*Metals*
Brass	copper and zinc
Bronze	copper and tin
Sterling silver	silver and copper
Steel	iron and carbon
Stainless steel	iron, chromium, nickel, and carbon

The best way to become familiar with alloys is to make one yourself.

A. Ship Alloy

What You Need

Bismuth (Bi)
Cadmium (Cd)
Lead (Pb)
Tin (Sn)

Beaker, 250 ml
Evaporating dish
Stirring rod

Asbestos pad
Balance
Burner
Matches
Paper towel
Ring
Ring stand
Safety goggles
Tongs

What to Do

a. Put 5 grams of Bi and 1.3 grams of Cd in an evaporating dish.
b. Heat gently over a burner until the metals melt. Stir continuously.
c. Add 1.3 grams of Sn and 2.5 grams of Pb. Continue stirring until all metals have melted.
d. Remove the evaporating dish from the burner and pour the melted alloy onto an asbestos pad.
e. When the alloy has cooled and solidified, examine it.

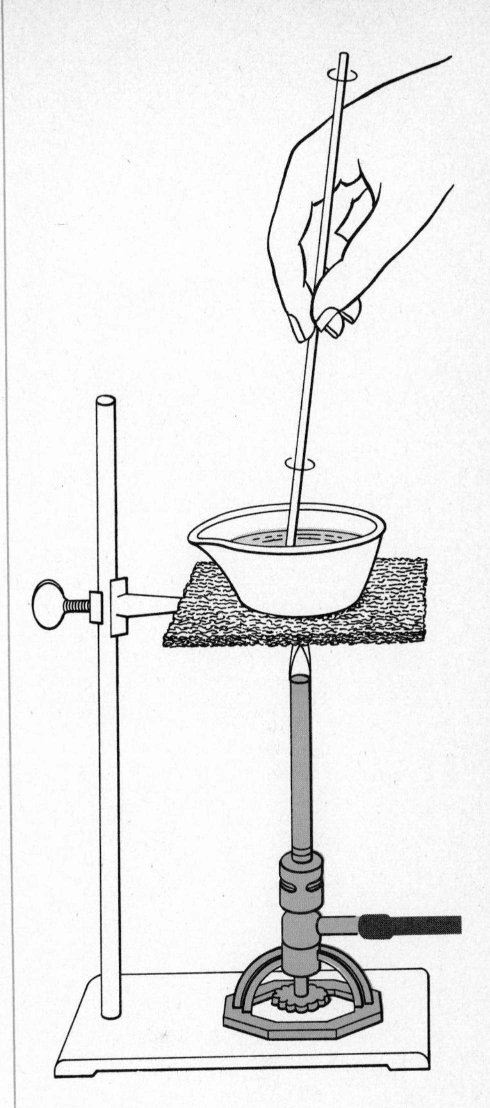

The alloy you have just made is called Wood's metal.

f. Put your Wood's metal in a beaker of water. Slowly heat it and observe.

1. What happens?

2. Does your alloy have a high or low melting point?

3. How would you determine the melting point of your alloy? Try it.

Alloys with low melting points are very useful in electric fuses and automatic sprinkler valves. Heat from an overloaded circuit or fire causes the alloy to melt. In electric fuses, this melting breaks the circuit. In a sprinkler system, the low melting point alloy acts as a plug. When it melts, the water is released.

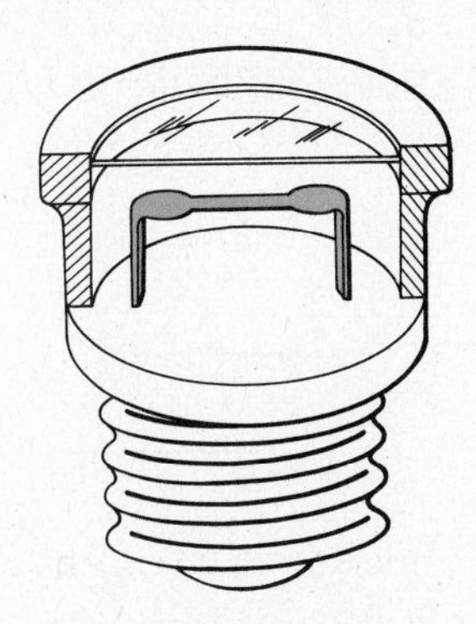

Bismuth is used to lower the melting point of alloys. Cadmium enables the alloy to stay pliable after it has cooled.

4. Design an alloy, using Bi, Cd, Pb, and Sn, that would have a higher melting point and be more brittle than the alloy you made.

Many thousands of people make their living in the metal industries.

Even the old village blacksmith was a metallurgist. He heated and hammered steel into horseshoes. You can get an idea of how he worked by doing the next activity.

B. Under the Spreading Chestnut Tree

When metals are heated to temperatures high enough to change their hardness, we say that they are *heat treated.*

What You Need

Nails, large
(16-penny; $6\frac{1}{2}$ cm), 3

Beaker,400 ml

Anvil
Asbestos pad
Burner
Hammer
Matches
Ring stand
Safety goggles
Tongs

What to Do

a. Place one nail on the anvil.

b. Place the second nail on the first in a crossed position.
c. Grip the end of the top nail tightly with your tongs.
d. With your hammer, strike the top nail with a sharp, hard blow, right at the junction of the two nails.

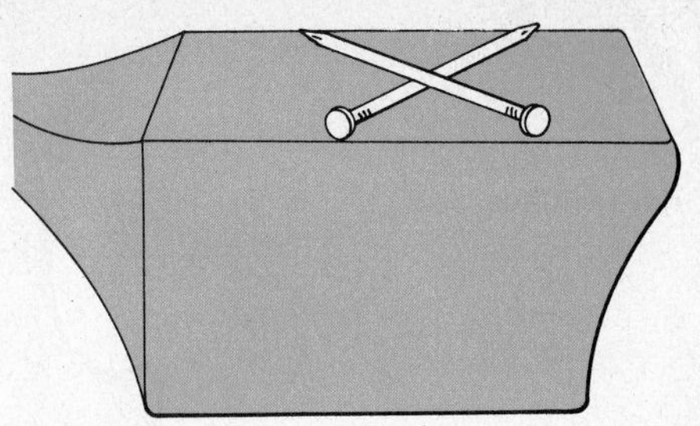

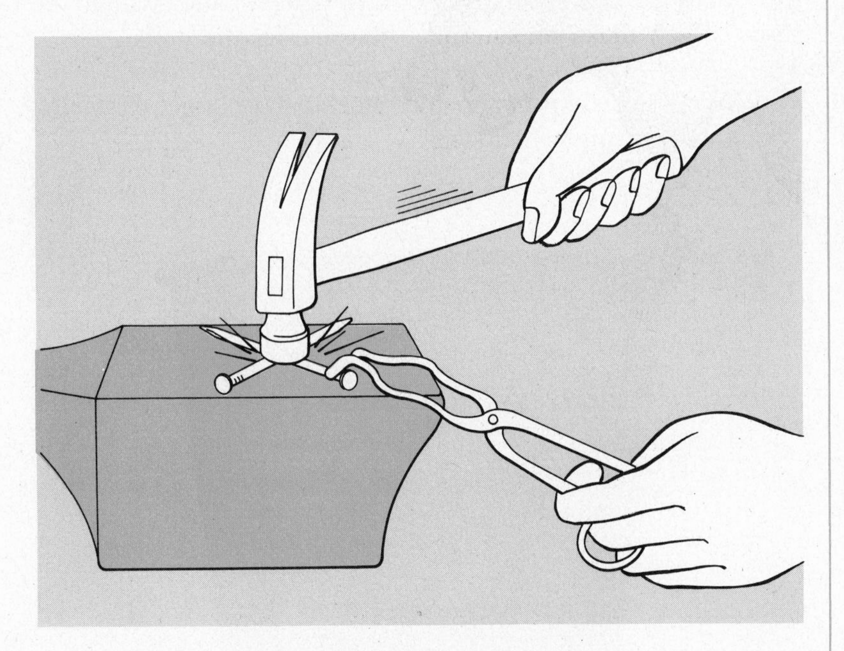

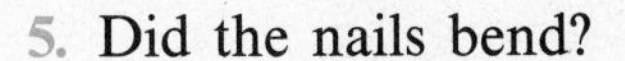

5. Did the nails bend?

e. Using your tongs, heat the center of the top nail to a bright red for 5 minutes.
f. Slowly raise the nail in the flame until it cools. Then let it cool to room temperature in the air. The cooling can be done by resting the nail on an asbestos pad.
g. Repeat steps **b** through **d.**

6. What happened to the heated nail?

Heating and slow cooling of steel in air will *anneal* or soften it. Annealing prevents brittleness.

h. Using a third nail, repeat step **e.**
i. Quickly remove the nail from the heat and immediately submerge it in a beaker filled with water.
j. Repeat steps **b** through **d** with the heated nail.

7. What happened to the nail?

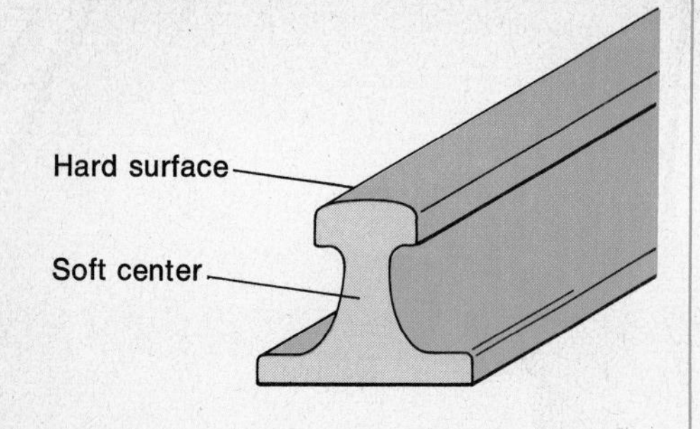

Heating and quickly cooling steel in this way is called *quenching*. Steel which needs a hard or tough surface and a soft center is quenched.

All metals have a grain structure. In general, the smaller the grains which make up the metal, the harder the metal.

8. Would you expect annealed steel to have larger or smaller grains than quenched steel?

C. Zinc or Zwim

What You Need

Galvanized steel strip, 1 cm × 5 cm
Hydrochloric acid (HCl), 0.1 *M*
Beaker, 50 ml
Graduated cylinder, 100 ml
Ruler, metric
Safety goggles
Tongs

What to Do

a. Place 40 ml of HC1 in the beaker.
b. With the tongs, grip the galvanized steel strip at one end.
c. Dip the strip into the acid, leaving it there for about 10 seconds.

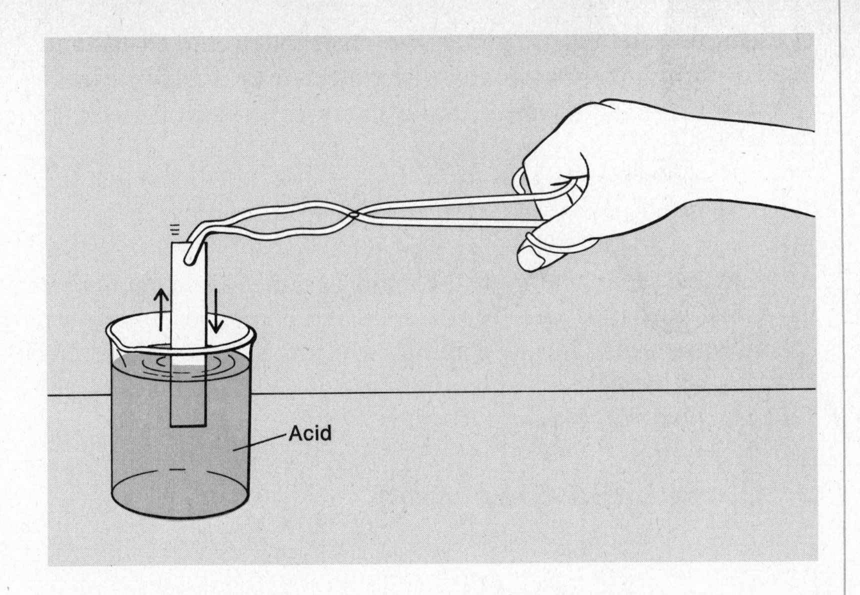

d. Remove the strip and wash it in running water.

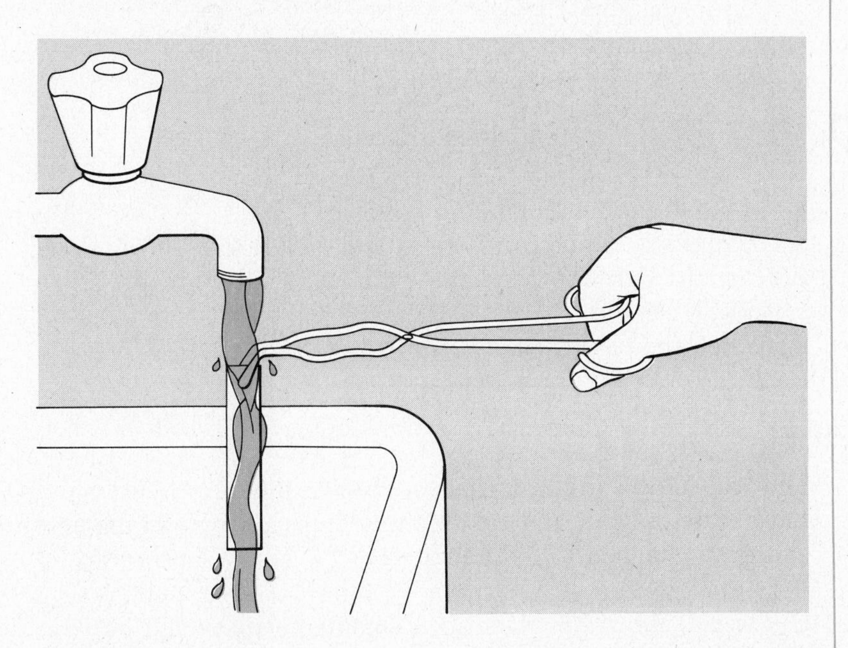

e. Repeat steps **b** through **d** until a definite grain structure appears.

9. Draw a picture of the grain structure.

10. Measure several grains and record the measurements in your notebook.

Galvanized steel is steel which has been rolled into sheets and coated with zinc. Zinc protects the steel from rusting. Steel's crystal structure is so small that a microscope must be used to see it.

11. Is steel harder or softer than zinc?

As you have seen, heating and cooling of a metal can change its hardness. A change in hardness indicates a change in grain or crystal structure. During cooling, when the crystal structure changes, an increase in the temperature of the metal takes place. This is called *recalescence.* Try it.

D. Glow, Little Glowworm

What You Need

Burner
Matches
Razor blade, single edge
Safety goggles
Tongs

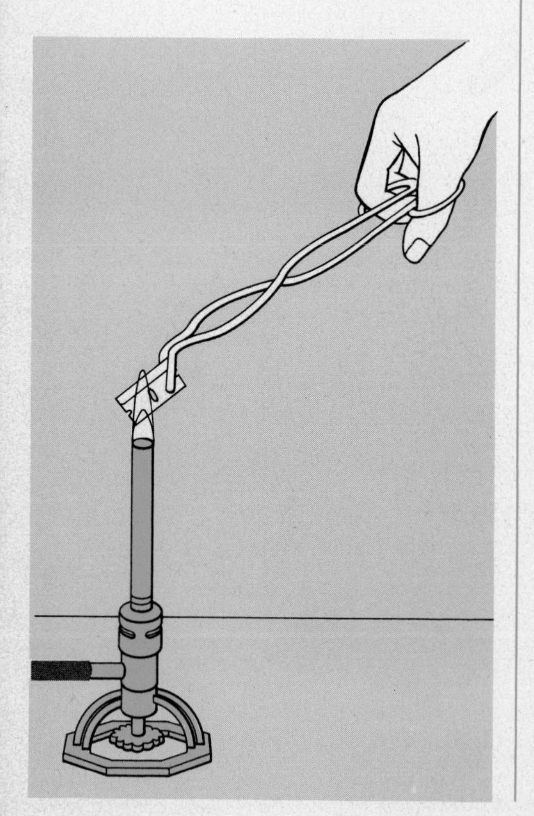

What to Do

a. Make the room as dark as possible.
b. Hold a razor blade with tongs and heat it to a glowing red.
c. Turn the burner off and allow the blade to cool in air. Try to hold it steady.
d. Watch the color of the blade closely as it cools.

12. Record your observations.

Sudden changes in the crystal structure of the razor blade cause the release of heat. You are able to see this, as some of the heat energy is changed to light energy.

e. Discard the razor blade in the container provided by your teacher.

The information you have learned about metals can be used in a number of practical ways. You now understand why a metal has certain properties. You can make metal objects by utilizing the various properties of different metals. A knowledge of metals plays an important part in your everyday life.

Cookbook Chemistry

INVESTIGATION 5

You have already had some experience with the chemistry of food, in Unit 4; but Food Technology is a very large field. Consumer needs have been changing—often dramatically—in recent years. People now are looking for food products that are more versatile, easier to prepare, and more nutritious. Science must continue to develop new products that will meet these needs.

General Foods and The Kraft Kitchens

Chemists in the food industry have the responsibility for maintaining quality control. They also develop new methods to prepare and preserve foods. Some of these new methods include freeze drying, dehydration, and radiation. Chemists must also test for the different constituents in foods.

A. The Caffeine Kick

Caffeine is an "upper." In medicine it is used as a cardiac and respiratory stimulant. But too much can cause nausea, nervousness, and insomnia. There is a lot of concern about caffeine in coffee. But what about other common drinks? Do you think there's caffeine in them, too?

What You Need

Chloroform ($CHCl_3$)	Beakers, 250 ml, 2	Asbestos pad
Cola	Beaker, 400 ml	Balance
Lead(II) acetate [$Pb(C_2H_3O_2)_2$], 0.5 *M* solution	Funnel	Clay triangle
Tea leaves	Graduated cylinder, 100 ml	Filter paper
	Separatory funnel	Matches
	Watch glass	Ring
		Ring stand
		Safety goggles
		Weighing paper

What to Do

a. Put 25 grams of tea leaves in a 400 ml beaker. Add 250 ml of water.
b. Boil 10 minutes.
c. Filter the liquid into a clean beaker.
d. Add 10 ml of $Pb(C_2H_3O_2)_2$ solution.

Lead acetate precipitates out a substance called tannic acid or tannin. Tannin is used in making paper, ink, and leather. In medicine it is used as an antidote for poisons and as a dressing for burns.

e. Filter the solution again into a clean beaker. Allow the filtrate to cool.
f. When cool, pour the filtrate into a separatory funnel. The stopcock of the funnel must be closed.
g. Work in the hood. Add 10 ml of chloroform to the separatory funnel. **Caution: chloroform is flammable and poisonous.**
h. Hold the stopper tightly in place and turn the funnel upside down. Open the stopcock and shake gently for 2 minutes.
i. Close the stopcock and turn the funnel right side up. Carefully loosen the stopper.
j. Place the separatory funnel in the ring stand. Let the liquid settle.

1. Why does the liquid settle into two layers?

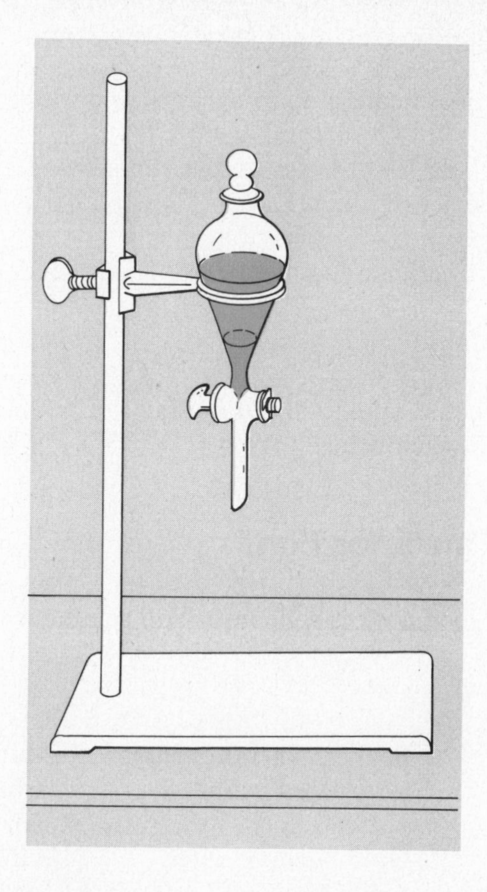

2. One layer in the funnel is water. What is the other layer?

k. Carefully open the stopcock and draw off the bottom layer into a watch glass.

The bottom layer is the chloroform. It contains the dissolved caffeine. The chloroform *extracts* the caffeine from the tea.

3. Is caffeine more soluble in water or in chloroform? How do you know?

l. Set the sample aside under the hood to evaporate to dryness. **Caution: Keep chloroform away from flames; it is highly combustible.**

The white crystal which remains is caffeine.

Photo courtesy of Paul Child

m. Clean the separatory funnel.
n. Get 50 ml of cola in a clean 250 ml beaker.
o. Heat to boiling.

4. What does heating do to the carbon dioxide in the cola?

p. Pour the cola into the separatory funnel.
q. Repeat steps **g** through **l.**

5. Is there caffeine in cola?

Try other drinks.

6. How would you go about finding the actual amounts of caffeine in each drink? (Try it.)

B. This One's for Fun

In reality, a good cook is a chemist. You can judge for yourself.

What You Need

Glucose ($C_6H_{12}O_6$), 3 *M* solution
4-hydroxy-3-methoxybenzaldehyde
Protein pellets
Sodium bicarbonate ($NaHCO_3$)
Soybean oil, hydrogenated
Sucrose ($C_{12}H_{22}O_{11}$)

Beaker, 400 ml or larger
Graduated cylinder, 25 ml
Graduated cylinder, 100 ml
Stirring rod
Thermometer, 0–200°C

Aluminum foil
Asbestos pad
Balance
Burner
Matches
Ring
Ring stand
Safety goggles
Tongs
Weighing paper

What to Do

Make certain all glassware and equipment are perfectly clean.

a. Lightly coat a 35-cm square of aluminum foil with hydrogenated soybean oil. Set it aside.
b. Place 240 grams of $C_6H_{12}O_6$ solution in a large beaker.
c. Add 500 grams of $C_{12}H_{22}O_{11}$ and 120 ml of water.
d. Heat carefully over a burner, while stirring, until the mixture reaches 110°C.

e. Add 340 grams of protein pellets and stir. Use tongs to hold the beaker.

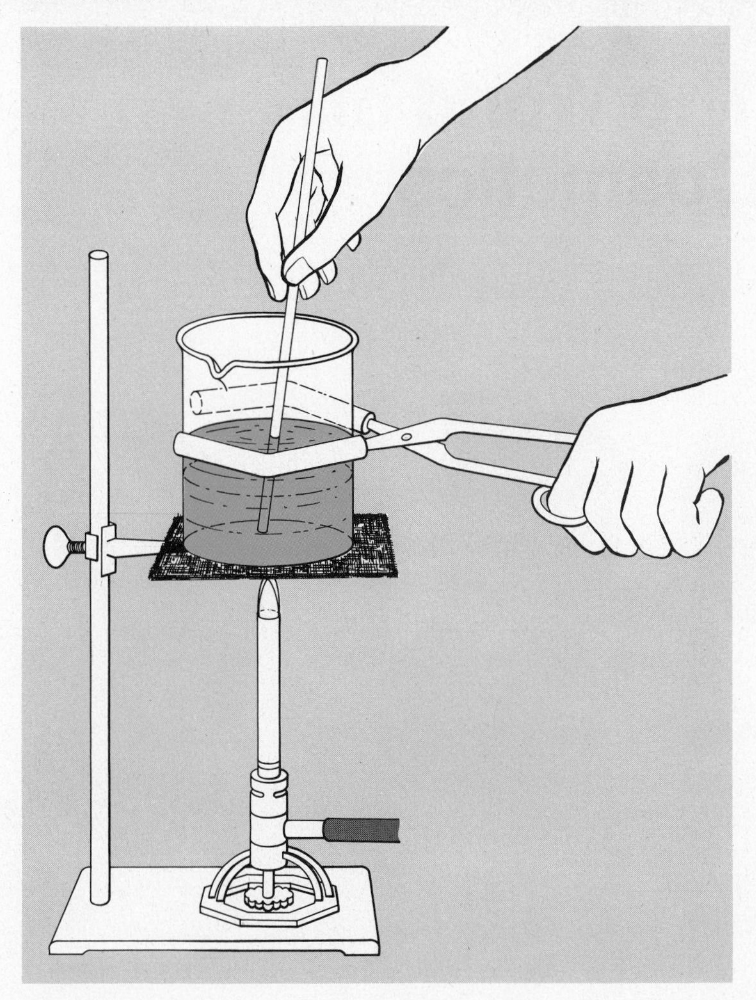

f. Heat the mixture to 150°C and remove the beaker from the burner.

g. Add to the beaker, while stirring, 10 ml of soybean oil, 2 ml of 4-hydroxy-3-methoxy-benzaldehyde, and 5 grams of $NaHCO_3$.

h. Pour the mixture onto the foil. Spread as thinly as possible. Clean out the beaker with a stirring rod.

i. Allow the mixture to cool.

7. Name your mixture.

j. Now for the best part—eat and enjoy!

INVESTIGATION 6

The World of Cosmetics

Lida Moser/dpi Michael Heron/Monkmeyer

Vivienne/dpi Georges Marry/Rapho United Nations (from Belgian Govt. Information Cent

Each civilization or culture has had its own ideas about what is beautiful.

Cosmetics are used to enhance beauty. Cosmetics are as old as civilization. The first recorded use is found in the Bible, which describes "anointing the head and body with oil." Perfumes, rouges, lipsticks, hair dyes, creams, and powders were used thousands of years ago. The most famous person of ancient times associated with cosmetics was the Queen of Egypt, Cleopatra.

During ancient times, it was the physician's job to mix and supply cosmetics. Later, tradesmen took over this job. Today, the cosmetics industry has sales of billions of dollars per year. Both men and women use cosmetics. Almost as many cosmetics are sold for men as for women.

Cosmetics have become among the most heavily advertised products in the world. Pick up almost any popular magazine or newspaper; watch television; or look at billboards. You will see cosmetics being advertised.

Today cosmetics can make you smell nice or not smell; add color to your skin or take color away; put oils into your skin or take oils away; give your hair body or take body away from your hair.

The actual cost of most cosmetics is relatively low. The 99¢ jar of face cream at the local discount store can cost $10.00 or more at "Ye Beauty Boutique." What is the difference? For the answer, we must look at the packaging and advertising, and also the income of the customer. Some products claim to do great or wondrous things because of rare ingredients. Medical evidence shows, however, that there is little absorption of chemicals through the skin. This means that a skin cream without an exotic ingredient might do the same job as one with. The one who really benefits from these rare ingredients is the manufacturer.

There are no magic cosmetics that can perform miracles. Cosmetics can be used to bring out or cover up a person's physical features. But we are more or less stuck with the faces and bodies we were born with.

FUNKY WINKERBEAN By Tom Batiuk

Funky Winkerbean by Tom Batiuk, Courtesy of Field Newspaper Syndicate

In this investigation, you will make some familiar cosmetics. You will make a hand lotion, shaving lotion, and bubble bath. In the process, you will have an opportunity to review quite a bit of chemistry.

A. Softer Than Soft

Almost everyone washes his or her hands several times a day. This removes the natural oils from the skin. There is a booming business in making products to put the oil back into your skin. Here's how to make your own special brand.

What You Need

Boric acid (H_3BO_3)
Glycerine ($C_3H_8O_3$)
Lanolin
Perfume (optional)
Petroleum jelly

Beaker, 100 ml
Bottle, wide mouth
Graduated cylinder, 10 ml
Stirring rod

Asbestos pad
Balance
Burner
Matches
Safety goggles
Spatula
Stopper for bottle, solid
Weighing paper

What to Do

a. Clean all glassware thoroughly.
b. Get 2 grams of boric acid and 12 grams of glycerine.
c. Mix them together in a clean 100 ml beaker.
d. Heat the mixture gently over a burner until the boric acid dissolves. Stir continuously.
e. Allow the mixture to cool. Add 12 grams of lanolin and 28 grams of petroleum jelly. Stir.
f. If desired, 5 ml of perfume may be added.
g. Stir the mixture and pour into a clean bottle. Stopper.

This mixture should make a fairly good hand lotion. A good hand lotion should replace the natural oils in the skin. It should leave the skin feeling soft and smooth and protect it against chapping during cold weather.

Hand lotions usually have several ingredients. Table 1 lists the typical ingredients.

TABLE 1: Ingredients of Hand Lotion

Ingredient	*Purpose*
Petroleum jelly	non-reactive base
Lanolin and glycerine	replace lost oils
Boric acid	antiseptic
Perfume	aroma
Emulsifying agents	allow oil and water to mix

1. Is the hand lotion you just prepared a compound or a mixture? Why?

B. Something for the Male Contingent

What You Need

Glycerine ($C_3H_8O_3$)
Isopropyl alcohol (C_3H_7OH)
Menthol ($C_{10}H_{19}OH$)
Water (H_2O)

Beaker, 250 ml
Bottle, wide mouth
Graduated cylinder, 100 ml
Stirring rod

Balance
Safety goggles
Spatula
Stopper for bottle, solid
Weighing paper

What to Do

a. Clean all glassware thoroughly.
b. Get the following:
2.75 grams of glycerine
0.5 gram of menthol
32.5 ml of isopropyl alcohol
30 ml of water
c. Mix the ingredients together in a clean beaker.
d. Pour the mixture into a bottle.
e. Pour a small amount on the back of your hand. Rub it in.

2. What happens?

This solution is effective as an after-shave lotion. A good after-shave lotion should dry quickly and leave a pleasant odor. The alcohol tends to dry out the skin.

The formulas for these cosmetics are only starting points. They can be changed to meet specific needs.

3. How could you change the hand lotion formula to make it better for very dry skin?

4. How could you change the after-shave lotion so that it would dry even more quickly?

C. B is for Bubbles

A bubble bath is not only for fun. It can also soften the water, and it smells nice, too.

What You Need

Castile soap, powdered
Cream of tartar
Sodium carbonate (Na_2CO_3)
Sodium sulfate (Na_2SO_4)
Bottle, wide mouth
Mortar and pestle
Balance
Spatula
Weighing paper

What to Do

a. Place the following in a mortar:
9 grams of Na_2CO_3
12 grams of cream of tartar
2 grams of Na_2SO_4
2 grams of powdered Castile soap

b. Grind all the ingredients together until fine and evenly mixed.

c. Pour the mixture into a bottle.

5. Does your bubble bath foam more in hot or cold water? Hard or soft water?

Hopefully this investigation gave you some beautiful insights.

INVESTIGATION

7

Seventh Inning Stretch

The plastics industry has a widespread impact on today's society. There are so many different kinds of plastic that it would be difficult to name them all. They come as fibers, like rayon, nylon, and Dacron; as films, like cellophane, Mylar, and polyethylene; and as solids, like Teflon, urethane foam, and polystyrene. Plastics are chemically related to several other compounds you have already studied. They are organic compounds which are synthetically joined together in long chains.

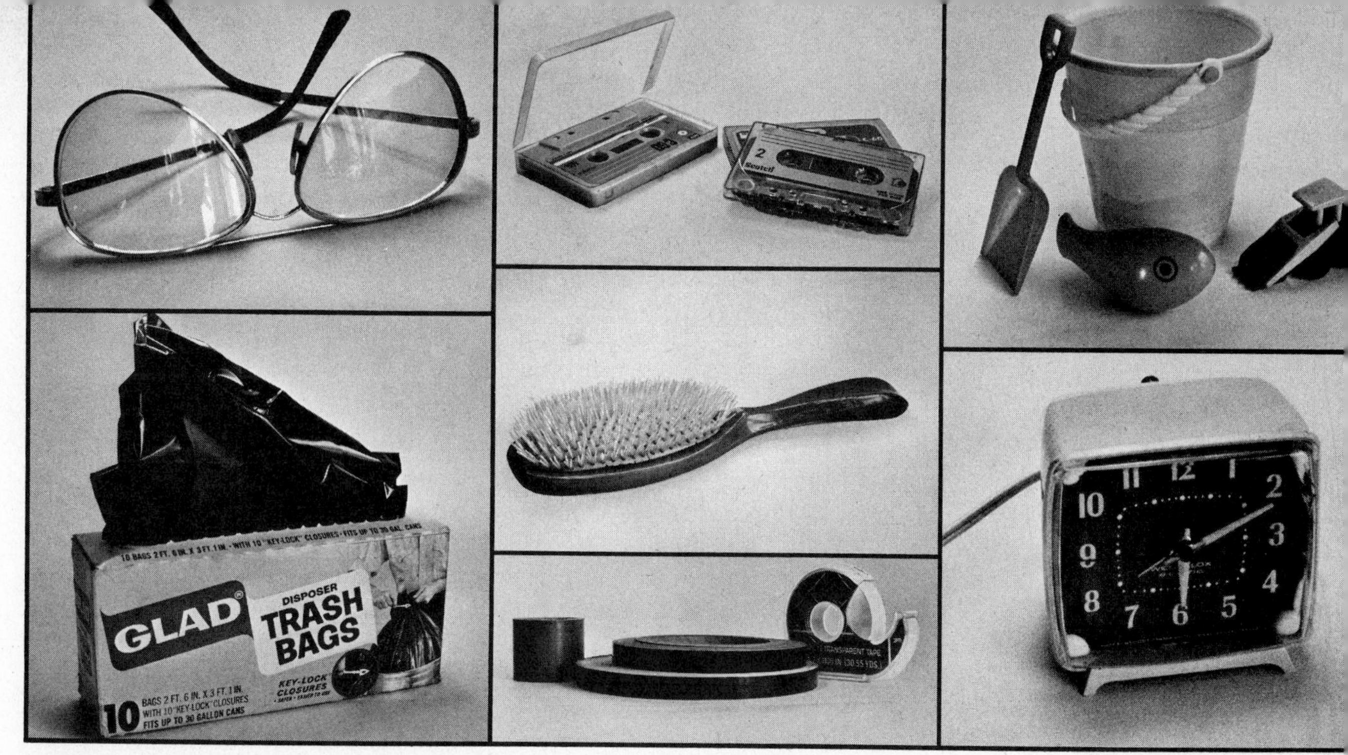

Cynara

The spectacular rise of the plastics industry has been due to the creative power of chemists. An enormous number of everyday items are made out of plastics. It is not surprising, then, that plastics chemists are in great demand. What do plastics chemists do? They discover and test plastics. So can you.

A. Repeat After Me, Me, me, me . . . . . .

What You Need

- Formaldehyde (HCHO), 37%
- Litmus paper, red
- Resorcinol [$C_6H_4(OH)_2$]
- Sodium hydroxide (NaOH), 6 *M* solution
- Beaker, 50 ml
- Beaker, 250 ml
- Graduated cylinder, 10 ml
- Stirring rod
- Thermometer
- Asbestos pad
- Balance
- Burner
- Copper wire
- Matches
- Ring
- Ring stand
- Safety goggles
- Spatula
- Stopper, 1-hole
- Test tube clamp
- Tongs
- Weighing paper

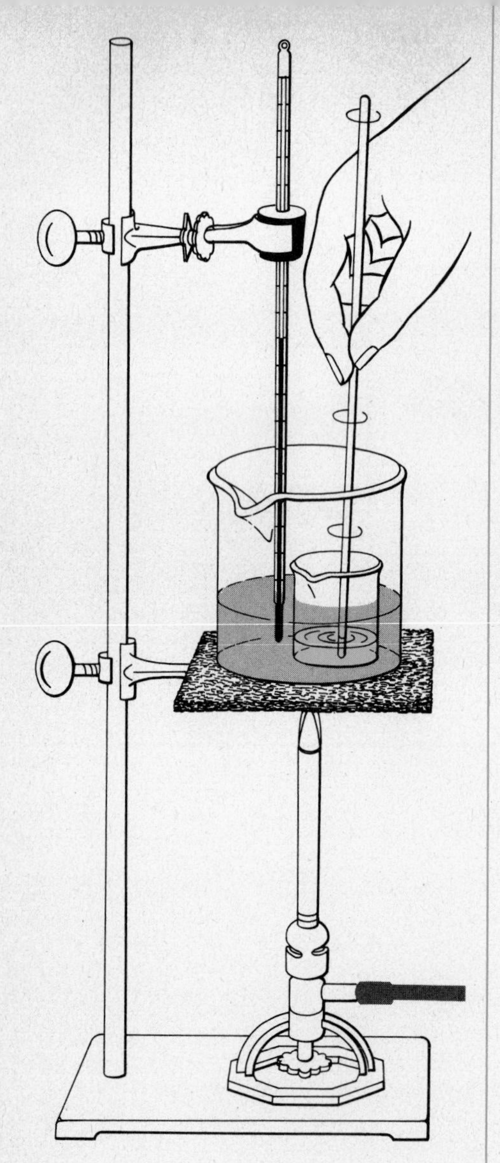

What to Do

a. Place 2 grams of resorcinol in a 50 ml beaker.
b. Add 3 ml of formaldehyde.
c. Test the solution with red litmus paper.

1. Is the solution acidic or basic?

d. Add NaOH, drop by drop, until the solution turns red litmus blue.

2. Is the solution now acidic or basic?

e. Set up a water bath. Heat it to 50°C.
f. Place the small beaker in the water bath as shown.
g. Keep the water bath at 50°C. Stir the mixture until all the resorcinol crystals have dissolved.
h. Remove the stirring rod and in its place use a loop of copper wire.

i. Heat the water bath to 70°C. Do not overheat. Keep this temperature for about 10 minutes.

3. What happens?

The product should be reddish-brown in color.
j. Use tongs to remove the small beaker from the water bath. Allow it to cool.

k. When the product has cooled, use the wire to pull it from the beaker.

4. Examine and describe the product.

When small organic molecules link together to form long chains, the process is called *polymerization.* The products are called *polymers.* For example, the gas ethylene (ethene, $H_2C{=}CH_2$), when heated under pressure, polymerizes to form a plastic called polyethylene. Polyethylene consists of CH_2 units linked together in an endless chain:

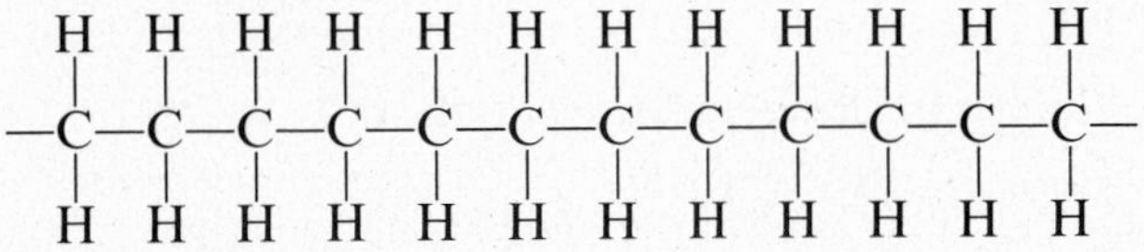

```
  H  H  H  H  H  H  H  H  H  H  H  H
  |  |  |  |  |  |  |  |  |  |  |  |
 —C—C—C—C—C—C—C—C—C—C—C—C—
  |  |  |  |  |  |  |  |  |  |  |  |
  H  H  H  H  H  H  H  H  H  H  H  H
```

Ethylene is called the *monomer* of polyethylene. A monomer is the basic unit of a polymer. Two or more different monomers may be mixed to form a polymer. This is what you did. You mixed formaldehyde and resorcinol to make a type of plastic—a polymer. Many substances you are already familiar with, such as plexiglass, nylon, rayon, and Dacron, are polymers. Not all polymers are plastics, though. Rubber, starch, and cellulose are all naturally-occurring polymers.

The two basic types of plastics are *thermoplastic* and *thermosetting.* Thermoplastic compounds may be softened by heat. Most plastics are thermoplastic. Thermosetting compounds undergo a chemical change when heated and cannot be softened. Plastic ovenware is thermosetting. In the next activity you can test which type of plastic you just made.

B. Get into Shape

What You Need

Plastic samples

Beaker, 250 ml
Graduated cylinder, 100 ml

Asbestos pad
Burner
Forceps
Matches
Ring
Ring stand
Safety goggles

What to Do

a. Get 100 ml of water in a 250 ml beaker and heat to boiling.
b. Using forceps, place a sample of plastic into the boiling water for 1 minute.
c. Remove the plastic from the water. Test it immediately to see if it bends.
d. Notice how long the sample stays hot after being removed from the water.
e. Repeat steps **b** through **d** with other samples of plastic.

5. Describe any changes that might have occurred in each plastic.

6. Which samples were thermoplastic?

7. Which samples were thermosetting?

8. Which type of plastic did you prepare in Part A?

9. Write up your results in the form of a report.

C. Have a Ball!

The polymer rubber is made up of the monomer isoprene, which looks like this:

```
               H
               |
  H        H—C—H        H
   \           |        /
    C==C—C==C
   /           |        \
  H            H        H
```

You can make synthetic rubber, which has many of the properties of the real thing, by doing the following:

What You Need

Ethylene dichloride (CH_2ClCH_2Cl)
Magnesium hydroxide [$Mg(OH)_2$]
Sodium hydroxide (NaOH)
Sulfur (S)

Beaker, 250 ml
Graduated cylinder, 100 ml
Stirring rod
Thermometer

Asbestos pad
Balance
Burner
Matches
Ring
Ring stand
Safety goggles
Spatula
Stopper, 1-hole
Test tube clamp
Weighing paper

What to Do

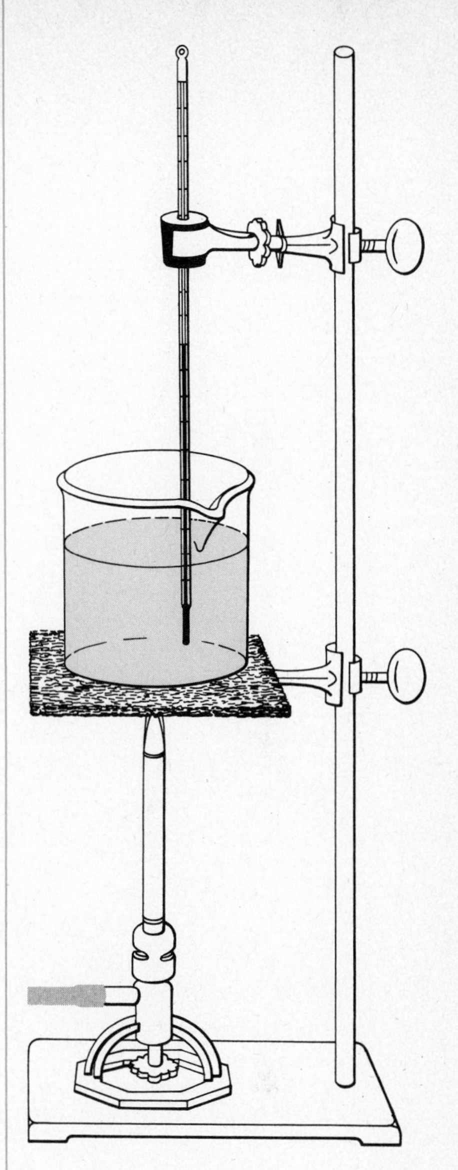

a. Dissolve 10 grams of NaOH in 150 ml of water in a 250 ml beaker.
b. Heat the solution until it boils.
c. Slowly add, while stirring, 20 grams of sulfur.
d. Continue to boil and stir for 10 minutes.
e. Allow the solution to cool.
f. Add a few grams of $Mg(OH)_2$ to the liquid and reheat the solution to 70°C.
g. Slowly add, while stirring constantly, 30 ml of ethylene dichloride to the beaker.
h. Keep the temperature between 70° and 80°C.
i. Continue the heating and stirring for 15 minutes.
j. Turn off the burner and allow the beaker to cool.
k. When cool, pour off the liquid.
l. Rinse the rubber left in the beaker with 50 ml portions of water until the water is almost colorless. Squeeze the rubber with a spatula while rinsing.

After rinsing the rubber, you may want to run tests on it to determine how well it stretches and bounces. Wash your hands after handling it.

D. What's in a Name?

What You Need

A sense of humor

What to Do

10. Name the following: a. (hexagonal ring with Fe at each of its six corners)

Fe
Fe Fe
Fe Fe
Fe

b. HI—HOAg c. $Ba(Na)_2$ d. $Au—H_2O$

e. CH_2O f. $Co(Fe)_2$

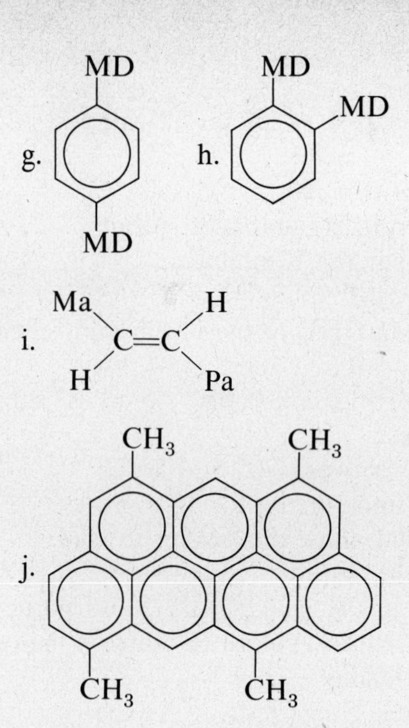
MD
MD
g.
MD
MD
h.
MD
Ma
H
i.
C=C
H
Pa
CH3
CH3
j.
CH3
CH3

INVESTIGATION 8

Clothing Makes the Man . . . Or Woman

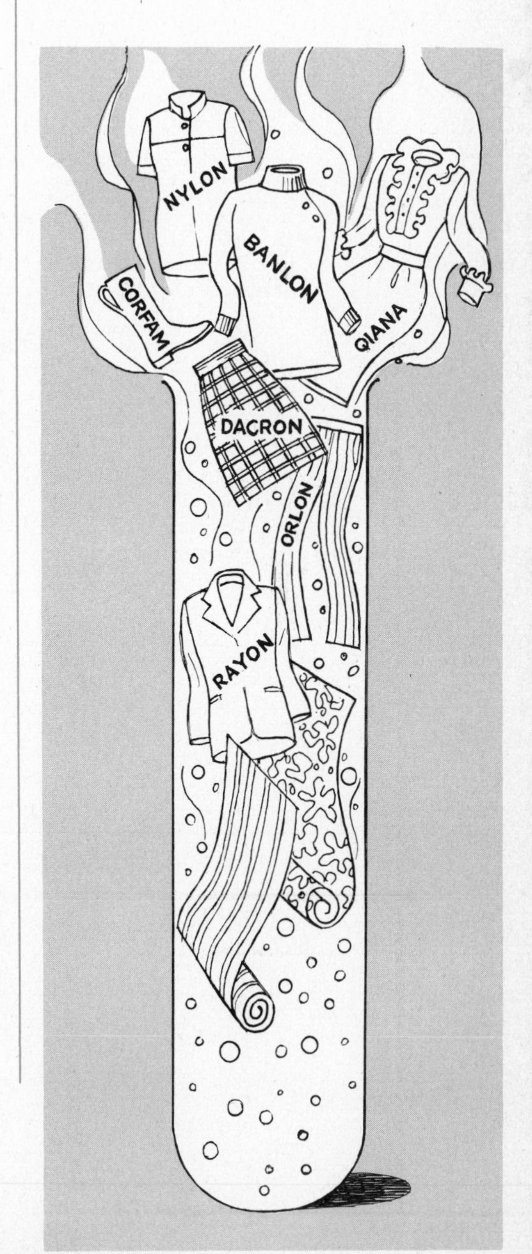

Have you ever had a piece of clothing fade, rip, shred, scorch, or dissolve in bleach? If so, you can realize the importance of the textile chemist.

The textile industry manufactures the fabrics used for clothing and furnishings. It is one of the largest industries in the country. Chemistry is a part of the textile industry in many different ways. Textile mills hire chemists to check the processes by which fabrics are made. Clothing manufacturers hire chemists to test their materials for purity and durability. Many companies employ chemists to invent new fibers. Nylon, Qiana, Orlon, Arnel, Dacron, and Banlon were all created in a test tube within the last forty years.

These fibers are called *synthetic* fibers. *Natural* fibers like cotton, wool, silk, and linen have been known for thousands of years. Can you tell natural from synthetic fibers? Can you tell what fibers make up a particular fabric? This is the place to learn something about textiles.

A. The Great Put-On

The old method of identifying fabrics was simply to look and feel. Today it is almost impossible to identify all fabrics that easily. We have synthetic fur that would fool a mink. Some suede and leather has never seen a cow. And the silkiest fabrics are not silk at all. A knowledge of chemistry can help you uncover the true qualities of fabrics.

What You Need

- Cotton samples
- Lead acetate paper
- Litmus paper, blue
- Litmus paper, red
- Polyester samples
- Wool samples
- Stirring rod
- Test tubes, Pyrex, 3
- Asbestos pad
- Burner
- Matches
- Safety goggles
- Scissors
- Test tube holder
- Test tube rack
- Tongs
- Wax pencil

What to Do

Natural fibers can come from plants or animals. Burning can tell you which. Natural plant fibers burn with an odor of burning paper. Natural animal fibers have the odor of burning feathers. Many synthetic fibers form round beads when burned.

a. Hold a few strands of cotton in your tongs. Ignite the fibers with a match.

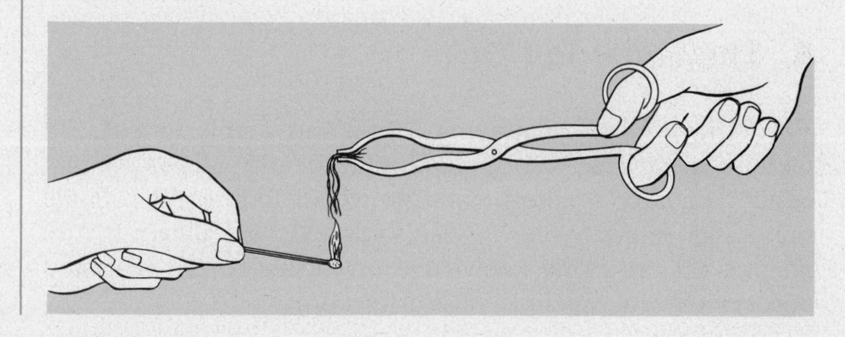

b. Note the rate of burning, the amount of residue, and carefully note the odor.
c. Repeat steps **a** and **b** with samples of wool and polyester.

1. Which fiber comes from plants?
2. Which comes from animals?

d. Label a test tube for each of your three fabric samples.
e. Cut up and place a small sample of the proper fabric in each test tube.
f. While heating each tube gently, hold strips of moistened blue litmus, red litmus, and lead acetate paper at the mouth.

3. Organize your results in a table similar to the one below.

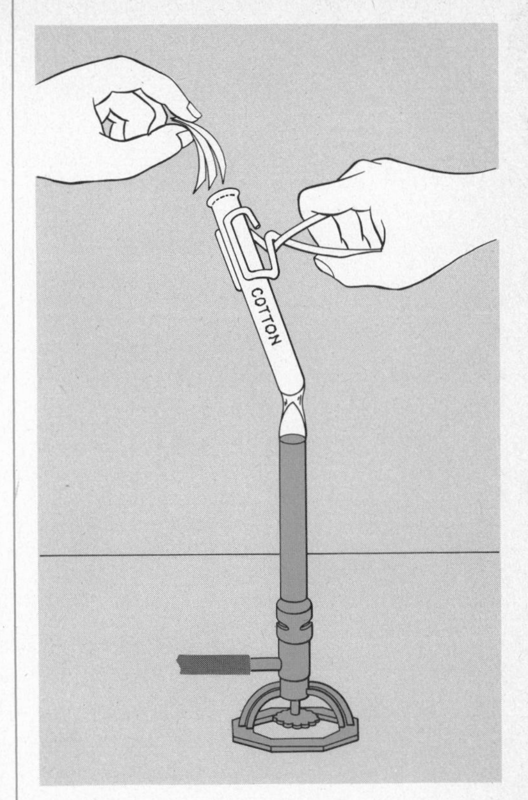

TABLE 1: Tests for Fabrics

Sample	*Burning odor*	*Source*	*Reaction with litmus paper*	*Reaction with lead acetate paper*
Cotton				
Wool				
Polyester				

A yellow-brown to dark color on lead acetate paper indicates the presence of sulfides. The darker the color, the higher the percentage of sulfides.

4. Which fiber contains the most sulfides?
5. What gas might cause red litmus to turn blue?
6. Which fibers contain nitrogen? How do you know?

B. Singing in the Rain

How rainproof is that raincoat of yours? Raincoats are rated not on how much water they let through (penetration), but on their relative repellency—their resistance to external wetting of the fabric. The fabric is rated by comparing its wetted pattern with pictures on a standard rating chart developed by the American Association of Textile Colorists and Chemists (AATCC).

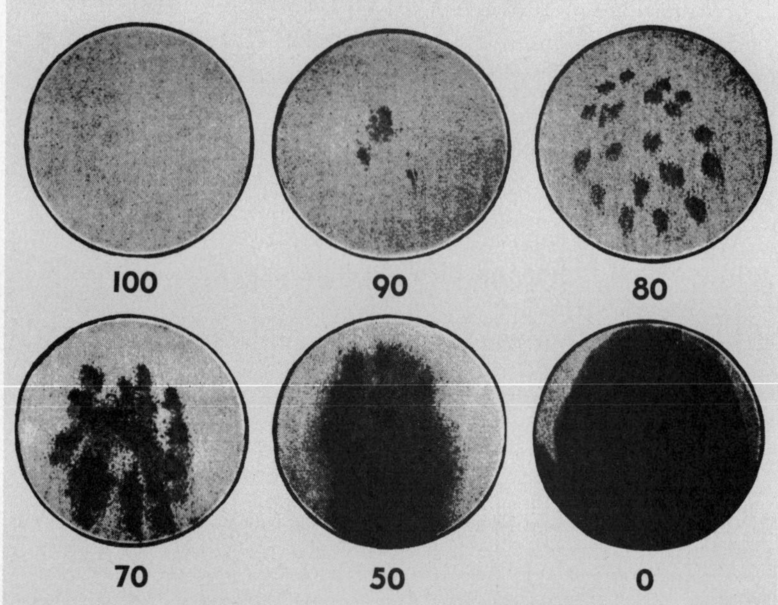

AATCC

What You Need

- Fabric samples, to fit embroidery hoop
- Water repellent sprays
- Funnel, 8 cm
- Graduated cylinder, 100 ml
- Embroidery hoop, 21 cm
- Ring
- Ring stand
- Rubber tubing, 10 cm
- Ruler, metric
- Safety goggles
- Scissors
- Spray head, type used to wet clothes during ironing

What to Do

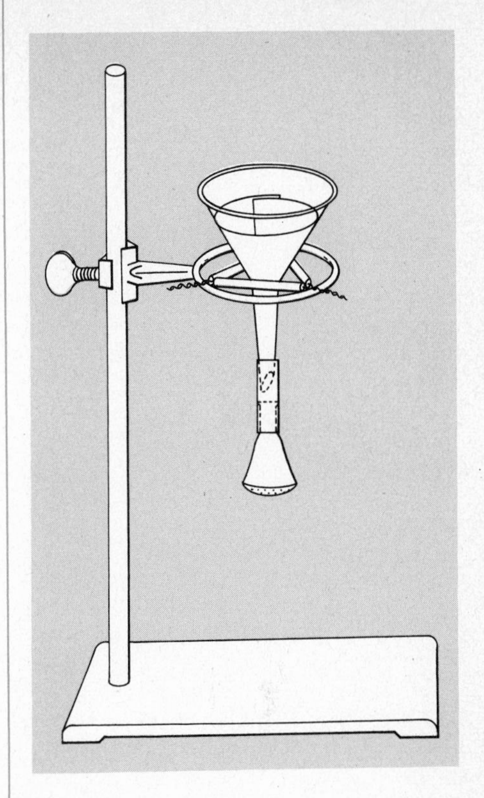

a. Assemble the repellency tester, in the sink.
b. Cut two pieces, each 20 × 20 cm, from a piece of fabric.
c. Following the label directions, spray one sample of the fabric with commercial water repellent. Let dry.
d. Place one fabric sample in the embroidery hoop.
e. Place the hoop under the tester so that the fabric is uppermost and in such a position that the center of the spray pattern coincides with the center of the hoop.
f. Pour 100 ml of water into the funnel of the tester and permit it to spray on the fabric.
g. When the spray stops, turn the fabric upside down and gently shake off any water drops.
h. Determine the water resistance of the fabric by comparing it to the Spray Test Rating Chart.
i. Repeat the experiment using the other sample.
j. Repeat the test with different fabrics and different water repellents.
k. Compare the efficiency of a heavy coat of water repellent vs. a thin coat.

7. Write a report explaining the significance of the test results.

8. Which of the brands you have tested is more effective?

9. Is a heavy coat of repellent better than a thin coat?

C. Blood, Tears, and Perspiration

Here is another method from the AATCC. This method is intended for determining the fastness of colored textiles to the effects of water and perspiration. A *color-fast* fabric is one in which the dye does not "bleed" or come out. The method is applicable to dyed or printed textile yarns and fabrics and to the testing of dyestuffs which are used on textiles.

The perspiration of different people can vary from acidic to basic. Therefore you will test fabrics with two perspiration solutions—one acidic and the other basic.

What You Need

Acid solution (acidic perspiration)
- Sodium chloride (NaCl), 1 gram
- Lactic acid ($C_3H_6O_3$), USP 85%, 0.5 gram
- Disodium hydrogen phosphate (Na_2HPO_4), 0.5 gram
- Add enough distilled water to make 100 ml of solution

Alkaline solution (basic perspiration)
- Sodium chloride (NaCl), 1 gram
- Ammonium carbonate [$(NH_4)_2CO_3$], 1 gram
- Disodium hydrogen phosphate (Na_2HPO_4), 0.5 gram
- Add enough distilled water to make 100 ml of solution

Fabric samples
Water (H_2O), distilled
Flasks, 250 ml
Cotton balls
Scissors

What to Do

a. Lay a piece of colored or dyed fabric flat on the desk.
b. Soak a cotton ball with distilled water and gently squeeze out the excess.
c. Lightly rub the wet cotton ball on the fabric. Use gentle strokes, going only in one direction.

10. Does the dye come off with water?

d. Repeat steps **a** thru **c** using first the acid and then the alkaline perspiration solutions. Remember to use a fresh part of the fabric sample and a fresh cotton ball for each test.

11. Construct a table showing the results. Rate the fabrics according to colorfastness to water and perspiration.

Unsatisfactory perspiration fastness may be due to bleeding of the color, or it may be due to a chemical change in the dyed material. It should be noted that in some cases an objectionable change in color may occur with no apparent bleeding. On the other hand, there may be bleeding with no apparent change in color; or there may be both bleeding and change in color.

The test you have just performed to determine the colorfastness of fabric is similar to the one used by chemists in the clothing and fabric industry. The chemists and chemical technicians of each industry have formed their own organizations. These organizations share as well as test developmental information.

The official name of the test you performed in Part C is "Colorfastness to Perspiration, AATCC Method 15-1962." According to the *official method,* the test fabric is thoroughly wet with the perspiration solution and then wrung out. The fabric is then sandwiched between two pieces of the same but undyed material. This sandwich is placed under a standard pressure in a small vertical press. The press is then placed in an oven at 38° ± 1°C for 6 hours.

At the end of the required time, the press is removed from the oven. The dyed fabric is examined to determine if the color has changed in any way or has transferred to the undyed fabric.

12. What is the purpose of the elevated temperature during the test?

13. Why put samples under pressure during the test?

Hopefully you now have a greater appreciation for the role of the textile chemist.

AATCC

AATCC Perspiration Tester

INVESTIGATION

9

No Pain, No Strain!

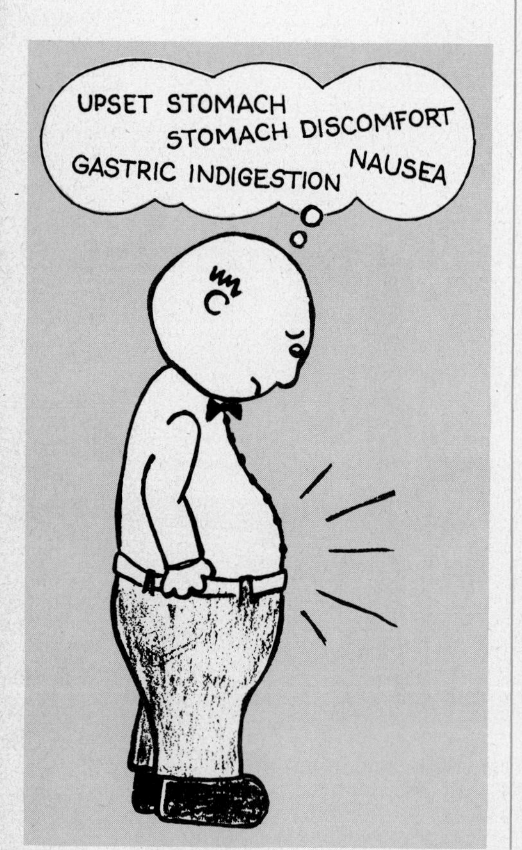

Kathy Wujciak

In our society, a pain can be relieved by making a quick trip to the drugstore. When your head aches, you take aspirin; when your stomach aches, you take an antacid. Taking medication can relieve the discomfort, but it may not cure the problem. A pain is your body's way of telling you that something is wrong. Just relieving the pain may not cure the source of the problem.

Many chemists are involved in making medicines designed to cure the various ills that plague humankind. New medicines are being synthesized all the time in order to help in the battle against disease and suffering. Before a new medical product is allowed on the market, it must undergo extensive testing in both animals and humans. Chemists in the health industry spend years testing a single product.

Obviously, such a testing program is beyond the scope of this course. You can, however, try your hand at some of the shorter testing procedures and at making a medical product familiar to all of you.

A. Don't Get Upset

Commercial products known as antacids are advertised as a means of relieving stomach upset. Your stomach is supposed to be acidic, but it can be too acidic. An antacid is a weak base used to neutralize excess acid in your stomach.

In this activity, you will test different antacids to see which will neutralize the most acid.

What You Need

Antacid tablets
Congo red indicator
Hydrochloric acid (HCl), 0.1 *M*
Buret
Dropper
Flask, 250 ml
Graduated cylinder, 100 ml
Mortar and pestle
Stirring rod
Balance
Buret clamp
Paper, white
Ring stand
Safety goggles
Weighing paper

What to Do

1. Copy Table 1 into your notebook.

TABLE 1: Comparison of Antacids

	Tablet 1	*Tablet 2*	*Tablet 3*
Brand name			
Mass of tablet			
Final volume of acid			
Initial volume of acid			
Volume of acid neutralized			

a. Find the mass of an antacid tablet.

2. Record the name of the tablet and the mass in Table 1 under *Tablet 1.*

b. Crush the tablet in a clean mortar with a clean pestle.

c. Add 50 ml of water to the mortar. Stir to dissolve the tablet.
d. Pour the contents of the mortar into your flask.
e. Rinse the mortar with 50 ml more water.
f. Pour the contents into the flask.
g. Add 5–8 drops of Congo red indicator to the flask.

3. What color is the solution in the flask?

h. Fill your buret with 0.1 *M* HCl.

Since stomach acid is about 0.1 *M* HCl, you will use this concentration to titrate.

4. Record the volume of hydrochloric acid in the buret before titrating.

i. Titrate with HCl until the contents of the flask turn a light blue.

5. Record the volume of acid in the buret in Table 1.

6. What is the volume of acid used in this titration?

j. Repeat steps **a** through **i** for other tablets.

7. Which antacid is the most expensive per tablet?

8. Which antacid neutralized the most acid?

9. Which antacid is the least expensive per tablet?

10. Which antacid neutralized the least acid?

It is not healthy to take too many antacid tablets or to take tablets that are very strong. Your stomach is supposed to be acidic. If you neutralize too much acid, your body supplies more acid to the stomach. Therefore, two much antacid causes more acid to be produced in your stomach. Hydrochloric acid in the stomach is used to digest protein, coagulate the casein in milk, and kill bacteria which enter the stomach.

B. The Laugh Is on You

Some pain relievers are very useful. The use of anesthetics has done much to minimize human suffering. Through the use of anesthetics, people can have a major operation without experiencing any pain during surgery.

In the next two activities, you will make pain relievers.

What You Need

Ammonium nitrate (NH_4NO_3)

Bottles, wide mouth, 4
Flask, 250 ml
Glass bends, right angle, 2
Glass plates, 2

Balance
Burner
Matches
Ring stand
Rubber tubing
Safety goggles
Stopper, 1-hole
Stopper, 2-hole
Test tube clamp
Trough, water
Weighing paper
Wood splints

What to Do

a. Place 12 grams of NH_4NO_3 in a flask.
b. Clamp the flask to a ring stand and set up the apparatus as shown in the illustration.

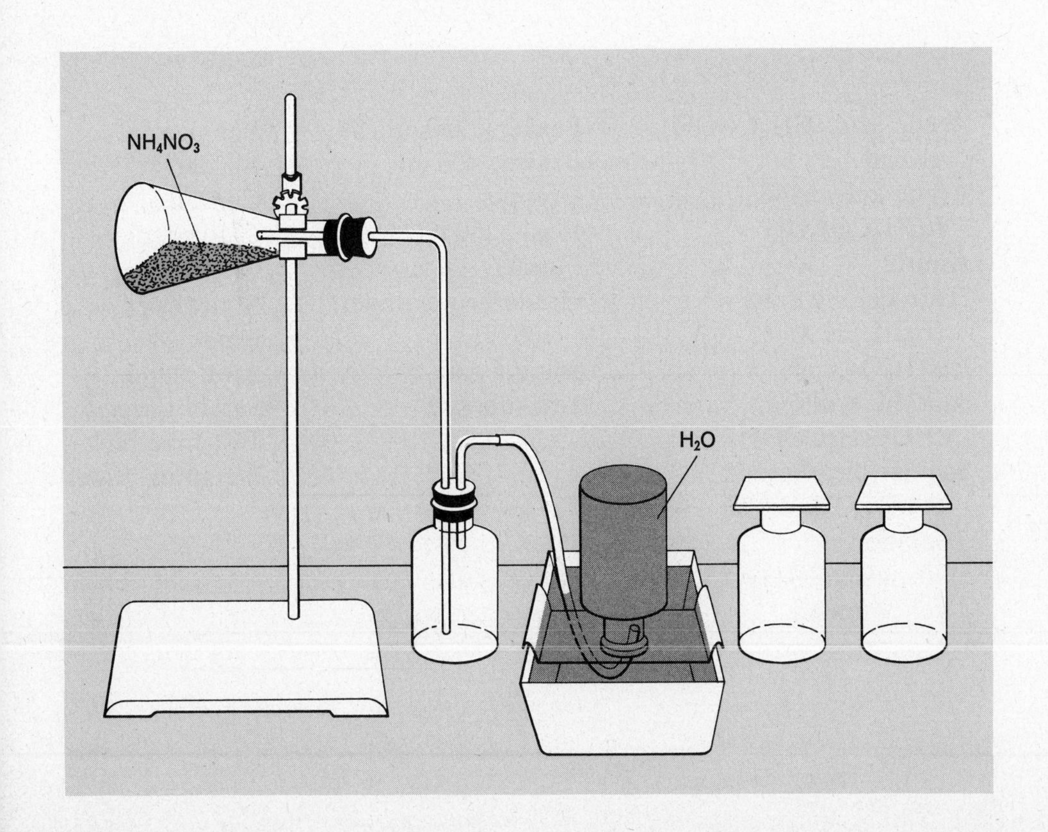

c. Have your instructor check your apparatus.
d. Heat the flask gently. **Do not heat it too much. If a brown-colored gas forms in the flask, remove the heat and allow to cool.**

11. The first bottle of gas collected should be discarded. Why?

e. Collect two more bottles of gas.
f. Use one bottle of gas to check the odor.

12. Describe the odor.

g. Thrust a lighted splint into the other bottle.

13. Describe what happens.

The gas you collected is known as "Laughing Gas." Laughing gas (N_2O) is used as an anesthetic.

C. Who's Aspirin Is Better?

Another useful pain reliever is aspirin, acetylsalicylic acid. Aspirin will relieve pain quickly all over the body, but there can be some side effects. Aspirin can cause ulcers in the stomach; and some people are allergic to it.

What You Need

Acetic acid (CH_3COOH), glacial
Acetic anhydride [$(CH_3CO)_2O$]
Aspirin
Hydrochloric acid (HCl), 0.1 *M*
Ice (H_2O)
Salicylic acid (HOC_6H_4COOH)
Sodium bicarbonate ($NaHCO_3$), 0.5 *M* solution
Sulfuric acid (H_2SO_4), concentrated

Beakers, 250 ml, 2
Beaker, 400 ml
Dropper
Flask, 250 ml
Funnel
Graduated cylinder, 100 ml
Stirring rod
Test tubes, 2

Balance
Burner
Can lid
Clay triangle
Filter paper
Matches
Ring
Ring stand
Safety goggles
Test tube rack
Weighing paper

What to Do

a. Get 5 grams of salicylic acid in a 250 ml flask.
b. Add 10 ml of acetic anhydride to the flask.
c. Stir for 2–3 minutes.
d. Add 10 ml of glacial acetic acid.
e. Add 2–4 drops of concentrated H_2SO_4.
f. Put 100 ml of water in a 250 ml beaker.
g. Add the contents of the flask to the beaker.
h. Place some crushed ice in the 400 ml beaker.
i. Cover the ice with water.
j. Place the 250 ml beaker inside the 400 ml beaker.
k. After the crystals have formed, filter.
l. Let the aspirin dry.

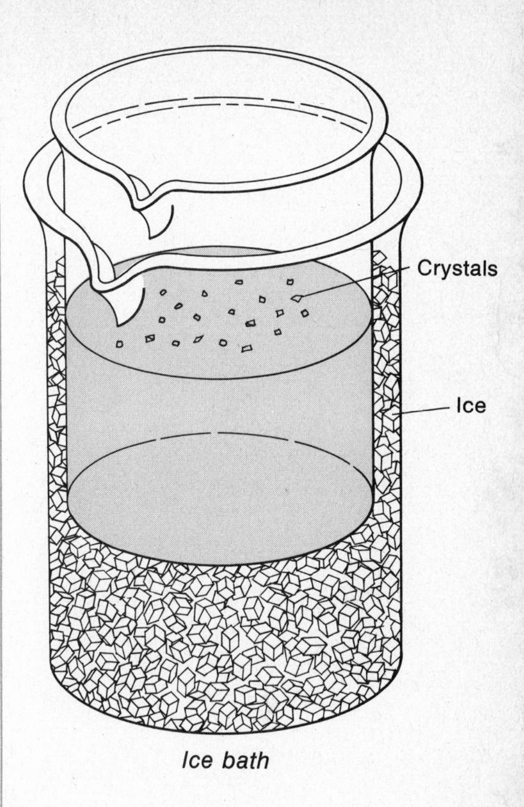

Ice bath

Now compare your product with some commercially produced aspirin.

14. Copy Table 2 into your notebook.

TABLE 2: Comparison of Aspirins

	Melting	*Solubility in water*	*Solubility in $NaHCO_3$*	*Solubility in HCl*
Laboratory aspirin				
Commercial aspirin				

m. Place small, equal samples of each aspirin about 2 cm apart on a can lid.
n. Heat the samples equally.

15. Do both samples melt at about the same time? Record your observations in Table 2.

o. Perform the other tests in Table 2.

16. Compare the properties of your product and the commercial aspirin. Do they react the same or do they react differently?

17. Do you believe your product is aspirin?

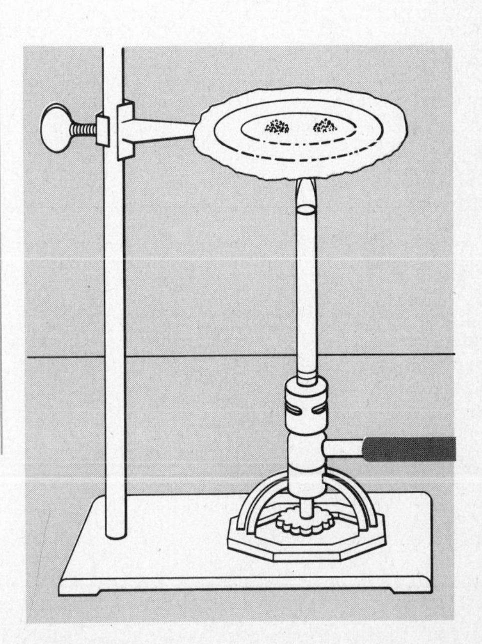

18. Compare the rate at which different commercial aspirins dissolve in 0.1 M HCl. (This is about the strength of acid in your stomach.)

Is the difference between the rates of dissolving large enough to warrant buying one particular brand? You can also consider the cost per tablet the next time you purchase aspirin.

As you have seen, there are many factors to be considered when purchasing medication. Some drugs are too weak, some too strong; and some may give you a bad reaction. It is also important to compare costs, especially if different products contain the same ingredients. Have a happy and healthy life!

INVESTIGATION 10

Paint Your Wagon

Today over 80,000 people work in more than 1,600 companies which manufacture paints, stains, varnishes, lacquers, enamels, and other coatings. If a structure, instrument, or tool is not given a protective coating, it will soon rust, rot, decay, corrode, mildew, or fade. Coatings are also decorative. The same paint that keeps your car from rusting gives it that shiny color. Houses, furniture, dishes, and tennis racquets all look nicer because of coatings put on them. Chemists research and develop a variety of coating products which go on everything from motorcycles to space capsules.

Cynara

A. Paint the Town

In ancient times the range of paints available was very limited. Today paint is a complex and important part of the giant chemical industry. Many colors are readily available. You can make a few in the lab.

What You Need

Copper(II) sulfate ($CuSO_4$)
Iron(III) oxide (Fe_2O_3)
Lead(II) nitrate [$Pb(NO_3)_2$]
Paint medium (30 ml clear gum + 10 ml H_2O + 1 drop liquid detergent)
Sodium bicarbonate ($NaHCO_3$)
Sodium carbonate (Na_2CO_3)

Beakers, 250 ml, 3
Beaker, 400 ml
Dropper
Graduated cylinder, 100 ml
Mortar and pestle
Stirring rod

Asbestos pad
Balance
Burner
Matches
Paper towels
Ring
Ring stand
Rubber policeman
Safety goggles
Spatula
Weighing paper

What to Do

a. Warm 50 ml of water in a 400 ml beaker. Dissolve 5 grams of Na_2CO_3 in it and let it cool.
b. Warm 75 ml of water in a 250 ml beaker. Dissolve 16 grams of $Pb(NO_3)_2$ in it.
c. Pour the $Pb(NO_3)_2$ solution slowly, with steady stirring, into the Na_2CO_3 solution.

1. What happens?

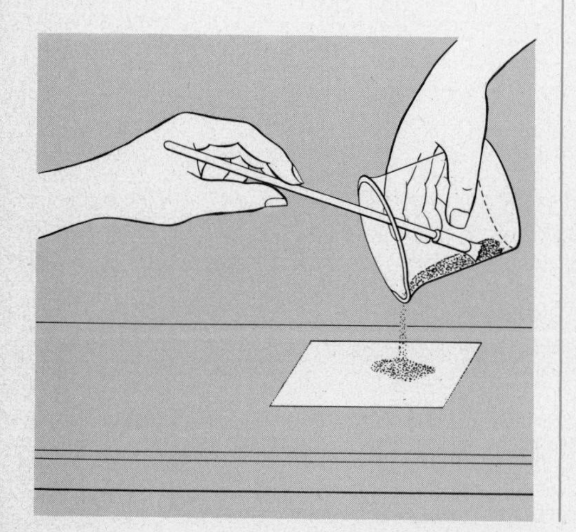

d. Allow the contents of the beaker to settle.
e. Carefully decant the clear, top layer of liquid. Discard it.
f. Add more water and stir. Allow to settle again and pour away the clear, top liquid.
g. Repeat step **f** four more times. Then spread the wet powder out on paper towels to dry.

You have just made a pigment, the powder from which a paint is made.

h. When your pigment is dry, mix it into a smooth paste with a few drops of paint medium.

You now have a paint. If it dries out, it can be softened again by adding a little lukewarm water.

i. Dissolve 15 grams of $CuSO_4$ in 100 ml of water in a 250 ml beaker.
j. In another 250 ml beaker, dissolve 5 grams of $NaHCO_3$ in 50 ml of water.
k. Add the $CuSO_4$ solution, while stirring, to the $NaHCO_3$ solution.

2. What happens?

3. What gas is given off?

l. Wash the pale green precipitate as you washed your white pigment.
m. When dry, make into a paint using a few drops of paint medium.
n. For a shade of red-brown, finely grind 7 grams of Fe_2O_3 in a mortar and pestle. Then mix with a few drops of paint medium.

This red-brown pigment is one of the most ancient ones known. It was used by the earliest painters. The table below lists the compounds used in making various pigments.

TABLE 1: Pigments

Color	*Compound*
White	$Pb(NO_3)_2$ and Na_2CO_3
White	TiO_2
Yellow	$Pb(NO_3)_2$ and K_2CrO_4
Red-brown	Fe_2O_3
Dark brown	PbO_2
Light green	$CuSO_4$ and $NaHCO_3$
Dark green	$(NH_4)_2Cr_2O_7$
Blue	$CuCO_3$ and $Cu(OH)_2$
Silver	Al

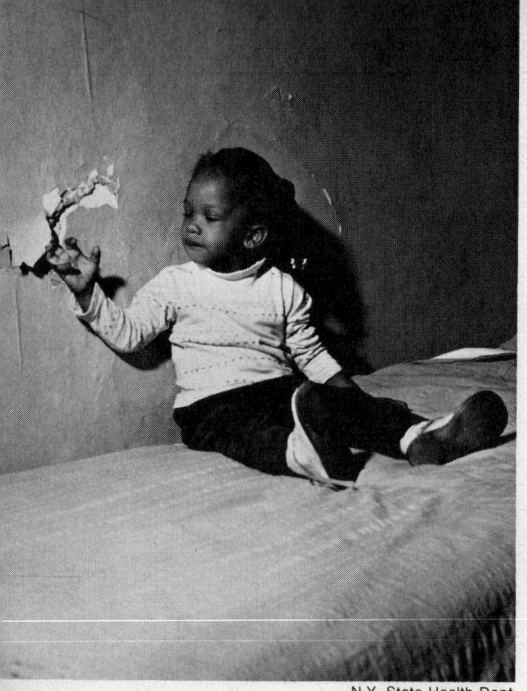
N.Y. State Health Dept.

Nowadays we are using more and more plastic and other synthetic coatings. These are not paints in the conventional sense, but they are made by the paint manufacturers. Synthetic coatings are extremely durable and can bond to (cover) almost anything. One reason synthetic coverings are used more now is that many common paints provide a very real danger to people, especially babies.

B. The Kiss of Death

Paint with lead in it is very dangerous. Young children can become victims of lead poisoning. They eat paint containing lead from toys, cribs, and furniture. A recent Federal law now prohibits the use of lead paints on children's furniture and toys, and in interior house paints. However, in older houses and apartment buildings, lead paint is still found. Lead poisoning from paint is still a very real health problem in many areas. Beware of paints on second-hand furniture.

Here is how a paint chemist might test for the presence of lead in paint. Try it on some paint samples or chips from your home. You can also test the paints you made in Part A.

What You Need

Acetic acid (CH_3COOH), 3 *M*
Ammonium hydroxide (NH_4OH), 3 *M* solution
Litmus paper, blue
Litmus paper, red
Nitric acid (HNO_3), 3 *M*
Paint samples
Potassium dichromate ($K_2Cr_2O_7$), 0.01 *M* solution
Sulfuric acid (H_2SO_4), 3 *M*

Beakers, 250 ml, 3
Dropper
Funnel
Graduated cylinder, 25 ml
Stirring rod

Asbestos pad
Balance
Burner
Clay triangle
Filter paper
Matches
Ring
Ring stand
Safety goggles
Weighing paper

What to Do

a. Put 1 gram of a paint sample into a beaker.
b. Add 25 ml of HNO_3.

c. While stirring, heat to boiling.
d. Carefully filter the solution into a clean beaker.
e. Add NH_4OH slowly to the filtrate until the solution is neutral.

4. How will you know when the solution is neutral?

f. Add acetic acid, drop by drop, to the filtrate until the solution just turns blue litmus paper red.
g. Place 10 ml of the solution into a clean beaker.
h. Add 1 ml of $K_2Cr_2O_7$.

5. What happens?

A yellow precipitate indicates the presence of lead.

i. To confirm the presence of lead, add 10 drops of H_2SO_4. The precipitate should not dissolve.

If you found evidence of lead in the paints in your home, and if there are young children around, contact your local public health service. They will recommend what to do.

Today new paint and coating products appear on the market every week. More than 80% of the paints and coatings now in use were unheard of as little as ten years ago. Chemists devise these coatings and test them for safety. You may want to invent some different paints, or test for other dangers in commercial coatings.

INVESTIGATION 11

The Plating Game

Have you ever admired the gleam of a chrome bumper, or the shine of a silver dish or a gold ring? If so, you are a prime supporter of one of the larger industries in the United States. It is the plating industry, which employs many thousands of laboratory technicians.

The plating technician must be able to:

a. analyze chemically the plating solutions (also known as baths);

Metal Finishing

b. formulate new solutions;
c. make additions to old solutions so that the used or spent plating baths can be reused; and
d. test the quality of the plate.

Plating is the process of putting a metallic coating on the metal surface. When electric current is used, the process is called *electroplating.* When current is not used, it is called *electroless plating.*

Plating is used to improve the appearance of materials and to protect against our old enemy, corrosion. If the bumper on your car was made of only iron or steel, it would rust out in a short time. By electroplating a combination of nickel and chromium onto the bumper, we give it a high long-lasting shine and also protect it from corrosion.

Technically, electroplating is the coating of an object with a thin layer of some metal, using electrolytic deposition. Oxidation-reduction reactions, with which you struggled in Unit 3, are the backbone of the plating industry. To refresh your memory, oxidation means a loss of electrons. Any substance that loses electrons becomes oxidized. Any substance that gains electrons becomes reduced. The process of gaining electrons is called reduction.

Look at this reaction:

$$2\ Na + Cl_2 \longrightarrow 2\ Na^+Cl^-$$

1. Which element is oxidized?
2. Which element is reduced?

You have already ranked several metals in the order of their activity. As you will recall, an iron nail placed in a copper sulfate solution is plated, or coated, with a layer of copper. In electroplating, the same principles apply. Electroplating uses:

a. a source of electric current;
b. a solution of a salt of the plating metal;
c. a piece of the plating metal; and
d. the object being plated.

The object to be plated is attached to the negative terminal (or *cathode*) of the power source. A strip of plating metal is attached to the positive terminal (or *anode*). The object to be plated and the strip of metal are placed in a solution of a salt of the plating metal. When the current is turned on, metal ions from the solution are deposited on the object to be plated. These ions are replaced by metal atoms from the metal strip.

A. A Set Fit for a King

You can start your own electroplating industry right here in the laboratory.

What You Need

Copper (Cu) strip
Copper(II) sulfate ($CuSO_4$)
Methyl alcohol (CH_3OH)
Stainless steel spoon
Steel wool soap pad
Sulfuric acid (H_2SO_4), 2 *M*
Beaker, 250 ml
Beaker, 400 ml
Graduated cylinder, 100 ml
Stirring rod
Alligator clips, 4
Balance
Dry cell, $1\frac{1}{2}$ volt
Insulated wire, ends stripped, 2
Safety goggles
Spatula
Weighing paper

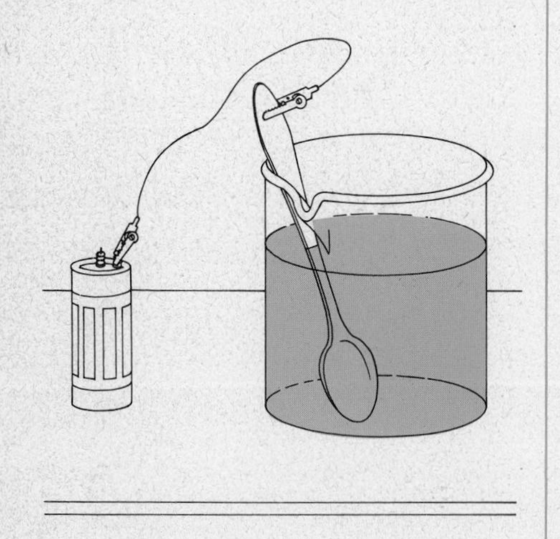

- **a.** Clean your spoon thoroughly with the steel wool soap pad. From this point on, handle the spoon only by the tip of the handle.
- **b.** Rinse the spoon in running tap water.
- **c.** Place the spoon in a beaker of alcohol, keeping it there until ready to use.
- **d.** Prepare 250 ml of a 0.2 *M* solution of $CuSO_4$ in a 400 ml beaker. To this solution, add 10 ml of dilute H_2SO_4.
- **e.** Use the alligator clips to attach a wire from the negative terminal of the dry cell to the handle of the spoon.
- **f.** Place the spoon in the beaker of copper sulfate solution.
- **g.** Attach a wire from the positive terminal of the dry cell to the copper strip. Use alligator clips.
- **h.** Place the copper strip in the beaker of copper sulfate solution. The spoon and the copper strip should not touch each other.

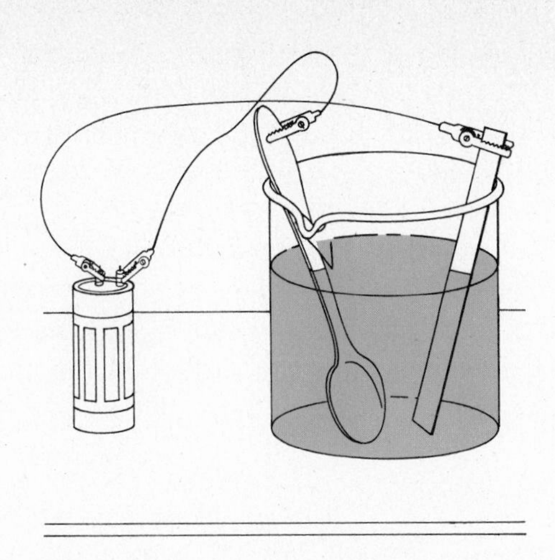

A simple electroplating cell.

i. After 15 seconds, remove the spoon from the copper sulfate solution.
j. Disconnect the wires from the power source.
k. Wash the spoon in cool running water.

3. What has happen to the spoon?

The chemical reactions that have occurred are oxidation-reduction reactions. The reaction on the spoon was:

$$Cu^{2+} + 2e^- \longrightarrow Cu$$

4. Where did the Cu^{2+} ions come from?

While this reaction was taking place, the atoms of copper in the copper strip were losing electrons:

$$Cu \longrightarrow Cu^{2+} + 2e^-$$

B. Color Me Gold

The plater uses his talents to beautify as well as to protect metal. A good and graphic example of this is *anodizing* aluminum metal. Because this metal has low density, is relatively strong, and is plentiful, it has become widely used for toys, decorations, airplanes, and buildings.

Aluminum oxidizes much more slowly than iron.

The Aluminum Association

5. What is oxidation in iron commonly called?

Aluminum oxidizes slowly and the products of the oxidation help protect it from further oxidation.

$$4\ Al + 3\ O_2 \longrightarrow 2\ Al_2O_3$$

Anodizing is a form of plating which colors the thin layer of aluminum oxide which coats the surface of the aluminum.

What You Need

Aluminum (Al) sheeting, 3 cm × 8 cm
Aluminum (Al) sheeting, 10 cm × 12 cm
Ammonium oxalate $[(NH_4)_2C_2O_4]$, 1 *M* solution
Detergent
Iron(III) chloride ($FeCl_3$), 1 *M* solution
Methyl alcohol (CH_3OH)
Sulfuric acid (H_2SO_4), 2 *M*

Beakers, 250 ml, 2
Beaker, 400 ml
Graduated cylinder, 100 ml
Stirring rod

Alligator clips, 4
Asbestos pad
Burner
Forceps
Insulated wire, ends stripped, 2
Matches
Power supply, 6 V. D.C.
Ring
Ring stand
Safety goggles

What to Do

a. Wrap the larger piece of aluminum around the 400 ml beaker to make a cylinder.
b. Fit it into the 400 ml beaker so that it acts as a liner.
c. Add 300 ml of 2 *M* H_2SO_4 to the beaker.
d. Using alligator clips, attach a wire between the large piece of aluminum and the negative terminal of the power source.

6. Will this piece of aluminum act as the cathode or the anode?

e. Thoroughly clean the smaller piece of aluminum with detergent, rinsing it thoroughly in cool running water. Handle this piece of aluminum only by the edges. Place it in alcohol until ready for use.

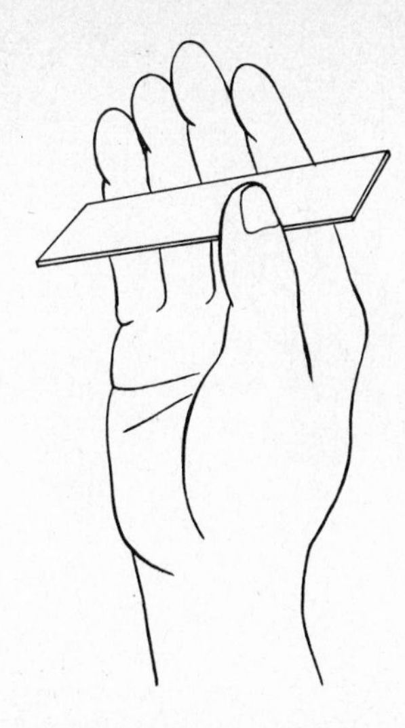

f. Using alligator clips, attach a wire to the smaller aluminum strip and loop it over the ring so that the strip hangs in the beaker of acid. **Caution: The two pieces of aluminum should not touch each other.**

g. Attach the smaller strip of aluminum to the positive pole of the power source. Within a few seconds a stream of bubbles will be seen rising from the cylinder of aluminum. This indicates all is well.

h. Allow the current to flow at least 10 minutes.

i. As you disconnect the center piece of aluminum from the positive pole, quickly remove it from the beaker.

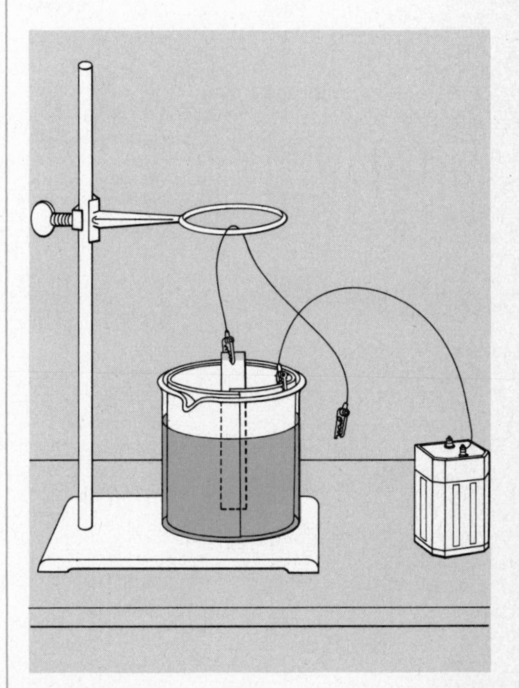

Remember, this piece should be handled only by the edges.

j. Thoroughly rinse this aluminum under a stream of running tap water.

k. Using your forceps, immerse this piece of aluminum into a solution of equal parts of 1 M $FeCl_3$ and 1 M $(NH_4)_2C_2O_4$. Allow the strip of aluminum to remain in the solution for 15 seconds.

l. Remove and rinse in running water.

Note the color.

m. To seal in this color, immerse the anodized plate into boiling water for 1–2 minutes.

Anodizing is a graphic example of how a protective coating can be made decorative.

INVESTIGATION

12

Water, Water, Everywhere

Almost everyone accepts water as a plentiful and commonplace substance. But there's more to water than meets the eye. Natural waters have many substances dissolved and suspended in them. On top of these, we have dumped sewage, poisonous gases, pesticides, and industrial wastes. Then we have wrung our hands over the fact that the waters have become contaminated. Our industrial progress can be traced by studying our exploitation of the earth's water.

Recently the Federal government has stepped into the picture and passed legislation to control water pollution. Technicians are now employed by many different industries to test the quality of water. Their jobs include the following tasks:

a. They see if the quality of the water agrees with the established standard.
b. They identify and interpret the effects of pollution on life in the water.
c. They predict the effects of pollution on people and their use of the water.
d. They predict how pollution can be controlled.

You will examine several of the tests used to determine water quality. You will also test several different water samples and draw conclusions about their quality.

A. C is for Chlorides

Chloride ions are found in all natural waters. They occur over a wide range of concentrations. Streams along the Atlantic Coast, the eastern Gulf Coast, and the Pacific Northwest normally contain low levels of chloride. The exceptions to this are in tidal regions and in polluted waters. Streams in desert and other dry areas usually contain very high levels of chloride. This is due to the high rate of evaporation in hot places. Additional chloride ions are added to natural waters by water softening processes.

In the following activity you will perform a *qualitative test* for chlorides. You will find out only *if* chlorides are present in a sample, not *how much*. A test to find out how much of a substance is present is called a *quantitative test*. In Parts B and C you will perform quantitative tests on water.

What You Need

Nitric acid (HNO_3), 0.5 *M*
Silver nitrate ($AgNO_3$), 0.025 *M* solution
Water samples
Dropper
Graduated cylinder, 10 ml
Stirring rod
Test tubes
Safety goggles
Test tube rack

What to Do

1. Set up a table in your notebook similar to the one below.

TABLE 1: Analysis of Water for Chlorides

Source	Results

a. Pour 10 ml of a water sample into a **clean** test tube.
b. Add 1 drop of HNO_3 and stir.
c. Add 1 ml of $AgNO_3$ solution and stir. A positive test for chlorides is indicated by a white precipitate.

2. Record the result of the test in your table.

d. Test other water samples.

3. Enter your results in the table.

B. How Much is That Chloride in the Water?

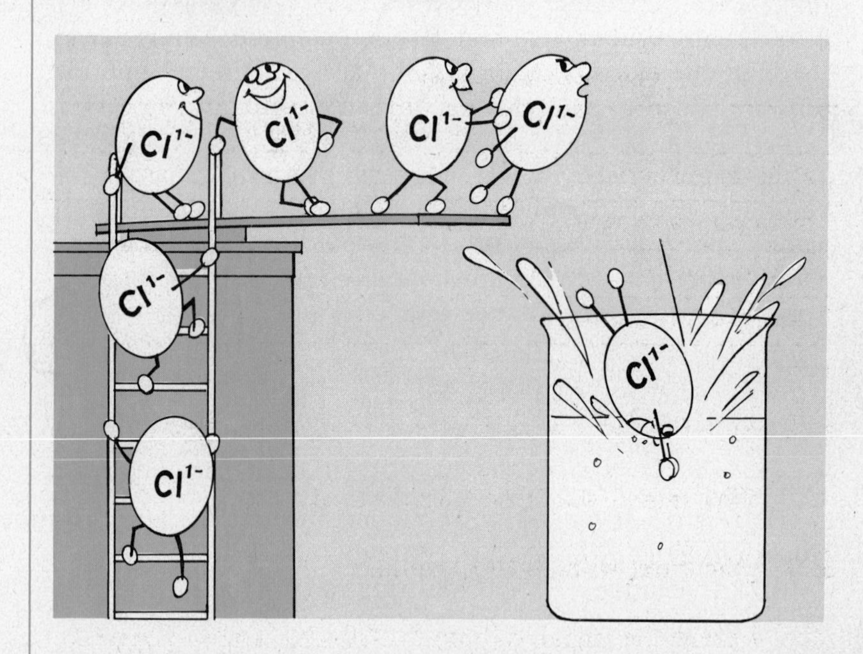

You have just tested water samples to see if they contained chloride ions. You can now test different samples to see which contain the most chloride ions. Silver nitrate reacts with chloride to form a precipitate, silver chloride:

$$Ag^{1+}NO_3^{1-} + Cl^{1-} \longrightarrow AgCl\downarrow + NO_3^{1-}$$

Potassium chromate is the indicator you will use. It turns a rust color in the presence of silver nitrate. The silver nitrate reacts first with chloride ions. When all the chloride ions have reacted, the potassium chromate reacts and turns a rust color. The rust color tells you that the reaction between silver nitrate and chloride is complete.

What You Need

Potassium chromate (K_2CrO_4) indicator
Silver nitrate ($AgNO_3$), 0.025 *M* solution
Water (H_2O), distilled
Water samples
Beaker, 100 ml
Beakers, 250 ml, 2
Buret
Funnel
Graduated cylinder, 25 ml
Stirring rod
Buret clamp
Paper, white
Ring stand
Safety goggles

What to Do

4. Copy Table 2 into your notebook.

TABLE 2: Ppm of Chloride Ions

	Sample 1	*Sample 2*	*Sample 3*
Final volume of $AgNO_3$			
Initial volume of $AgNO_3$			
Volume of $AgNO_3$ reacted			
Ppm of chloride ions			

a. Clean all the glassware and rinse in distilled water.

5. Why should you rinse with distilled water?

b. Fill the buret with $AgNO_3$ solution.

6. Record the initial volume in the table under *Sample 1.*

c. Put 50 ml of Water Sample 1 in the 100 ml beaker.
d. Add 1 ml of K_2CrO_4 indicator solution.
e. Slowly titrate the sample with $AgNO_3$ until the solution turns and stays a rust color. While titrating, constantly stir the solution.

The reacting chloride ions clump together and need to be separated. Stirring well separates them.

f. After the solution turns rust colored, wait 2–3 minutes to make sure the reaction is complete. If the rust color disappears, add 1–2 drops more $AgNO_3$ from the buret and stir. Repeat until the color stays for 2–3 minutes.

7. Record the final volume of silver nitrate.

8. Calculate and record the volume of silver nitrate used in the reaction.

The concentration of impurities in water is given in *parts per million,* or *ppm.* Parts per million is equivalent to milligrams per liter (prove this to yourself). To get the ppm of chloride, use the following formula:

$$\text{Chloride (ppm)} = \frac{V \times M \times 71{,}000}{S}$$

where V = ml of $AgNO_3$ reacted
M = molarity of $AgNO_3$
S = ml of Water Sample

9. Calculate and record the ppm of chloride in your water sample.

g. Repeat steps **a** through **f** with other water samples.

10. Calculate and record the results for each sample. Compare your data with those of your classmates.

C. Pure as Mountain Dew?

Besides chlorides, a great many substances are soluble, at least slightly, in water. Therefore, a wide variation in the chemical composition of natural waters can be expected. The list below shows substances which have been found in natural waters.

Acids: carbonic, hydrochloric, and sulfuric
Aluminum: oxide and sulfate
Calcium: bicarbonate, carbonate, chloride, phosphate, and sulfate

Iron: carbonate, oxide, and sulfate
Lithium: bicarbonate, carbonate, chloride, and sulfate
Magnesium: bicarbonate, carbonate, chloride, and sulfate
Potassium: bicarbonate, carbonate, chloride, phosphate, and sulfate
Silicon: usually found as silica
Sodium: bicarbonate, carbonate, chloride, and sulfate

As you can see from the list, everything but the kitchen sink can be found in water—and sometimes even that!

While so many solid materials can be dissolved in water, the compounds of calcium and magnesium are especially important. Water which has few calcium or magnesium ions is called "soft." High concentrations of calcium and magnesium ions cause water to be "hard." Hard water leaves solid deposits in pipes and boilers which are costly to remove. Also, the calcium and magnesium ions form a scum with soap which hinders washing. Soapsuds do not form well in hard water.

The total solids dissolved in water include more than the calcium and magnesium ions used to determine the hardness. But if you find the total solids dissolved in a water sample, you will have a fairly good indication of how hard or soft the water is. Table 3 shows the degrees of softness and hardness of water.

TABLE 3

Class	*Total dissolved solids, in ppm*
Soft	0–60
Moderately hard	61–120
Hard	121–180
Saline	over 180

If you had a 100 ml water sample with a hardness of 250 ppm, could you detect it?

$$\begin{aligned} 250 \text{ ppm} &= 250 \text{ mg in 1 liter} \\ &= 250 \text{ mg in 1000 ml} \\ &= 25 \text{ mg in 100 ml} \\ &= 0.0025 \text{ g in 100 ml} \end{aligned}$$

Grant Heilman

11. Can your balance measure 0.0025 grams? If so, you can do the following experiment.
12. If not, how could you detect 250 ppm using the balance you have?

What You Need

Water samples

Beakers, 250 ml, 2
Dropper
Funnel
Graduated cylinder, 100 ml

Asbestos pad
Balance
Boiling chip
Burner
Clay triangle
Filter paper
Forceps
Matches
Ring
Ring stand
Safety goggles

What to Do

13. Copy Table 4 into your notebook.

TABLE 4

Mass of beaker + boiling chip + residue	____ g
Mass of beaker + boiling chip	____ g
Mass of residue from 100 ml	____ g
Mass of residue from 100 ml	____ mg
Mass of residue from 1000 ml	____ mg
Concentration of residue	____ mg/liter

a. Thoroughly shake the container holding your water sample. Filter 110 ml of it into a 250 ml beaker.

14. Why should you filter the sample?

b. Clean and dry another 250 ml beaker.
c. Add a clean, dry boiling chip.
d. Find the mass of the beaker plus boiling chip as accurately as possible.

15. Record the mass in Table 4.

e. Add exactly 100 ml of filtered water sample to this beaker.

16. How could you add 100 ml of water without measuring its volume?

f. Carefully evaporate the water to dryness over a burner. Be careful not to heat it so much that it splatters.
g. When the water has entirely evaporated, turn off the burner. Allow the beaker to cool.
h. When the beaker has cooled completely, find the mass.

17. Record this in Table 4.

18. Complete the table.

19. What is the concentration, in ppm, of the total solids dissolved in your water sample?

20. Would this water sample be classified as soft, moderately hard, hard, or saline?

21. What might account for your getting results which seem too high?

i. Repeat the procedure with other water samples. You might try samples from wells, rivers, lakes, ponds, small streams, and your town water supply.

22. Prepare a table to record your data. Be sure to indicate the source of each water sample. Call this Table 5.

23. What is the relationship between the origin of the water sample and its total dissolved solids?

Now this is my idea of hard water!

D. Why All the Fuss?

What You Need

Soap solution
Sodium borate ($Na_2B_4O_7$)
Sodium carbonate (Na_2CO_3)
Tri-sodium phosphate (Na_3PO_4)
Water (H_2O), distilled
Water samples
Dropper
Graduated cylinder, 25 ml
Stirring rod
Test tubes, 3
Safety goggles
Spatula
Stoppers, solid, 3
Test tube rack
Wax pencil

What to Do

a. Label two test tubes, "Hard" and "Soft."
b. Place 10 ml of distilled water into the "Soft" tube. Place 10 ml of hard water into the other tube.
c. Add 15 drops of soap solution to each tube.
d. Stopper and shake each tube vigorously.

24. Describe the results.

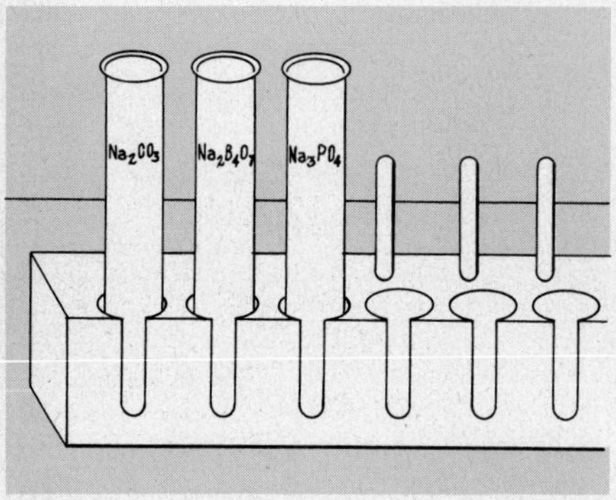

e. Place 10 ml of hard water in each of three clean test tubes. Label the tubes as shown at the left.
f. With a spatula, add a small amount of the proper compound to each tube.
g. Stir each tube until the sodium salts have dissolved.
h. Add 5 drops of soap solution to each tube.
i. Stopper and shake each tube ten times.

25. Describe the results.

26. Compare with the results you recorded in question 24. What differences did you see?

These are only a few of the many tests that can be done on water.

27. Study the data in your tables. Which water samples would you consider the most polluted? Why?

28. Which of these tests could be used to show pollution from mine drainage?

Paper Moon

INVESTIGATION 13

How could we get by without paper? Paper is the material on which newspapers, manuscripts, books, magazines, and record album covers are printed. In a world concerned with money, there are paper checks, money, stocks and bonds. What industry would be able to exist without paper?

Scientists have found ways to make paper flameproof, waterproof, and resistant to weather. In Japan, paper is used for housing.

Consulate General of Japan, N.Y.

Paper is a very useful item. As a paper chemist, you would test paper for its absorption (ability to pick up liquid), strength, bending quality, and other properties. You would analyze it to see how it was made, what it contains, and how it stands up under various conditions.

A. The Quicker-Picker-Upper

Did you ever wonder what lies behind those commercials of harassed housewives mopping up spills from a kitchen counter or a floor? It's a conscientious laboratory technician performing a quantitative test. This test will measure quantitatively the amount of water absorbed by different brands of paper towels.

What You Need

Paper towel samples
Beaker, 400 ml
Buret
Funnel
Buret clamp
Ring stand
Rubber bands
Ruler, metric
Safety goggles
Scissors
Timer (watch) with sweep second hand

What to Do

a. Cut 10 × 10 cm samples of different brands of paper towel.
b. Label each sample in one corner with its brand name.
c. Attach a paper towel sample to a large beaker with a rubber band, as shown below.

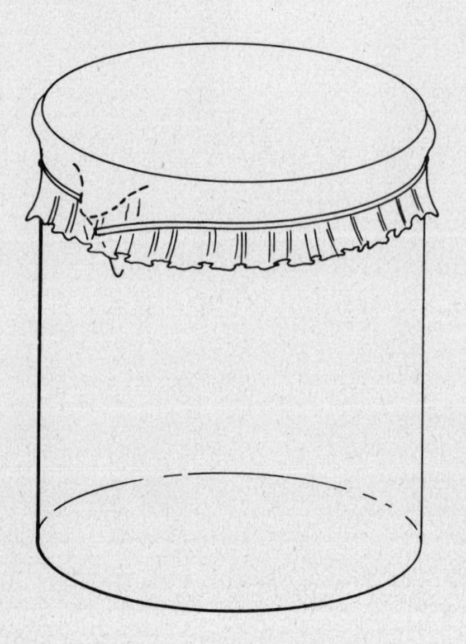

d. Fill a buret with water.
e. Place the paper towel sample and beaker under the tip of the buret. The buret should be centered over the sample. The tip of the buret should be approximately 1 cm above the surface of the paper.
f. Note the time and immediately place 0.5 ml of water from the buret on the center of the paper towel sample.
g. After 30 seconds have elapsed, measure the longest and shortest diameters of the stain.
h. Average the results. This is an approximation of the size of the stain.
i. Make at least 5 tests on each kind of paper. Average the results.
j. Repeat the tests with samples of other brands of paper towel.

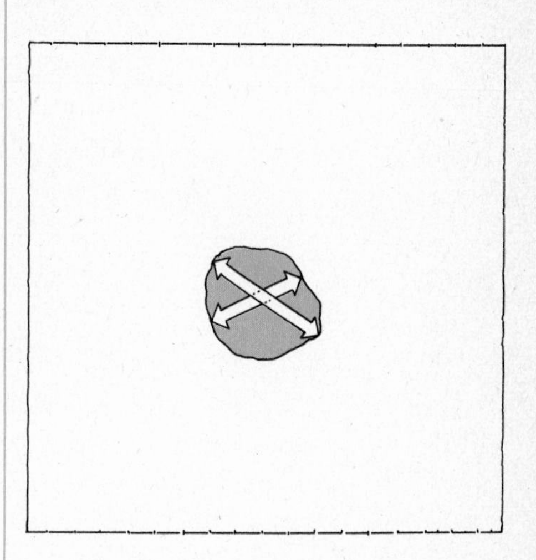

1. Record all the data.

2. Write a report comparing the different samples of paper you tested.

3. Devise at least one other test method that could be used to test the efficiency of different brands of paper towels.

If your teacher agrees, try out your method.

This type of paper testing is of great interest to government agencies which check for truth in advertising.

4. Which brand of paper towels did you find to be most absorbent?

5. Do your results agree with those of your classmates? If not, why not?

6. Which brand of paper towels is the most economical?

B. Before It Goes up in Smoke

It is said that nothing in this world is permanent. Paper is one of the least permanent materials we use in our daily lives. It wrinkles, it rots, it tears, and it burns. Yet most of our important documents are written on paper. Chemists try to develop processes to make paper more permanent. Here is one method of making paper fireproof.

What You Need

Ammonium sulfate $[(NH_4)_2SO_4]$
Borax ($Na_2B_4O_7$)
Boric acid (H_3BO_3)
Paper samples

Beaker, 100 ml
Graduated cylinder, 100 ml
Stirring rod
Thermometer
Watch glass

Asbestos pad
Balance
Burner
Matches
Ring
Ring stand
Safety goggles
Spatula
Tongs
Weighing paper

What to Do

a. Put 4 grams of $(NH_4)_2SO_4$, 1.5 grams of H_3BO_3, and 0.9 grams of $Na_2B_4O_7$ in a 100 ml beaker.
b. Add 50 ml of water and stir to dissolve all chemicals.
c. Heat to 50°C and hold at this temperature while you use the solution.
d. Dip one of your paper samples into the hot solution. Allow the paper to absorb as much solution as it will.
e. Allow the paper to dry thoroughly.
f. Place the paper on a watch glass.
g. Try to ignite the paper with a match.

7. What happens? Why?

8. To what uses could fireproof paper be put?

The study of paper chemistry could take a lifetime. The paper chemist is interested not only in testing, evaluating, and developing paper products, but also in preserving the paper already in existence. Paper has been used to record the written word for more than two thousand years. A great deal of recent research has been spent in trying to devise methods to preserve important documents and manuscripts. Remember, few things are more powerful than the written word.

INVESTIGATION 14

Let's Clean Up This Mess

Soaps and detergents are everywhere. They are used to wash hands, dishes, floors, walls, cars, and shirts. Indeed, there are few objects that do not require cleaning. In Part A you can actually make soap.

DeWys

A. Let's Lather Up

A soap is a sodium or potassium salt of a long chain fatty acid. When fat and sodium hydroxide are mixed together, a chemical reaction takes place.

What You Need

Coconut or cottonseed oil
Ethyl alcohol (CH_3CH_2OH)
Sodium chloride ($NaCl$)
Sodium hydroxide ($NaOH$), 8 *M* solution

Beaker, 250 ml
Evaporating dish, 200 ml
Graduated cylinder, 25 ml
Graduated cylinder, 100 ml
Stirring rod

Asbestos pad
Burner
Matches
Paper towels
Ring
Ring stand
Safety goggles

What to Do

a. Place 10 ml of your oil into an evaporating dish.
b. Add 10 ml of ethyl alcohol and stir.
c. Heat gently. **Keep wet paper towels handy in case the alcohol catches fire.**
d. Slowly and carefully add 10 ml of NaOH solution.
e. While heating, stir the mixture constantly until the odor of alcohol is gone.

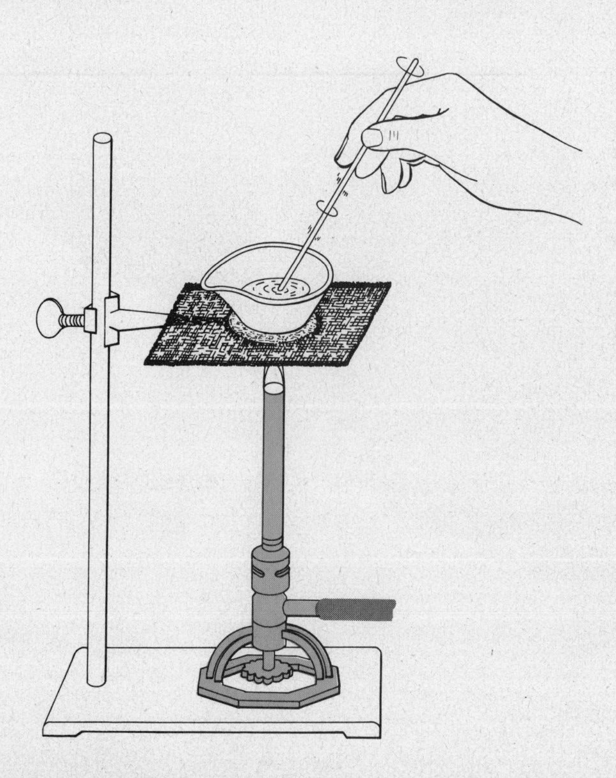

f. Add 20 ml of water and stir until thoroughly mixed.
g. Allow to cool.
h. Make up 100 ml of a saturated solution of NaCl in a 250 ml beaker.
i. Pour the cooled mixture slowly, while stirring, into the beaker.
j. Carefully pour off the top liquid layer. What remains is your soap. Save it for Parts B and C.

The reaction you have just completed is:

$$\underset{\text{Fat}}{C_3H_5(C_{17}H_{35}COO)_3} + \underset{\text{Lye}}{3\ NaOH} \longrightarrow \underset{\text{Glycerine}}{C_3H_5(OH)_3} + \underset{\text{Soap}}{3\ C_{17}H_{35}COONa}$$

1. How would you test to see if your soap is acidic or basic? What would you expect it to be?

Before 1930 the only household laundering agent was soap. Soap worked fairly well on all of the fibers used at that time. It removed the dirt and kept the "whites" white. But soap does not work well in hard water. It forms an insoluble deposit which is difficult to remove.

In the 1930's chemists began adding chemicals to soap in order to solve the "hard water" problem. Thus was born the chemistry of *detergents.*

B. Soap vs. Detergent

What You Need

Detergent, liquid
Soap from Part A
Water (H_2O), distilled
Water (H_2O), hard

Dropper
Graduated cylinder, 10 ml
Test tubes, 4

Safety goggles
Stoppers, solid, 4
Test tube rack
Wax pencil

What to Do

a. Label two test tubes **A** and **B.**
b. Place equal-sized pieces of your soap from Part A in each tube.
c. To tube **A** add 5 ml of distilled water.
d. To tube **B** add 5 ml of hard water.
e. Stopper both tubes and shake for 30 seconds.

2. In which tube did suds form?

3. Describe what soap does to hard water.

f. Repeat steps **a** through **e,** this time using the liquid detergent in step **b.**

4. In which tube did suds form?

5. Would soap or detergent be more effective in hard water?

6. How does soap compare with detergent when used in combination with your school tap water?

7. Does the area in which you live have hard or soft water? Explain why.

C. The Great Lift Off

With what you now know about soaps and detergents, see if you can explain how cleaners remove oil stains from cloth.

What You Need

Detergent, liquid
Oil
Soap from Part A

Dropper
Graduated cylinder, 10 ml
Test tubes, 3

Safety goggles
Stoppers, solid, 3
Test tube rack
Wax pencil

What to Do

a. Label three test tubes as shown below.

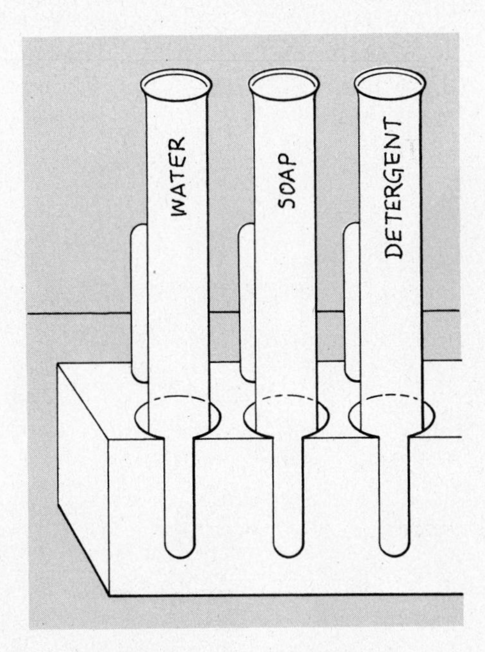

b. Place 2 drops of oil in each tube.
c. Put 5 ml of water in each of the tubes.
d. Stopper and shake each for 30 seconds.

8. Describe the contents of each tube.

e. To the tube marked "Soap" add 1 drop of your soap from Part A.
f. To the tube marked "Detergent" add 1 drop of detergent.
g. Stopper and shake all three tubes for 30 seconds.

9. Describe the action of soap and detergent on your oil/water mixture.

10. Which—soap or detergent—will better aid in removing oil and grease from clothing?

Detergents have many advantages over soap, but they also present problems. Early detergents could not wash cottons, so phosphates were added to give extra cleaning power. Many of the chemicals used would not *degrade* or go back into the earth and water as elements. Different compounds which were *bio-*

degradable had to be found. Biodegradable compounds are those which are decomposed by bacteria found in soil and water. Phosphates are not biodegradable. Phosphates are a problem because they stimulate the growth of algae and water plants. These plants clog streams and waterways. They also remove the oxygen from the water, killing the fish.

Wide World Photos

Chemists, who invented detergents, now have the job of solving this type of pollution.

INVESTIGATION 15

The Final Curtain

This is it. You have completed a year's study of chemistry. Some of what you learned you will never use and will soon forget. Some of what you have learned can be useful to you in your everyday life. That will be remembered. If you become a laboratory technician in one of the many fields using chemistry, most of what you have learned will be used—and therefore should not be forgotten.

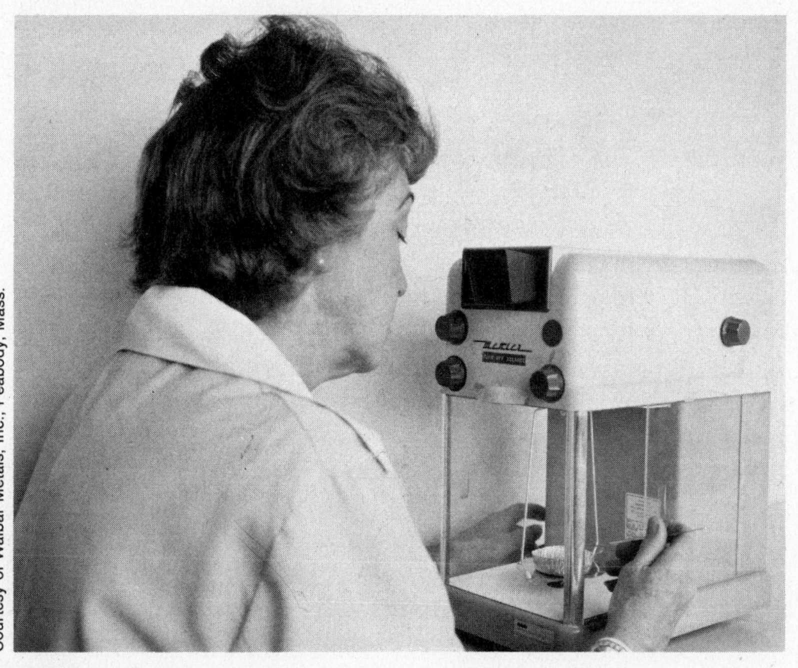

Courtesy of Walbar Metals, Inc., Peabody, Mass.

Chris Reeberg/dpi

As you have discovered this year, chemistry has given you many advantages in life that your parents did not have. Further advances will probably give your children even more advantages. Today chemistry helps us to be healthier, to eat better, to have more entertainment, to travel farther and faster, to live in better housing, and to wear better clothing. The list of advantages could go on and on.

There is, however, a price to pay for all of this. We have polluted our planet beyond belief. The air we breathe is filled with impurities which our activities have put into it. The waters of the earth are polluted with chemicals and oils. Some foods we eat contain almost as much artificial material as natural material.

Chemistry has also played an important part in making wars bigger and more spectacular. The ancient Chinese invented gunpowder, but more recent advances in chemistry gave the world the atomic bomb.

What does all this mean? Knowledge brings increased responsibility. In order to make intelligent decisions in life, or in the laboratory, you must have as many facts at your disposal as possible. It is up to you to put all that you have learned to work for you—whether in everyday life or in your chosen profession.

Wide World Photos

Let's go back to the beginning of the school year. Have your views changed in the past year?

Gaze into your crystal ball and take a look at the future of chemistry. How do you think the world of the future will look in the areas of:

1. food?
2. shelter?
3. clothing?
4. health?
5. recreation?
6. transportation?
7. communication?

Well this is it—not the end of a course, but the beginning of a future. Good luck!

Home Activities

Investigation 1

1. Discuss with a nurse or physician some of the diseases directly related to abnormal blood chemistry. Write a report on your findings.
2. Explain the difference between molar, normal, and percentage solutions.
3. After discussing with local doctors or nurses the importance of chemistry in the field of medicine, write a report on the topic.

Investigation 2

1. Write a report discussing one or more of the questions asked in the first paragraph of this investigation.
2. Design and perform an experiment to show that water expands when it freezes.
3. Draw a diagram showing how crude oil is refined. Name as many by-products from the refining of crude oil as you can.

Investigation 3

1. With your new knowledge of photography, produce a display on one of the following topics:
   a) pollution
   b) traffic problems

c) a tour of the school
d) a chemistry experiment

2. Determine the chemistry of the fast-developing systems pioneered by Polaroid.

Investigation 4

1. By looking through the Sunday paper want ads, determine the types and number of positions open which require a knowledge of metals or metallurgy.

2. Define the following terms:
   a) ferrous alloy
   b) non-ferrous alloy
   c) stainless steel
   d) brass
   e) pewter

Investigation 5

1. Write a recipe for a cake, converting all the amounts to metrics and all of the common ingredients to their chemical names.

2. What chemical principals are involved in making rock candy?

3. Make a list of all the chemicals found in your kitchen.

Investigation 6

1. Using ingredients found in the home, what types of cosmetics could you concoct?

2. Would a mixture of salt and cream of tartar make a good tooth powder? Why?

3. Go into various stores in your neighborhood and list all the brands and prices of cold creams. List the ingredients found in them and determine what, if any, is the difference.

4. What is the major or active chemical in:
   a) nail polish remover?
   b) perfume?
   c) after-shave lotion?
   d) anti-perspirant?

Investigation 7

1. List the plastic items you can find in your:
   a) home
   b) car
   c) schoolroom

2. In the manufacture of plastics, is environmental pollution a problem? If so, how?

Investigation 8

1. Explain how a water repellant works.
2. What are the various types of dyes used in the clothing industry? Choose one and explain the chemical reactions involved.

Investigation 9

1. Write a report outlining the chemist's role in identifying various drugs.
2. How does the chemist regulate the quality of drugs used in medicine?

Investigation 10

1. Go to a local paint or hardware store and make a list of the different paint bases that are sold.
2. Investigate and report on how artists of the 15th, 16th, and 17th centuries formulated and mixed their paints.
3. Devise a test procedure to show how paint withstands scrubbing. Soap, water, and scrub brushes are allowed.

Investigation 11

1. The disposal of effluents, or waste plating solutions, has become a major concern of environmentalists. Explain why and discuss possible solutions.
2. Using the principle of electrolytic plating, discuss chemical milling.

Investigation 12

1. What causes some water to pit aluminum pots and pans?
2. How pure is rain water? Collect some and find out.
3. Name some other impurities or pollutants in water and devise a method for testing them.

Investigation 13

1. Describe the process whereby a tree is converted to paper.
2. Describe the process whereby paper is recycled.

Investigation 14

1. Why are detergents containing phosphates considered pollutants?
2. Will soaps "clean" better than detergents? Why?
3. Some soaps contain corn meal and pumice. Why?

INDEX

C

D

E

F

G

H

I

K

L

M

N

O

P

Q

R

S

T

U

V

W

Z